Application Development Systems

The Inside Story of Multinational Product Development

Edited by Tosiyasu L. Kunii

With 242 Figures

Springer-Verlag
Tokyo Berlin Heidelberg New York
London Paris

Dr. Tosiyasu L. Kunii
Professor and Director
Kunii Laboratory of Computer Science
Department of Information Science
Faculty of Science
The University of Tokyo

Cover Design: Hardware installation diagram of CrossoverNet at the Kunii Laboratory of Computer Science. (See the paper "An Overview of CrossoverNet" in this volume.)

ISBN 4-431-70017-X Springer-Verlag Tokyo Berlin Heidelberg New York
ISBN 3-540-70017-X Springer-Verlag Berlin Heidelberg New York Tokyo
ISBN 0-387-70017-X Springer-Verlag New York Heidelberg Berlin Tokyo

Library of Congress Cataloging-in-Publication Data

Application development systems.
Revised papers originally presented at 1983
1984 & 1985 IBM computer science symposia.
Bibliography: p.
1. System design.   2. Computer software—Development.
I. Kunii, Tosiyasu.   II. International Business Machines Corporation.
QA76.9.S88A67 1986   004.2′1   86-14625
ISBN 0-387-70017-X (U.S.)

All rights reserved. No part of this publication may be reproduced or transmitted in any form or by any means, electronic or mechanical, including photocopy, recording, or any information storage and retrieval system, without permission in writing from the publisher.

© Springer-Verlag Tokyo 1986

Printed in Japan
Printing and Binding: Sanbi Printing, Tokyo

Preface

Applications are the parts of computer systems which directly satisfy users' requirements for information processing. Effective development of applications is the key to the success of any computer-related project. This is why the dominant part of investment in computer systems development is dedicated to applications. However, little work has been published so far on how to develop applications effectively.

"Application Development Systems" directly resolves this omission by presenting basic approaches, both theoretical, e.g., methodologies and frameworks, and practical, e.g., application development tools and environments, to effective development of applications. Many world famous cases of successful application development in the USA, Europe, and Japan are also presented. From this book, the reader will gradually gain an insight into such successful cases as those of General Motors, Toyota, IBM, Yamaha, Nippon Steel, Mitsui Shipbuilding, Mitsui Bank, and Xerox.

This book has taken over 4 years to compile and edit. Indeed, without the contribution of many leading figures in academic fields and industry and the support of IBM Japan and IBM World Trade, this task could not have been achieved. Three consecutive IBM Computer Science Symposia in 1983, 1984, and 1985 were dedicated to the theme of application development. All the contributions from these symposia were reviewed and revised several times before being made into this volume.

Tosiyasu L. Kunii
Editor

Contents

Application Development Systems

FORMAL: A Forms-Oriented and Visual-Directed Application Development System for Non-Programmers
(N.C. Shu) 2

Intelligent Support for Querying Application Databases
(R. Bayer) 27

NIL: A Very High Level Programming Language and Environment
(R. Strom, S. Yemini) 40

DF Language: A Program Development Tool Applying a Concept of Data Flow
(H. Ishida, M. Ohnishi) 52

A Document Management System
(Y. Kambayashi) 65

An Overview of CrossoverNet: A Local Area Network for Integrating Distributed Audio Visual and Computer Systems
(T.L. Kunii, Y. Shirota) 78

Human Interface Development Systems

Issues in the Design of Human-Computer Interfaces
(J. Nievergelt) 114

CAVIAR: A Case Study in Specification
(B. Flinn, I.H. Sørensen) 126

Measurement of SQL: Problems and Progress
(P. Reisner) 165

Human Factors in Electronic Musical Instruments
(Y. Mochida) 188

Incorporating Human Engineering in Motor Vehicle Design
(S. Tsuchiya) 196

CAD/CAM/CAE Development Systems

From Solid Modeling to Finite Element Analysis
(D.A. Field) 220

PC Based Customized VLSI
(T. Nonaka) 250

Computer Aided Robotics
(H. Matsuka, T. Yoshida) 266

Computer Aided Engineering: Will There Be Any ... Ever?
(R.E. Miller, Jr., R.P. Dube) 283

Very Large Application Systems Development

Development of the Banking System in Japan
(E. Ueda) 308

Large-Scale Computer Systems in the Steel Industry
(Y. Inoue) 334

Large Systems Software Implementation
(R.A. Radice) 354

Names and Addresses of Contributors 375
Subject Index 377

Application Development Systems

FORMAL
A Forms-Oriented and Visual-Directed Application Development System for Non-Programmers

Nan C. Shu
IBM Los Angeles Scientific Center

I. INTRODUCTION

Stimulated by the rapid decline of computing costs and the desire for productivity improvement, demands for computerized applications have increased sharply in recent years. We have now reached a point that we can no longer hope to satisfy the tremendous demands simply by producing more professional programmers to honor the requests. As T. J. Freiser, president of John Diebold & Associates, pointed out, "Information is an integral part of end user's everyday activity. It is not realistic that to get this information in the future end users will place a formal request and be placed on a priority list" (Chabrow 1984).

To solve this problem, an apparent approach is to provide easy-to-learn and easy-to-use tools with which end users can create and develop their own applications. It is not surprising that many "user friendly" facilities have appeared in the market place. Icons, pointing devices, and user friendly menus have taken the pain out of learning and remembering commands. Word processors, query facilities and spread sheet programs have become important tools for text preparation, simple inquiries and financial analysis. Further more, some of the application generators have made it possible for a user to choose a particular application from a set of templates and customize the chosen application to suit his/her purpose.

Nevertheless, to most non-DP professionals, the usefulness of a computer is bounded by the usefulness of the "canned" application software available on the computer. Predesigned packages written for a mass audience seldom give every user all the capabilities that he/she needs. Those who wish to use the computer to do something beyond the capabilities of the "canned programs" discover that they have to "program".

Learning to program, unfortunately, is a time consuming and oftentimes frustrating endeavor. Moreover, even after the skill is learned, writing and testing a program is still a time-consuming and detail-intensive chore. Many non-DP professionals stay away simply because the time and effort required for programming often outweighs the potential benefits of using the computer as a tool to solve their problems, particularly when the problem to be solved is not of routine nature. Programming has the tendency to lead to what has been termed "analysis paralysis". "The means become the ends as you forget what you wanted to get out of the computer and become wrapped up in the process of getting it out." (Brown and Sefton 1984).

Thus, the real bottleneck in access to computing is in the **ease of programming.** The challenge is to bring computing capabilities, **usefully and simply,** to people without special training.

At the same time, the computing environment is undergoing an evolution. From the centralized computing at the time-shared mainframes, many corporations are moving toward distributed computing on families of workstations and mainframes, interconnected by networks within a user community. The need to obtain data from various disparate systems meant that the transformation and restructuring of data will play an increasingly important role in applications.

To address these issues, an experimental application development system is being built at the IBM Los Angeles Scientific Center. The primary goal of this project is to develop a very high level facility for data manipulation so that people with little or no DP training can be induced to computerize their own applications. A secondary and more distant goal is to explore the usage of this experimental system as a mechanism for sharing data among different systems.

II. FORMS-ORIENTED APPROACH

To accomplish our goal, we are taking a forms oriented approach. There are several reasons for taking this approach. First, after close examination of the non-programmers' infomation processing needs, we concluded that much of the data manipulation can be naturally expressed or thought of as forms processing.

Second, we concur with many others that forms are the most natural interface between a user and data (see Lefkovitz et al. 1979, for example). The last few years have seen a vast amount of work focusing on "forms" in the interest of office automation and office information systems. For instance, forms have been used as templates for documents which are logical images of business paper forms (Hammer et al. 1977; Ellis and Nutt 1980), and later generalized as templates for messages which include not only the paper forms and text, but also graphics (Kitagawa et al. 1984) and voice utterances (Tsichritzis 1982). Form editors have been implemented to facilitate the entering and querying of data via computerized forms (Bernal 1982; Kitagawa et al.1984; Rowe and Shoens 1982; Yao et al. 1984). Query languages and database systems have been extended to deal with forms (deJong 1980; Luo and Yao 1981; Stonebraker et al.1983; Zloof 1982). The flow of information via forms and messages and the control of such happenings have formed the framework for the modeling and analysis of office activities (Baumann and Coop 1980; Cook 1980; Ellis 1979; Nutt and Ricci 1981; Tsichritzis 1982) and the automation of office procedures (Lin et al. 1983; Lum et al. 1982; Shu et al. 1982; Zisman 1977).

Third, a forms-oriented language offers an opportunity to drastically alter the nature of programming. Traditionally, programming is a specialized arduous task requiring detailed textual instructions which must adhere to strict syntactical rules. Both the structre of the traditional programming languages and the data objects manipulated by them are geared toward the internal (computer) representations. Perceptions and convenience of non-DP persons are sacrificed or ignored for the sake of machine efficiency. However, to encourage end user programming, end user convenience must be considered.

It is our belief that the concept of "form" is not only a convenient representation of data objects, but also a convenient representation of program structure. A forms oriented language is more akin to the user's point of view than the computer's.

The concept of form has been proposed as a programming vehicle in the past. Embley (1980) has investigated the extent to which the features and characteristics of ordinary forms can be exploited to provide a basis for a nonprocedural programming system, and has concluded that it is promising.

The QBE/OBE (Zloof 1981, 1982) is a two-dimensional table-based language which has received wide recognition as a user-friendly means to retrieve, update, define and control the relational data base. Its data restructuring and data manipulation capabilities, however, are limited.

The records processing facility of the Xerox's STAR system uses forms for data definition (Purvy et al.1983). However, for data manipulation, the user is expected to use CUSP, STAR's customer programming language. As pointed out by Purvy et al.(1983), "The complexity of user applications is essentially unbounded, which makes some sort of programming language virtually mandatory." Unfortunately, what is envisioned is: "Eventually, CUSP will become a full programming language, with procedures, variables, parameters, and a programming environment."

In this article, we demonstrate that it is not only feasible but also practical to use a visual-directed forms-oriented language as a programming language for data processing of fairly complex nature, in a simple, natural manner.

The exerimental system is implemented on VM/CMS. The user's familiarity with forms is exploited in three aspects: A form data model, a forms-oriented language, and the familiar way of filling in forms for the user/system communication. These three aspects are elaborated in the following sections.

III. "FORMS" as VISUAL REPRESENTATION of DATA

A "form" data model was first suggested in Shu et al.(1975) as a two-dimensional representation of hierarchical data. It was intended at that time as a visual aid to assist the users in understanding the high level operations performed on their data. The nested table (NT) model and the high level algebraic operations for manipulating NTs reported by Kitagawa et al.(1984) are similar to the form data model and the form (or CONVERT) operations reported by Shu et al.(1975) and by Housel and Shu (1976). It is interesting to note that even though both the NT operations and the CONVERT operations are very high level, they are still procedural. For complicated manipulations, users of these languages still have to specify the operations in a step by step manner.

In Shu et al.(1982), the form data model was refined to be a structured representation, corresponding to an abstraction of conventional paper forms. This structured form is a fundamental concept that permeates our system and enables us to lift the data manipulation language from high level explicit operations (Shu et al.1975; Housel and Shu 1976; Kitagawa et al.1984) to truly non-procedural specifications. The natural tendency of people to draw pictures or images of the information processes that they are attempting to perform is exploited in this approach.

A few definitions are in order. A **form** is a named collection of instances (or records) having the same data structure. Components of a form can be any combination of fields and groups (or subforms). A field is the smallest unit of data that can be referenced in a user's application. A group (or subform) is a sequence of one or more fields and/or subordinate groups. Groups can be either repeating or non-repeating. A non-repeating group is a convenient way to refer to a collection of consecutive fields (e.g. DATE as a non-repeating group over MONTH, DAY and YEAR). A repeating group may have multiple instances and can have, recursively, another repeating group as its subordinate component.

A **form heading** is a stylized two-dimensional description of data structure. In a form heading, the form name is placed on the top line. The names of the first level components (i.e. fields or groups) are shown in columns under the form name. The names of components of groups, in turn, are placed under the associated group names. Parentheses are used to denote repeating groups. A double line signals the end of a form heading. To visualize the data, form instances can be displayed under the heading. As an example, Figure 1 shows the form heading and a few sample instances of the PERSON form.

The form heading represents a convenient way to precisely describe a hierarchical structure of arbitrary depth and width. A relational table is simply a flat form of no repeating groups, i.e. a one-level hierarchy.

The compactness of the form heading allows us to visualize many instances at a time. This property is advantageous in the application development environment. Unlike the office environment where instances of forms are sent, received, and handled one at a time, data processing applications require the handling of a collection of instances.

It is interesting to note that Luo and Yao (1981) adopted this form data model as the basis for the user interface of FOBE (Form Operations By Example), because they consider it to be "a user-friendly model of data". We have employed the form data model as a fundamental concept underlying our work, not only because it is a simple concept for the user to grasp, but also because the preciseness of the form headings enables us to generate code for automatic data restructuring. This last point should become clear in the next section.

IV. A FORMS ORIENTED PROGRAMMING LANGUAGE

Based on the observation that most of the data processing activities of non-programmers can be expressed as forms processing, we have provided the users with a forms oriented programming language. We call this language **FORMAL,** an acronym for Forms ORiented MAnipulation Language. Essentially, we ask the user to think of his/her data processing activity as a form process (or a series of form processes). In general, each form process takes one or two forms as input and produces another form as output. As an example, consider an application to provide product information organized by vendor, given information organized by product. This application is illustrated in Fig. 2 as a form process.

```
I-------------------------------------------------------------------------|
I                              (PERSON)                                   |
I-------------------------------------------------------------------------|
IENO|DNO|  NAME  |PHONE|JC|   (KIDS)    |        (SCHOOL)        |SEX|LOC|
I   |   |        |     |  |-------------|------------------------|   |   |
I   |   |        |     |  |KNAME   |AGE| SNAME   |   (ENROLL)    |   |   |
I   |   |        |     |  |        |   |         |---------------|   |   |
I   |   |        |     |  |        |   |         |YEARIN |YEAROUT|   |   |
I=========================================================================|
I05 |D1 |SMITH   |5555 |05|JOHN    |02 |PRINCETON|1966   |1970   |F  |SF |
I   |   |        |     |  |MARY    |04 |         |1972   |1976   |   |   |
I-------------------------------------------------------------------------|
I05 |D1 |SMITH   |5555 |05|JANE    |01 |         |       |       |F  |SF |
I-------------------------------------------------------------------------|
I07 |D1 |JONES   |5555 |05|DICK    |07 |SJS      |1960   |1965   |F  |SF |
I   |   |        |     |  |JANE    |04 |-------------------------|   |   |
I   |   |        |     |  |        |   |BERKELEY |1965   |1969   |   |   |
I-------------------------------------------------------------------------|
I11 |D1 |ENGEL   |2568 |05|        |   |UCLA     |1970   |1974   |F  |LA |
I-------------------------------------------------------------------------|
I12 |D1 |DURAN   |7610 |05|MARY    |08 |         |       |       |M  |SF |
I   |   |        |     |  |BOB     |10 |         |       |       |   |   |
I   |   |        |     |  |JOHN    |12 |         |       |       |   |   |
I-------------------------------------------------------------------------|
I18 |D1 |LEWIS   |6673 |05|HIM     |02 |STANFORD |1940   |1944   |M  |LA |
I   |   |        |     |  |HER     |05 |         |       |       |   |   |
I-------------------------------------------------------------------------|
I19 |D1 |HOPE    |3150 |07|MARYLOU |10 |         |       |       |M  |SJ |
I   |   |        |     |  |MARYANN |07 |         |       |       |   |   |
I-------------------------------------------------------------------------|
I02 |D2 |GREEN   |1111 |01|RON     |15 |         |       |       |M  |SF |
I   |   |        |     |  |DAVID   |04 |         |       |       |   |   |
I-------------------------------------------------------------------------|
I20 |D2 |CHU     |3348 |10|CHARLIE |06 |HONGKONG |1962   |1966   |F  |LA |
I   |   |        |     |  |CHRIS   |09 |-------------------------|   |   |
I   |   |        |     |  |BONNIE  |04 |STANFORD |1967   |1969   |   |   |
I   |   |        |     |  |        |   |         |1972   |1975   |   |   |
I-------------------------------------------------------------------------|
I21 |D2 |DWAN    |3535 |12|        |   |USC      |1970   |1974   |F  |SJ |
I-------------------------------------------------------------------------|
I          .                                                              |
I          .                                                              |
I          .                                                              |
I-------------------------------------------------------------------------|
```

Fig. 1.   Form heading and a few instances of PERSON form

(PRODUCT)							
PROD_NO	PNAME	TYPE	(SUPPLIER)	(STORAGE)		PRICE
			VNAME	BIN_NO	LOC	
====	====	====	====	====	====	====
110	PIPE	PVC	AQUA	B1	SJC	0.79
			CHEMTRON	B2	SJC	
				B3	SFO	
120	PIPE	STEEL	ABC	B4	SFO	4.10
			CHEMTRON			
210	VALVE	STEEL	AQUA	B5	SJC	0.45
			ABC	B6	SFO	
			CHEMTRON			
221	VALVE	COPPER	ABC	B7	SJC	1.25
			CHEMTRON	B8	SFO	
			ROBINSON			

↓

(VENDPROD)			
VNAME	(PROD)		
	PROD_NO	TYPE	PNAME
====	====	====	====
ABC	120	STEEL	PIPE
	210	STEEL	VALVE
	221	COPPER	VALVE
AQUA	110	PVC	PIPE
	210	STEEL	VALVE
CHEMTRON	110	PVC	PIPE
	120	STEEL	PIPE
	210	STEEL	VALVE
	221	COPPER	VALVE
ROBINSON	221	COPPER	VALVE

Fig. 2.   Example of a Form Process

A **form specification** is a concise and non-procedural description of a form or a form process. Its purpose is either to DEFINE a form or to depict a process to CREATE a new form from existing forms. The result of entering a DEFINE form specification is an entry in the catalog of form descriptions. When the defined form is needed later as a source for the creation of a new form, relevant information about this form can be extracted from the catalog. Entering a CREATE form specification, on the other hand, causes not only an entry in the catalog of form descriptions for the target form, but also the generation of a tailored program to produce the desired output.

In either case, a form specification is a two-dimensional description consisting of:

(1) A title line, indicating whether the specification is used to DEFINE a form, or to depict a named process aimed to CREATE a new form from existing form(s).

(2) A form heading of the form being DEFINEd or the target form to be CREATEd.

(3) Under the form heading, properties of data and/or process can be specified in any row order. Properties of data include DATA_TYPEs and optional "KEY" field(s). Properties of process include:

SOURCE of data;
Fields to be MATCHed when an output instance is constructed from two input sources;
CONDITIONs for selecting instances from the input(s);
ORDERing of instances within a form or within a repeating group.

(4) An "END" line to signal the end of a specification.

In most cases, an application can be described completely by a form specification. In some cases, however, more than one specification may be required. For instance, the current implementation limits the number of input forms to two for each form process. Thus, if the creation of a desired output requires data from three input forms, two specifications (representing two steps) would be necessary. One or more form specifications grouped as a unit is called a **FORMAL program.**

Examples of FORMAL programs are shown in figures 3 through 10. The title line, the form heading and the "END" line need no further explanations. Other elements of a form specification are discussed below.

DATA_TYPE describes the type of data of a field. Fixed or variable length character strings and numbers with or without a decimal point are acceptable data types. For an example, see Fig. 3(a) where a PROJECT form is defined. A few instances of PROJECT are shown in Fig. 3(b).

Note that the description of the DATA_TYPE is mandatory only when the purpose of the specification is to DEFINE a form. For a form process, DATA_TYPE need only be stated for the derived (new) fields or when a change in data type is desired. Normally, data type of an output field is inherited from its corresponding source field.

```
DEFINE PROJECT
           I------------------------------------------|
           I                  (PROJECT)               I
           I------------------------------------------|
           I         |       |       (PROJ)           |
           I  DNO    |  MGR  |------------------------|
           I        |       | PJNO  |  (EQUIP)       | COST  |
           I        |       |       |----------------|       |
           I        |       |       | NAME  | USAGE  |       |
==========I========|=======|=======|========|========|=======|
DATA_TYPE I CH(2) |CH(6)  |CH(4)  | CHV(8) | NUM(2) |NUM(8) |
----------I-------|-------|-------|--------|--------|-------|
KEY       I  Y    |       |  Y    |   Y    |        |       |
----------I-------|-------|-------|--------|--------|-------|
END
```

Fig. 3(a). Example of defining a form

```
I---------------------------------------------|
I               (PROJECT)                     |
I---------------------------------------------|
IDNO| MGR    |           (PROJ)               |
I   |        |--------------------------------|
I   |        |PJNO|   (EQUIP)     | COST      |
I   |        |    |---------------|           |
I   |        |    | NAME  |USAGE  |           |
I===========================================|
ID1 |112340| P11| ABCD  | 50   | 10000      |
I   |      |    | CDE   | 40   |            |
I   |      |----|-------|------|------------|
I   |      | P12| XXXX  | 60   | 5000       |
I----------------------------------------|
ID2 |204550| P12| XXXX  | 30   | 72500      |
I   |      |----|-------|------|------------|
I   |      | P22| ACE   | 50   | 54560      |
I   |      |    | CDE   | 20   |            |
I   |      |    | XYZ   | 25   |            |
I----------------------------------------|
ID3 |301302| P31| XYZ   | 40   | 12450      |
I   |      |    | ABC   | 75   |            |
I   |      |    | ABCDE | 25   |            |
I----------------------------------------|
ID5 |555555| P51| ABC   | 30   | 12345      |
I   |      |    | DUMMY | 50   |            |
I----------------------------------------|
IDX |      | PXX|  ?    |      |            |
I----------------------------------------|
```

Fig. 3(b) Example of PROJECT data

KEY denotes a field (or a collection of fields) whose value uniquely identifies an instance within a form or within a group. For the PROJECT form in Fig.3, DNO uniquely identifies the form instances. Within a given form instance, PJNO is the unique identifier of the repeating PROJ instances; and within each PROJ instance, NAME is the unique identifier of the EQUIP instances. When specified, KEY information is useful for the generation of optimized code.

SOURCE specifies how or where to obtain the relevant value for the operation. There are many ways to specify the sources:

1) An asterisk (i.e.'*') under one or more fields means that the value of these fields is supplied on-line at execution time.

2) A form name under one or more components specifies that the value of these components should be obtained from the corresponding components of the source form.

3) An expression involving arithmetic or string operations specifies the computation or derivation of a new value.

4) A varying assignment specifies case-by-case evaluation for a particular item.

5) Reference to a user program provides a mechanism to invoke a user function written in PL/I, COBOL or IBM 370 assembler language. We don't expect that the non-programmers will use this feature. But it did not require too much effort to make it available in our implemented system, so the "user hook" is there for those who have conventional programming skills.

As an example, a program (written in FORMAL) for the application in Fig. 2 is shown in Fig. 4. Note that we have simply presented the form heading of the desired output, VENDPROD, and specified PRODUCT form as its SOURCE. From the differences in the output and input form headings, the FORMAL compiler is able to generate code to actually perform the implied data restructuring.

```
VENDPROD:   CREATE VENDPROD

            I--------------------------------------------------------|
            I                        ( VENDPROD )                    |
            I---------------|----------------------------------------|
            I               |                (PROD)                  |
            I     VNAME     |---------------|----------|-------------|
            I               | PROD_NO       | TYPE     | PNAME       |
   =========I===============|===============|==========|=============|
   SOURCE  I     PRODUCT                                              |
   --------I--------------------------------------------------------|
   END
```

Fig. 4. A FORMAL program for the process shown in Fig.2.

SOURCE can also be used to specify how a new field is to be derived. Arithmetic and string functions can be performed on the extracted data before the resulting value is placed in the output. The example in Fig. 5 shows that (1) the values of PROD_NO, PNAME and TYPE are extracted directly from the corresponding components in the PRODUCT form; (2) VENDOR is obtained from VNAME of the PRODUCT form; and (3) PRICE, on the otherhand, is computed according to the note <1> of the FORMAL program. Note <1>, in turn, specifies a case by case assignment dependent on the successive tests. For each instance of PRODUCT input, the compiler generated code will first test whether PROD_NO equals 210. If the test succeeds, the user will be asked to supply a new value of PRICE to be placed in the corresponding PROD (output) instance. If the test failed (i.e. PROD_NO is not 210), then the next test (and its accompanying assignment) will be effective. That is, if an instance of PRODUCT.PRICE is less than 1.00, then the computed result of PRODUCT.PRICE times 1.05 will be assigned to the corresponding instance of PROD.PRICE, otherwise, PRODUCT.PRICE times 1.06 will be the value of PROD.PRICE.

```
PROD: CREATE PROD

         I-----------------------------------------------------------|
         I                        ( PROD )                           |
         I-----------------------------------------------------------|
         I         |         |       |              | (SUPPLIER)     |
         I PROD_NO | PNAME   | TYPE  | PRICE        |----------------|
         I         |         |       |              | VENDOR         |
=========I=========|=========|=======|==============|================|
SOURCE   I    PRODUCT                |     <1>      | PRODUCT.VNAME  |
---------I-----------------------------------------------------------|
ORDER    I    DES  |                 |              | ASC            |
---------I-----------------------------------------------------------|
<1>      I *  WHERE PRODUCT.PROD_NO EQ 210                           |
         I  PRODUCT.PRICE TIMES 1.05 WHERE PRODUCT.PRICE LT 1.00     |
         I  PRODUCT.PRICE TIMES 1.06 OTHERWISE                       |
---------I-----------------------------------------------------------|
END
```

Fig. 5. Example of a FORMAL program showing ORDER and case-by-case assignment

CONDITION is a means for stating the criteria for selecting instances from input(s) for processing. In most cases, conditions can be specified under column headings, and Boolean operations are implied. As a rule, conditions specified in two or more rows under the same field (column) are "ORed". Conditions specified for different fields are then "ANDed". For example, Fig. 6 produces a form (VPROD) having the same structure as that of VENDPROD (in Fig. 4). However, only those instances having VNAME not equal to 'ABC' and (TYPE equal to 'STEEL' or 'COPPER') will be selected from PRODUCT form, restructured according to the form heading of VPROD, and placed in the output.

```
VPROD:   CREATE VPROD

              I------------------------------------------------|
              I                    ( VPROD )                   |
              I---------------|--------------------------------|
              I               |           (PROD)               |
              I     VNAME     |--------------------------------|
              I               | PROD_NO  |  TYPE  |  PNAME     |
    =========I===============|==========|========|============|
    SOURCE   I        PRODUCT                                  |
    ---------I------------------------------------------------|
    CONDITION I  NE 'ABC'    |          |'STEEL' |             |
              I              |          |'COPPER'|             |
    ---------I------------------------------------------------|
    END
```

Fig. 6.   Example of a FORMAL program showing CONDITIONs
          stated under column (field) headings

There are occasions when we would like to base our selection on a field which will not appear in the output. Then a global condition (i.e. a condition not bounded by field separators) can be stated. Figure 7 shows an example.

```
VPROD1:   CREATE VPROD1

              I------------------------------------------------|
              I                    ( VPROD1)                   |
              I---------------|--------------------------------|
              I               |           (PROD)               |
              I     VNAME     |--------------------------------|
              I               | PROD_NO  |  TYPE  |  PNAME     |
    =========I===============|==========|========|============|
    SOURCE   I        PRODUCT                                  |
    ---------I------------------------------------------------|
    CONDITION I  PRODUCT.PRICE GE 1.00                         |
    ---------I------------------------------------------------|
    END
```

Fig. 7.   Example of a FORMAL program showing CONDITION
          stated "globally"

When a global condition is specified, Boolean operations (if any) must be stated explicitly. If a user is uncomfortable with explicit Boolean expressions, he/she can always write a two-step program: In the first step, selections are performed to produce a temporary form which has the same structure as the source form (this enables the specification of CONDITIONs under column headings). In the second step, the temporary form is specified as source for the desired output.

The **MATCH** specification is used to tie two input forms in a meaningful way. In Fig. 8, DINFO is created by taking restructured data from two different sources: PROJECT (Fig. 3) and PERSON (Fig. 1). In order to tie the project information of a particular department with the employee information of the same department, we match the department number (DNO). When a match is found, an instance of DINFO is produced.

```
DINFO:   CREATE DINFO

             I-------------------------------------------------------------|
             I                          ( DINFO )                          |
             I-------------------------------------------------------------|
             I DNO | MGR |   (PROJ)    |          ( EMPLOYEE )             |
             I     |     |-------------|------------------------------------|
             I     |     |PJNO|BUDGET|ENO | NAME |         (SCHOOL)        |
             I     |     |    |      |    |      |-------------------------|
             I     |     |    |      |    |      |SNAME |   (ENROLL)       |
             I     |     |    |      |    |      |      |------------------|
             I     |     |    |      |    |      |      |YEARIN|YEAROUT|
             ======I=============================================================|
    SOURCE   I          PROJECT          |              PERSON              |
    ---------I-------------------------------------------------------------|
    MATCH    I    PROJECT.DNO,   PERSON.DNO                                |
             I       ELSE   ERPROJ = PROJECT, ERPERSON = PERSON            |
    ---------I-------------------------------------------------------------|
    END
```

Fig. 8.   Example of creating a form from two source forms

Note that the matching is not restricted to only one field from each input. But the match fields must be paired. Neither is it necessary for the match fields to appear at the top level of input/output form(s). Figure 9 is an example to create a RESDIR form from DEP and DIRECTRY forms matched on DEP.ENO and DIRECTRY.ENO pair, where ENO is not at the top level of DEP or RESDIR. The form headings of the source forms are included in the program as comments. As in PL/I, a comment in a FORMAL program begins with "/*" and ends with "*/".

It is worthwhile to note that normally (as in the example of Fig.9) an instance of output will be produced only when a match of input instances is found. However, if the user chooses to treat the "no match" situation as an error, he/she may do so by using an optional ELSE clause which assigns un-matched instances of a specified form to a designated file. This option may be applied to one or both inputs. For the example in Fig. 8, un-matched instances of PROJECT are deposited in ERPROJ, whereas the un-matched instances of PERSON are deposited in ERPERSON.

Another option is to create output instances based on the "prevailing" form(s). This option is called "PREVAIL" option, and is used in Fig. 10. When specified, an instance of output will be produced for each instance in the prevailing form. In cases of no match, "NULL" values will be assigned to the missing values in the output instance.

```
/*   I-----------------------|     I--------------------------------|
     I     ( DIRECTRY )      |     I            ( DEP )             |
     I-----------------------|     I--------------------------------|
     I  ENO |  NAME | PHONE  |     I       |      |      |   (EMP)  |
     I=======================|     I  DNO  | MGR  | DIV  |----------|
                                   I       |      |      | ENO | JC |
                                   I================================|
*/

RESDIR: CREATE RESDIR
```

```
               I-----------------------------------------------------|
               I                    (RESDIR )                        |
               I-----------------------------------------------------|
               I                |              (EMP)                 |
               I                |------------------------------------|
               I      DNO       |    NAME    |    ENO    |   PHONE   |
==========I================|============|===========|===========|
ORDER     I      ASC       |    ASC     |           |           |
----------I----------------|------------|-----------|-----------|
SOURCE    I      DEP       |         DIRECTRY                   |
----------I-----------------------------------------------------|
MATCH     I      DEP.ENO,       DIRECTRY.ENO                    |
----------I-----------------------------------------------------|
END
```

Fig. 9   Example of a match field not at the top level

ORDER specifies the ordering of instances within a form or ordering of group instances within a parent instance. Order direction is represented by ASC (for ascending) or DES (for descending). When there is more than one significant sort field, the order direction can be followed by a digit (e.g. ASC 1) to indicate the relative significance of the associated sort fields. When associated with an input form, ORDER specifications provide information that is useful for the FORMAL compiler to generate optimized code. When specified for an output, it serves as a request for the desired ordering.

The fore-going discussion completes our informal description of the FORMAL language as implemented at the time of this writing. With a few simple concepts, we are able to provide some very powerful functions. Fig. 10 offers a comprehensive example. Using data available in PERSON form (Fig. 1) and PROJECT form (Fig. 3(b)), a DEPTMENT form is created. The program looks deceptively simple. But the transformation of the source data into the target data involves very extensive data restructuring operations. This should be evident when the resulting DEPTMENT form (shown in Fig. 11) is compared with its sources (PERSON and PROJECT forms in Fig.1 and Fig.3 respectively).

DEPTMENT: CREATE DEPTMENT

```
          I-----------------------------------------------------------|
          I                    ( DEPTMENT )                           |
          I-----------------------------------------------------------|
          I DNO|              (RESOURCE )              |    (PROJ)    |
          I    |---------------------------------------|--------------|
          I    |  JC  |            (EMPLOYEE)          | PJNO | BUDGET|
          I    |      |--------------------------------|      |       |
          I    |      |     |     |     |   (SCHOOL)   |      |       |
          I    |      |NAME |PHONE| LOC |--------------|      |       |
          I    |      |     |     |     |SNAME|(ENROLL)|      |       |
          I    |      |     |     |     |     |--------|      |       |
          I    |      |     |     |     |     |YEAROUT |      |       |
==========I===========================================================|
SOURCE    I         PERSON                            |PROJECT| < 1 > |
----------I-----------------------------------------------------------|
<1>       I PROJECT.PROJ.COST TIMES 1.5                               |
----------I-----------------------------------------------------------|
DATATYPE  I                                           | NUM(9)        |
----------I-----------------------------------------------------------|
MATCH     I   PERSON.DNO,   PROJECT.DNO                               |
          I   ELSE PERSON, PROJECT PREVAIL                            |
----------I-----------------------------------------------------------|
CONDITION I  |GE '05'|          | 'LA'|                               |
          I  |       |          | 'SJ'|                               |
----------I-----------------------------------------------------------|
ORDER     I ASC|  DES  | ASC|                                         |
----------I-----------------------------------------------------------|
END
```

Fig. 10 A comprehensive example

V. "FILLING IN FORMS" for USER-SYSTEM COMMUNICATION

Filling in forms is a familiar activity in everyone's daily life. We therefore use it as a means of communication between the user and our system. When a user types "FORMAL", a **Request Form** (shown in Fig. 12) shows up on the screen. At this time, the user may fill in a form/program name next to the chosen activity, and the system will honor his/her request. In case the user is not sure of the form/program name, he/she may enter an "*" instead, and the system will bring out a list of the names of the forms/programs available for the chosen activity.

We now describe the system's reponse when a specific form/program name is filled in for an activity.

```
I----------------------------------------------------------------|
I                        (DEPTMENT)                              |
I----------------------------------------------------------------|
IDNO|           (RESOURCE)              |       (PROJ)           |
I   |-----------------------------------|------------------------|
I   |JC|          (EMPLOYEE)            |PJNO|    BUDGET         |
I   |  |-----------------------------   |    |                   |
I   |  |  NAME |PHONE|LOC|   (SCHOOL)   |    |                   |
I   |  |       |     |   |------------- |    |                   |
I   |  |       |     |   | SNAME |(ENROLL)|  |                   |
I   |  |       |     |   |       |--------|  |                   |
I   |  |       |     |   |       |YEAROUT |  |                   |
I================================================================|
ID1 |07|HOPE   |3150 |SJ |SJS    |77     |P11 |      15000       |
I   |----------------------------------------|P12 |       7500   |
I   |05|ENGEL  |2568 |LA |UCLA   |74     |    |                  |
I   |  |LEWIS  |6673 |LA |STANFORD|44    |    |                  |
I----------------------------------------------------------------|
ID2 |18|MINOR  |2346 |SJ |UCD    |75     |P12 |     108750       |
I   |----------------------------------------|P22 |      81840   |
I   |12|ANDO   |3321 |SJ |SJS    |       |    |                  |
I   |  |       |     |   |BERKELEY|77    |    |                  |
I   |  |DWAN   |3535 |SJ |USC    |74     |    |                  |
I   |  |GHOSH  |7649 |SJ |       |       |    |                  |
I   |-----------------------------------------    |              |
I   |10|CHU    |3348 |LA |HONGKONG|66    |    |                  |
I   |  |       |     |   |STANFORD|69    |    |                  |
I   |  |       |     |   |        |75    |    |                  |
I----------------------------------------------------------------|
ID3 |08|PETER  |2046 |SJ |       |       |P31 |      18675       |
I   |-----------------------------------------                   |
I   |07|BROOKS |3392 |SJ |       |       |    |                  |
I----------------------------------------------------------------|
ID4 |16|WADE   |1429 |SJ |       |       |    |                  |
I   |-----------------------------------------                   |
I   |13|HUNTER |6666 |SJ |USC    |68     |    |                  |
I   |-----------------------------------------                   |
I   |10|KING   |6397 |SJ |UCD    |71     |    |                  |
I   |  |       |     |   |       |75     |    |                  |
I----------------------------------------------------------------|
ID5 |10|PARKS  |6967 |SJ |MIT    |66     |P51 |      18517       |
I   |  |       |     |   |BERKELEY|71    |    |                  |
I   |-----------------------------------------                   |
I   |08|CHOY   |6059 |SJ |MIT    |75     |    |                  |
I   |  |NEWCOMB|7619 |SJ |       |       |    |                  |
I----------------------------------------------------------------|
IDX |  |       |     |   |       |       |PXX |          0       |
I----------------------------------------------------------------|
```

Fig. 11   DEPTMENT form produced by program in Fig. 10

```
|------------------------------------------------------------------------|
| Welcome to FORMAL                                                      |
|   The following is a list of things that you may choose to do.         |
|                                                                        |
| If "GO ON" is shown at the bottom left corner,                         |
|     please enter the Form/Program NAME next to your choice.            |
|     (or enter a "*" to list those available for the chosen activity)   |
| If "GO ON" is not shown, press PF3 for the next step.                  |
|     (or issue command directly at the bottom "====>" line if you wish.)|
|                                                                        |
|    |-------------------------------------------------------------|     |
|    |         Things to do                 |  Form/Program name   |     |
|    |======================================|======================|     |
|    | Describe a Form                      |                      |     |
|    |--------------------------------------|----------------------|     |
|    | Describe a Form Process              |                      |     |
|    |--------------------------------------|----------------------|     |
|    | Look at or modify a description      |                      |     |
|    |--------------------------------------|----------------------|     |
|    | Translate into computer program      |                      |     |
|    |--------------------------------------|----------------------|     |
|    | Run the computer program             |                      |     |
|    |--------------------------------------|----------------------|     |
|    | Enter data into a new form           |                      |     |
|    |--------------------------------------|----------------------|     |
|    | Display data of a form               |                      |     |
|    |--------------------------------------|----------------------|     |
|    | Display data with summary            |                      |     |
|    |--------------------------------------|----------------------|     |
|    | Modify data in existing form         |                      |     |
|    |--------------------------------------|----------------------|     |
|    | Print the data                       |                      |     |
|    |--------------------------------------|----------------------|     |
|                                                                        |
|                                                                        |
| GO ON                                                                  |
| ====>                                                                  |
|------------------------------------------------------------------------|
```

Fig. 12   Request Form as a means to communicate with the system

If a name, say ABC, is entered in the column next to **"Describe a Form"**, the system checks whether a FORMAL program of the given name already exists. If the program exists, it will be presented to the user. If it does not exist, then a skeleton program for defining a form (shown in Fig. 13) appears on the screen. The user can use the skeleton program as a basis to complete a form specification.

In a similar manner, if a name is entered to **"Describe a Form Process"** and that particular name is not in the catalogue of FORMAL programs, a skeleton program for a form process (as shown in Fig. 14) appears on the screen.

```
| /** What you see is a skeleton for DEFINING a form.
|   Use editing facilities to expand it horizontally or vertically.
|   Fill in relevant information, then file it.                **/
|
| DEFINE xxxxxxx
|
|           I-----------------------------------------------------|
|           I                     (xxxxxxx)                       |
|           I-----------------------------------------------------|
|           I
|           I          (complete form heading here)
|           I
| =========I=======================================================|
| DATATYPE  I                                                      |
| ---------I-------------------------------------------------------|
| KEY       I                                                      |
| ---------I-------------------------------------------------------|
| ORDER     I                                                      |
| ---------I-------------------------------------------------------|
| END
```

Fig. 13   Skeleton program to DEFINE a form

```
| /** What you see is a skeleton for describing a form PROCESS.
|   Use editing facilities to expand it horizontally or vertically.
|   Fill in relevant information, then file it.                 **/
|
| xxxxxxx: CREATE xxxxxxx
|
|           I-----------------------------------------------------|
|           I                     (xxxxxxx)                       |
|           I-----------------------------------------------------|
|           I
|           I          ( complete form heading here )
|           I
| ==========I======================================================|
| DATATYPE  I  (for derived fields)                                |
| ----------I------------------------------------------------------|
| SOURCE    I                                                      |
| ----------I------------------------------------------------------|
| MATCH     I                                                      |
| ----------I------------------------------------------------------|
| CONDITION I                                                      |
| ----------I------------------------------------------------------|
| ORDER     I                                                      |
| ----------I------------------------------------------------------|
| END
```

Fig. 14   Skeleton program to specify a form process

The next three items of the "Things to do" (on the Request Form) are straight forward. In response to **"Look at or modify a description"** the system presents the requested FORMAL program on the screen in the edit mode. **"Translate into computer program"** causes the invocation of the FORMAL compiler. The result is the translation of the specification written in FORMAL into a machine manipulatable computer program. **"Run the computer program"** causes the execution of the translated program to produce the desired form. We could have presented the translation and execution in one step. But we choose to have them as separate steps because a translated program can be used repeatedly while the translation process need not be repeated.

Once the definition of a form has been translated, one can **"Enter data into a new form"**. The system's response to this request is the presentation of a form heading on the screen in edit mode. The user may then proceed to enter data under the heading. When the entering of data is completed, the form data is transformed into "internal data" stored in the computer, ready for accessing.

Both the data entered via our data entry facility and the data produced by the compiled form processing programs are stored in an "internal" format. When a form name is entered to **"Display data of a form"**, the internal data is transformed into formatted data and presented under its form heading.

If, along with the data, summary information (SUM, AVG, etc.) is desired, then the user may choose to **"Display data with summary"**. When a form name, say DEPTMENT, is entered for this purpose, a skeleton summary program (shown in Fig. 15) appears on the screen.

At this time, the user may fill in "Y" (meaning Yes) for the summary information desired. Furthermore, if a user is interested only in a subset of the data and/or wishes to have the instances presented in certain order, he/she may specify CONDITION and/or ORDER. Example of a completed summary program is shown in Fig. 16.

As soon as the user has filed the completed summary program, the system will perform the requested selection, sorting, and computation. If the form is not "flat", the aggregation will be performed for successive levels up the hierarchical path, until the "grand" amounts are computed. The results are recorded in a summary form and presented to the user. Fig. 17 shows excerpts of the summary form resulting from the summary program for DEPTMENT (Fig. 16). It can be seen that SUM, AVG, MAX and MIN are computed for the PROJ.BUDGET within its parent (in this case, it is computed for each DEPTMENT instance), followed by the grand values for all instances in the DEPTMENT form. As requested, within each DNO, RESOURCE instances are in the ascending order of JC, and PROJ instances are in ascending order of BUDGET. The "internal data" of the DEPTMENT form is not affected by the summarizing process. Thus many forms of summary analysis may be performed in succession.

The next item on the "Things to do" list is to **"Modify data in existing form"**. Originally, we contemplated the provision of UPDATE, INSERT and DELETE (in addition to DEFINE and CREATE) in specifications. However, we have decided not to provide the explicit operations, mainly because we believe that the data maintenance is a function of a database management system. When the application development system is integrated with a data base management

```
      I----------------------------------------------------------------|
      I                         (DEPTMENT)                             |
      I----------------------------------------------------------------|
      I DNO|           (RESOURCE)              |      (PROJ)           |
      I    |-------------------------------------|--------------------|
      I    | JC|         (EMPLOYEE)            |PJNO| BUDGET          |
      I    |   |-------------------------------|    |                 |
      I    |   | NAME |PHONE|LOC|   (SCHOOL)   |    |                 |
      I    |   |      |     |   |--------------|    |                 |
      I    |   |      |     |   | SNAME |(ENROLL)|  |                 |
      I    |   |      |     |   |       |--------|  |                 |
      I    |   |      |     |   |       | YEAROUT|  |                 |
=======I================================================================|
COUNT  I  |   |   |      |     |   |       |        |    |            |
-------I----------------------------------------------------------------|
SUM    I  |   |   |      |     |   |       |        |    |            |
-------I----------------------------------------------------------------|
AVG    I  |   |   |      |     |   |       |        |    |            |
-------I----------------------------------------------------------------|
MAX    I  |   |   |      |     |   |       |        |    |            |
-------I----------------------------------------------------------------|
MIN    I  |   |   |      |     |   |       |        |    |            |
-------I----------------------------------------------------------------|
CONDITION I |   |  |      |     |   |       |        |    |           |
-------I----------------------------------------------------------------|
ORDER  I  |   |   |      |     |   |       |        |    |            |
-------I----------------------------------------------------------------|
END
```

/* (NOTE) FILL IN Y (MEANING YES) FOR SUMMARY INFORMATION WHERE DESIRED.
         SPECIFY ORDER/CONDITION WHERE APPROPRIATE.
         FILE IT WHEN ALL DONE. */

Fig. 15. A skeleton summary program presented to the user when
         DEPTMENT is entered as form name next to "Display data
         with summary" in the Request Form

system, the users are expected to use the facilities of the controlling DBMS to update, insert and delete.

The users in our current environment can extract data from existing forms (assuming that they have been granted the right) <u>for their own use.</u> The extracted data, the data that they have entered with our data entry facility, and the data that they have created with their FORMAL programs become their private data. Update, insert, or delete of private data can be accomplished by normal editing. (This approach is similar to the record processing of STAR (Purvy 1983)). Thus, when the user enters a form name next to "Modify data in existing form", he/she will be presented with the formatted data in edit mode. After the edit session is completed, the system automatically transforms the edited version into "internal data" for later use.

```
        I-----------------------------------------------------------------|
        I                          (DEPTMENT)                             |
        I-----------------------------------------------------------------|
        I DNO|             (RESOURCE)              |     (PROJ)           |
        I    |--------------------------------------|---------------------|
        I    | JC|           (EMPLOYEE)            |PJNO| BUDGET         |
        I    |   |----------------------------------|    |                |
        I    |   | NAME |PHONE|LOC|    (SCHOOL)    |    |                |
        I    |   |      |     |   |-----------------|    |                |
        I    |   |      |     |   | SNAME |(ENROLL)|    |                |
        I    |   |      |     |   |       |--------|    |                |
        I    |   |      |     |   |       |YEAROUT |    |                |
========I=================================================================
COUNT   I    |   |      |     |   |       |        |    |                |
--------I-----------------------------------------------------------------|
SUM     I    |   |      |     |   |       |        |    |      Y         |
--------I-----------------------------------------------------------------|
AVG     I    |   |      |     |   |       |        |    |      Y         |
--------I-----------------------------------------------------------------|
MAX     I    |   |      |     |   |       |        |    |      Y         |
--------I-----------------------------------------------------------------|
MIN     I    |   |      |     |   |       |        |    |      Y         |
--------I-----------------------------------------------------------------|
CONDITION I  |   |      |     |   |       |        |    |                |
--------I-----------------------------------------------------------------|
ORDER   I  |ASC|      |     |   |       |        |    |     ASC         |
--------I-----------------------------------------------------------------|
END
```

/* (NOTE) FILL IN Y (MEANING YES) FOR SUMMARY INFORMATION WHERE DESIRED.
         SPECIFY ORDER/CONDITION WHERE APPROPRIATE.
         FILE IT WHEN ALL DONE. */

Fig. 16. A completed summary program

If the user enters a name next to **"Print the data"**, he/she will be asked whether summary information is desired. Depending on the answer, the system will go through either the process to "Display data of a form" or the process to "Display data with summary" as we described above. The resulting formatted form or summary form will then be sent to the printer. However, if the system has found that the creation date of the formatted/summary form is more recent than the associated "internal form", the process described above will be skipped. The formatted or summary form will be sent to the printer directly.

It should be pointed out that the use of "filling in forms" for user/system communication is intended to make the user feel at ease. The Request Form is open ended. What is shown in Fig. 12 reflects what the user can do with our current system at the time of this writing. Included in the Request Form, there is also a command ("====>") line. The experienced users who can remember the few commands (e.g. COMPILE, RUN) are welcome to enter their request directly at the command line, if they wish.

```
I------------------------------------------------------------------|
I                         (DEPTMENT)                               |
I------------------------------------------------------------------|
IDNO|         (RESOURCE)                    |      (PROJ)          |
I   |-----------------------------------------|                    |
I   |JC |          (EMPLOYEE)             |PJNO|    BUDGET         |
I   |   |-------------------------------------|                    |
I   |   | NAME  |PHONE|LOC|    (SCHOOL)   |   |                    |
I   |   |       |     |   |---------------|   |                    |
I   |   |       |     |   | SNAME |(ENROLL)|  |                    |
I   |   |       |     |   |       |--------|  |                    |
I   |   |       |     |   |       |YEAROUT|   |                    |
I==================================================================|
 ID1 |05 |ENGEL  |2568 |LA |UCLA    |74    |P12 |      7500        |
 I   |   |LEWIS  |6673 |LA |STANFORD|44    |P11 |     15000        |
 I SUM:                                          |     22500|
 I AVG:                                          |  11250.00|
 I MAX:                                          |     15000|
 I MIN:                                          |      7500|
 I   |------------------------------------------|              |
 I   |07 |HOPE   |3150 |SJ |SJS     |77    |    |                  |
I------------------------------------------------------------------|
 ID2 |10 |CHU    |3348 |LA |HONGKONG|66    |P22 |     81840        |
 I   |   |       |     |   |STANFORD|69    |P12 |    108750        |
 I   |   |       |     |   |        |75    |    |                  |
 I SUM:                                          |    190590|
 I AVG:                                          |  95295.00|
 I MAX:                                          |    108750|
 I MIN:                                          |     81840|
 I   |------------------------------------------|                  |
 I   |12 |ANDO   |3321 |SJ |SJS     |      |    |                  |
 I   |   |       |     |   |BERKELEY|77    |    |                  |
 I   |   |DWAN   |3535 |SJ |USC     |74    |    |                  |
 I   |   |GHOSH  |7649 |SJ |        |      |    |                  |
 I   |------------------------------------------|                  |
 I   |18 |MINOR  |2346 |SJ |UCD     |75    |    |                  |
 I------------------------------------------------------------------|
 I                           .                                     |
 I                           .                                     |
 I                           .                                     |
 I==================================================================|
SUM I   |   |       |     |   |        |      |    |    250282|
    I   |   |       |     |   |        |      |    |          |
AVG I   |   |       |     |   |        |      |    |  35754.57|
    I   |   |       |     |   |        |      |    |          |
MAX I   |   |       |     |   |        |      |    |    108750|
    I   |   |       |     |   |        |      |    |          |
MIN I   |   |       |     |   |        |      |    |      7500|
 I------------------------------------------------------------------|

     DATE:  08/08/85            TIME:  18:19:46
```

Fig. 17. A summary form resulted from the program in Fig. 16

VI. SUMMARY and DISCUSSION

FORMAL is a very high level language of a moderate extent of visual expressions, designed and implemented to allow the computerization of a fairly wide range of information processing tasks. The two-dimensional form headings and the forms-oriented program structure play an important role as a **means of programming.** At this point, it may be interesting to compare FORMAL with another two-dimensional language, the well known Query-By-Example (Zloof[12]).

Both the Query-By-Example (QBE) and FORMAL are non-procedural. Users do not tell the computer what steps to follow in order to achieve the results. They only tell the computer what they want done, but not how to do it.

Both QBE and FORMAL are two-dimensional programming languages. Skeleton tables (in the case of QBE) and form headings (in the case of FORMAL) are visual expressions. These visual expressions are designed as an integral part of the respective languages. However, tables of QBE are restricted to one-level flat tables while the form headings of FORMAL can accomodate multi-levels of nested and/or parallel (but related) tables.

The most important differences between QBE and FORMAL lie in the scope and emphasis. Because of FORMAL's ability to handle data structures much more complex than the relational tables underlying QBE, FORMAL has a larger problem domain - a broader scope of applicability - than QBE. In other words, FORMAL can handle more complex situations than QBE. As far as emphasis is concerned, QBE is a <u>relational database management system,</u> designed for interactive queries and data maintenance activities (such as insert, update and delete), while FORMAL is the heart of an <u>application development system,</u> designed for non-DP professionals to computerize their information processing (as opposed to data maintenance) applications.

One of the most important features of FORMAL is perhaps its ability to perform automatic data restructuring. Data restructuring, often an integral but non-trivial part of an application, is implied in the differences in the output and input form headings. For example, the specification in Fig. 4 implies an inversion of a hierarchical structure. The inversion is automatically performed by the compiler generated code. In a sense, "**what you sketch is what you get**".

Using the visual representation of an output form as a starting point, a user can make use of simple, familiar concepts (e.g. SOURCE, MATCH, CONDITION, and ORDER) to perform complex processing. For example, the program in Fig. 10 implies the following operations:

   1) "Projection", "restriction", and "outer-join" (to borrow the relational terminologies) of hierarchical data;
   2) "Stretching" of hierarchical levels along one branch of a tree;
   3) Derivation of new data by arithmetic operations;
   4) Sorting of instances within a form and sorting of group instances within parents.

The fact that the FORMAL compiler is able to generate code to accomplish all the above operations (implicit in the program specified in Fig. 10) demonstrates that it is not only feasible but also practical to use a forms oriented language as a **visual programming language** for fairly **complex data processing.** More work is required to bring further advances in that direction.

Earlier, we have discussed the role of the form data model in the visual representation of data. The current version of our system creates forms based on this model and stores them as CMS files. In this manner, the form data model now serves as an interface between the user's conception of data and the data stored as internal files. A natural extension would be the use of the form model as a **common interface** between the data stored in various databases of disparate systems on different machines. In that environment, FORMAL can serve as a means for mapping and restructuring of data extracted from various dis-similar sources. We have some thoughts on this, but more investigation is required.

Along another avenue, networks of advanced workstations connected to mainframes are emerging. It would be interesting to exploit the capabilities of advanced workstations and devise a scheme for dynamic distribution of the functions provided for application development among the workstations and the host computer. Many important issues would have to be addressed.

We have now reached a point where the system is usable for experimental purposes. We intend to gain experience with its use as a guide for our further efforts.

ACKNOWLEDGEMENT

Many concepts embodied in this work were initially kindled by an office procedure automation effort while the author was a member of the IBM San Jose Research Laboratory. The author is indebted to V.Y. Lum, D.M. Choy, F.C. Tung and C.L. Chang for their contributions to that effort. Implementation of the application development system is carried out at the IBM Los Angeles Scientific Center. The author is grateful for the management support; the encouragement from many colleagues and visitors; and the discussions with A.E. Schmalz, P.S. Newman, R.C. Summers, and J.G. Sakamoto.

REFERENCES

Baumann, L.S. and Coop, R.D.
  "Automated workflow control: A key to office productivity". Proceedings of the National Computer Conference, (1980) pp.549-554.

Bernal, M.
  "Interface concepts for electronic forms desigh and manipulation". Office Information Systems, N. Naffah (ed.), INRIA/North Holland Publishing Company, (1982) pp.505-519.

Brown, G.D. and Sefton, D.H.
  "The micro vs. the applications logjam". Datamation, (Jan. 1984) pp.96-104.

Chabrow, E.R.
"Computing Shifting From IM Shops To Users", Information Systems News, (Oct. 1, 1984) p.24.

Cook, C.L.
"Streamlining office procedures - An analysis using the information control net model". Proceedings of the National Computer Conference, (1980) pp.555-566.

deJong, S.P.
"The system for business automation (SBA): A unified application development system". Proceedings of the IFIP Congress 80, (1980) pp.469-474.

Ellis, C.A.
"Information control nets: A mathematical model of office information flow". Proceedings of the ACM Conference on Simulation, Modeling, and Measurement of Computing Systems, (Aug. 1979) pp.225-240.

Ellis, C.A. and Nutt, G.J.
"Office Information Systems and Computer Science". Computing Surveys, Vol.12, No.1, (1980) pp.27-60.

Embley, D.W.
"A forms-based non-procedural programming system". Technical Report, Dept. of Computer Science, University of Nebraska, (Oct. 1980).

Hammer, M., Howe, W.G., Kruskal, V.J. and Wladawsky, I.,
"A very high level programming language for data processing applications". Comm. of ACM, Vol.20, No.11, (Nov.1977) pp.832-840.

Housel, B.C. and SHU, N.C.
"A high-level data manipulation language for hierarchical data structures". Proceedings of the Conf. on Data Abstraction, Definition and Structure, (March 1976), pp.155-168.

Kitagawa, H., Gotoh, M., Misaki, S. and Azuma, M.
"Form Document Management System SPECDOQ - Its Architecture and Implementation". Proceedings of the Second ACM Conference on Office Information Systems, (June 1984) pp.132-142.

Lefkovitz, H.C. et al.,
"A status report on the activities of the CODASYL end user facilities committee (EUFC)". SIGMOD Record, Vol. 10, (2 and 3), (Aug. 1979).

Lin, W.K., Ries, D.R., Blaustein, B.T. and Chilenskas, B.M.
"Office procedures as a distributed database application". in Databases for Business and Office Applications, Proceedings of Annual Meeting, (May 1983) pp.102-107.

Lum, V.Y., Choy, D.M. and Shu, N.C.
"OPAS: An office procedure automation system." IBM Systems Journal, Vol21, No.3, (1982) pp.327-350.

Luo, D. and Yao, S.B.
  "Form operation by example - A language for office information processing". Proceedings of SIGMOD Conf. (June 1981) pp. 213-223.

Nutt, G.J. and Ricci, P.A.
  "Quinault: An Office Modeling System". Computer, (May 1981), pp.41-57.

Purvy, R., Farrell, J. and Klose, P.
  "The design of Star's records processing: Data processing for the noncomputer professional". ACM Transactions on Office Information Systems, Vol.1, No.1, (Jan. 1983), pp.3-24.

Rowe, L.A. and Shoens, K.
  "A form application development system." Proceedings of SIGMOD Conf. (June 1982) pp. 28-38.

Shu, N.C., Housel, B.C., and Lum, V.Y.
  "CONVERT: A high level translation definition language for data conversion". Communications of the ACM, Vol.18, No.10 (Oct. 1975) pp.557-567.

Shu, N.C., Lum, V.Y., Tung, F.C. and Chang, C.L.
  "Specification of forms processing and business procedures for office automation". IEEE Transactions on Software Engineering, Vol.SE-8, No.5 (Sept. 1982) pp.499-512.

Stonebraker, M., Stettner, H., Lynn, N., Kalash, J. and Guttman, A.
  "Document processing in a relational database system". ACM Transactions on Office Information Systems, Vol.1, No.2 (April 1983) pp.143-158.

Tsichritzis, D.
  "Form management". COMM. of ACM, Vol.25, NO.7, (July 1982) pp.453-478.

Yao, S.B., Hevner, A.R., Shi, Z. and Luo, D.
  "FORMANAGER: An Office Forms Management System". ACM Transactions on Office Information Systems, Vol.2, No.3 (July 1984)

Zisman, M.D.
  "Representation, specification and automation of office procedures". Ph.D. Thesis, Wharton School, University of Pennsylvania (1977).

Zloof, M.M.
  "QBE/OBE: A language for office and business automation". Computer (May 1981) pp.13-22.

Zloof, M.M.
  "Office-by-Example: A business language that unifies data and word processing and electronic mail". IBM Systems Journal, Vol.21, No.3 (1982) pp.272-304.

Intelligent Support for Querying Application Databases

Rudolf Bayer
Technical University Munich

1. Application Semantics, Database Schemata, and Schema Constraints

This paper presents a technique to support an application system by semantic knowledge about the data and the database schema of the application. In particular this semantic knowledge may be used to support querying the database in an intelligent way. The technique will be described for database schemata, which are defined according to the entity-relationship data model as used in the Cypress [5] database system, but it may of course be adapted to other data models as well. Semantic information about a database, in particular a database schema, should be augmented by strong types in order to be able to support the task of query formulation more effectively.

Our data model is basically relational, but in addition to relations and attributes of rather primitive types (string, integer, Boolean) it also knows hierarchies of types and of entity classes. The attributes of a relation are typed by either strong types or entity classes. We call entity classes also EntityTypes. CYPRESS knows hierarchies of entity classes (called "domains" in CYPRESS), but it does not have strong types. We assume that the members of a StrongType are defined statically, as e.g. the definition of weekdays in Pascal. In contrast the members of EntityTypes are created and destroyed dynamically. As a consequence we assume that the definition of a StrongType is kept as a closed description whereas the members of an EntityType are explicitly kept in a special one-place relation. We will use T as a mnemonic for StrongTypes and E as a mnemonic for EntityTypes.

The database schema we use from here on is basically that of CYPRESS extended by StrongTypes. The schema itself is described by relations, which is a standard technique in modern relational database systems as e.g. in System R [1]. But our schema contains much more and much more detailed information. The schema consists of 11 relations having attributes of the EntityTypes **Relation, Attribute, EntityType, StrongType, TypeDefinition.**We list the schema relations with their signatures and a brief description of their meaning.

E (e : EntityType)

The relation E contains a 1-tuple for every EntityType, namely the name of the EntityType. So E is initialized with the four 1-tuples:

Relation
Attribute
EntityType
StrongType
TypeDefinition

Now for each EntityType there is a 1-place relation containing exactly the names of the exisiting entities of that EntityType. Note that for EntityType that relation is E itself.

---

*Some of the research presented in this paper was performed while spending sabbatical time at XEROX PARC.

R (r : Relation)
    R contains all relation names, it is initialized as : E, R, A, T, AT, AE, AR, SE, ST, TD.

A (a : Attribute)
    A contains all attribute names.

T (t : StrongType)
    T contains the names of all explicitly defined StrongTypes, it is initially empty.

D (d : TypeDefinition)
    D contains the string defining a StrongType, initially empty.

AT (a : Attribute, t : StrongType)
    For each attribute, AT contains the name of the type of that attribute.

The following schema-relations are self-explanatory:

AE (a : Attribute, e : EntityType)

AR (a : Attribute, r : Relation)

SE (e1 : EntityType, e2 : EntityType)
    e1 is a subclass of e2.

ST (t1 : StrongType, t2 : StrongType)
    t1 is a subtype of t2.

TD (t : StrongType, d : TypeDefinition)

We summarize in tabular form the initialization of these 11 base relations making up our database schema:

E:		R :		A :	
	Relation		E		e
	Attribute		R		r
	EntityType		A		a
	StrongType		T		t
	TypeDefinition		D		d
			AT		e1
			AE		e2
			AR		t1
			SE		t2
			ST		
			TD		

T   = empty

D   = empty

AT   = empty

AE :   e   , EntityType
            r   , Relation
            a   , Attribute
            t   , StrongType

	d	, TypeDefinition
	e1	, EntityType
	e2	, EntityType
	t1	, StrongType
	t2	, StrongType

AR:
	e	, E
	r	, R
	a	, A
	t	, T
	d	, D
	a	, AT
	t	, AT
	a	, AE
	e	, AE
	a	, AR
	r	, AR
	e1	, SE
	e2	, SE
	t1	, ST
	t2	, ST
	t	, TD
	d	, TD

SE = empty,   **ST** = empty,   **TD** = empty

To clarify and to demonstrate the role of this database schema, let us assume that we define a new relation

SP (S#: SupplNumber, P#: PartNumber, QTY: Quantity)

with the relation name SP, attribute names S#, P#, QTY and the types SuplNumber, PartNumber, Quantity for the attributes resp. Then this definition would cause the following changes to the relations of the database schema:

insert   SP   into R

insert   S#
          P#
          QTY   into A

insert   SupplNumber
          PartNumber
          Quantity   into T

insert   S#   SP
          P#   SP
          QTY   SP   into AR

insert   S#   SupplNumber
          P#   PartNumber
          QTY   Quantity   into AT

If the strong types SupplNumber, PartNumber, and Quantity were not defined before, i.e. are not in the relation T, then they would also have to be defined and relations T, D, TD would have to be changed accordingly. Of course, the above changes are not made by the application user directly, they are performed by a special transaction, which handles schema definitions, i.e. accepts a definition like SP (S#: SupplNumber, P#: PartNumber, QTY: Quantity) and causes the changes to R, A, T, AR, and AT described above.

In the following we will express rules in the query language Q , for details see [3]. The rules of Q are basically Horn-clauses, but Q uses a different technique than PROLOG to identify and to unify variables, which is much more suitable for database applications.

For a database schema to be correct, the following integrity constraints must hold. Transactions updating the schema on behalf of declarations must preserve these integrity constraints and should be proven to do so. It has been observed in the literature on deductive database systems, that integrity constraints can be expressed conveniently as rules or Horn-clauses, and I will use this technique in the following. The reader should convince himself, that our initial schema obeys these constraints:

Integrity constraints for schema relations:

A (a)	←	AT (a , t)
T (t)	←	AT (a , t)
A (a)	←	AE (a , e)
E (e)	←	AE (a , e)
A (a)	←	AR (a , r)
R (r)	←	AR (a , r)
E (e1)	←	SE (e1 , e2)
E (e2)	←	SE (e1 , e2)
T (t1)	←	ST (t1 , t2)
T (t2)	←	ST (t1, t2)
T (t)	←	TD (t , d)
D (d)	←	TD (t , d)

Further integrity constraints should include:
"The transitive closure of SE is acyclic"

"The transitive closure of ST is acyclic"

$1 = |TD (t , d)|$ ← T (t)   there must be exactly one type definition for t

$1 < |AR (a , r)|$ ← R (r)   r must have at least one attribute

$1 < |AR (a , r)|$ ← A (a)   a must appear in at least one relation

Note: StrongTypes and EntityTypes are defined independently of relation definitions and before they can be used in relation definitions.

$1 = |AT (a , t)| + |AE (a , e)|$ ← A (a)   an attribute must have exactly one StrongType or one EntityType.

$1 = |ER(e , r)|$ ← E(e)
$1 = |AR(a , r)|$ ← ER(e , r)
    E(e)   ← ER(e , r)
    R(r)   ← ER(e , r)

2. Auxiliary Rules

Next we want to describe some auxiliary predicates needed for later query formulation. These predicates are expressed as Horn-clauses or rules, using the particular syntax of Q.

F (q : Query , r : Relation)
    This predicate should state that *the query q does not contain the relation r*. We do not define F in this paper.

RAA (r : Relation , a : Attribute, b : Attribute)
    Both attributes a and b appear in the same relation r.

        RAA (r , a , b)   ←   AR (a , r) AR (a ↑ b , r)

    The renaming a↑b of a into b is used in the query language Q to separate the two attributes a appearing on the right side and to prevent a join. Note that the join attritube on the right side of this rule is r.

TAA (a : Attibute , b : Attribute)
    Both attributes a and b have the same type.

        TAA (a , b) ←   AT (a , t) AT (a ↑ b , t)

EAA (a : Attribute , b : Attribute)
    Both attributes a and b have the same EntityType.

        EAA (a , b) ←   AE (a , e) AE (a ↑ b , e)

JAA (a : Attribute , b : Attribute)
    The two attributes a and b are join-compatible.

        JAA (a , b) ←   TAA (a , b) + EAA (a, b)

    This uses the union operatory + . Staying closer to Horn clauses the following two rules would be needed alternatively:

        JAA (a , b) ←   TAA (a, b)

        JAA (a , b) ←   EAA (a , b)

Supertypes and Superentities:

The transitive closure of the type and entity hierarchies are needed, in order to decide easily when two attrributes can be coerced to be of the same type and to become join compatible:

UT (t1 : StrongType , t2 : StrongType)
    t2 is a proper supertype of t1.

        UT (t1 , t2) ←   ST (t1 , t2)

        UT (t1 , t2) ←   UT (t1 , t2 ↑ t) ST (t1 ↑ t, t2)

This rule is a linear recursion, which is probably preferable to the following cascading recursion:

UT (t1 , t2) ←   UT (t1 , t2 ↑t) UT (t1 ↑t , t2)

Note that renaming t2 , t1 to t is used for serveral purposes here :

- to distinguish the two t2 ,
- to distinguish the two t1 ,
- to make the first t2 on the right side equal to the second t1 on the right side and to thus force an equijoin.

UE (e1 : EntityType, e2 : EntityType)
e2 is a proper superentity of e1.

UE (e1 , e2)   ←   SE (e1 , e2)

UE (e1 , e2)   ←   UE (e1 , e2 ↑ e) SE (e1 ↑ e , e2)

   or alternatively the cascading recursion:

UE (e1 , e2)   ←   UE (e1 , e2 ↑ e) UE (e1 ↑ e , e2)

Coercion Rules: Next we need some coercion rules telling us, when an attribute has a supertype or a superentity of the other and can be coerced to become join-compatible.

CE (a : Attribute , b : Attribute)
Both attributes can be made join compatible by coercing one of them.

CE (a , b) ←   AE (a , e) UE (e1 ↑ e , e2 ↑ f) AE (a ↑ b , e ↑ f)

CE (a , b) ←   AE (a , e) UE (e1 ↑ f , e2 ↑ e) AE (a ↑ b , e ↑ f)

CE (a , b) ←   AT (a , t) UT (t1 ↑ t , t2 ↑ u) AT (a ↑ b , t ↑ u)

CE (a , b) ←   AT (a , t) UT (t1 ↑ u , t2 ↑ t) AT (a ↑ b , t ↑ u)

Note that CE is almost an equivalence relation , but it is not reflexive. Reflexivity could be obtained by adding the rule:

CE (a , b) ←   TAA (a , b) + EAA (a , b)

which might be useful to simplify the whole rule set.

3. Rules for Query Formulation

We need a strong type for properly formed queries in Q, we simply call it Query. Query could be defined by a set of additional rules specifying formally the predicate that a string of characters is a syntactically correct query in Q. For now we omit these rules and refer the reader to [3] to learn how queries are formed syntactically correct.

Let us assume that in addition to the relation SP we have the following two relations:

S (S# : SupplNumber , SNAME : Name , STATUS : [1:20] , CITY : String)
P (P# : PartNumber , PNAME : Part , COLOR : Colors , WEIGHT : Real , CITY : String)

We want to define classes of queries involving at least two of the following : Attributes, EntityTypes, StrongTypes or Relations. For brevity we call a system providing intelligent support for querying a database an IQ-System. If the user of IQ wants to get information, e.g. about our Attribute SNAME and a StrongType COLOR, he would ask the IQ-system something like:

? SNAME , COLOR

Without knowing the database schema or being forced to formulate any detailed queries himself he should then be presented with a sequence of increasingly complex queries referring to SNAME and COLOR , from which he should choose the desired one for evaluation.

In general a class of queries -e.g. the class QAA of those queries involving two attributes- can be defined by a predicate over the database schema, i.e. more precisely over the 11 relations constituting the representation of our database schema. These predicates can of course again be defined as rules and will be described in the rule language Q.

In the following we will restrict ourselves to the special, but probably most important case of queries involving just two given concepts, any combination of Attributes, EntityTypes, StrongTypes or Relations. This will yield a reasonably simple experimental rule system to define the query classes. It can be generalized of course to describe classes of queries involving more than two concepts, but we conjecture that the rule system presented will already be very useful for practical applications.

The most important class of queries is QAA involving two attributes. We define it first:

QAA (a : Attribute , b : Attribute , q : Query)
This predicate states: *a and b are both involved in q.*

$\qquad$ QAA (a , b , r (a , b)) $\qquad\qquad$ ← $\qquad\qquad$ RAA (r , a , b)

This rule says : *if a and b are both attributes of relation r , then r (a , b) is a properly formed query involving both a and b.* Note that r (a , b) is a query formulated in Q , in particular it is the projection of r to a and b.

$\qquad$ QAA (a , b , q r (b)) ← $\qquad$ QAA (a , b , q) AR (a↑b , r) F (q , r)

If q is a query involving attributes a and b and if b is an attribute of r and if q is free of r , then we can form a new query namely q r (b) , a join of the result relation of q with relation r w.r. to join attribute b. The new query obviously also involves a and b. This is expressed by the left side QAA(a,b,qr(b)) of the rule, saying that the attributes a, b are involved in the query qr(b).

$\qquad$ QAA (a , c , q r (b , c)) $\qquad\qquad$ ← $\qquad\qquad$ QAA (a , b , q) RAA (r , a↑b , b↑c)
$\qquad\qquad\qquad\qquad\qquad\qquad\qquad\qquad\qquad\qquad\qquad$ F (q , r)

Here a join via attribute b of q and a of r is performed , if b and a are in fact the same attribute name. This is checked by performaing a join on attribute b of QAA and a on RAA, the join operation being expressed by renaming a as a↑b.

Note : The notation QAA (a, c, q r (b, c)) means that the expressions a, c, qr (b, c) are assigned to the attributes a, b, q of QAA resp., thus obtaining a new tuple -namely (a,c,qr(b,c)) - satisfying QAA. The notation could be considered as an abbreviation for:

$$QAA (a := a, b := c, q := q r (b, c))$$

QAA (a, c, q r (c↑b))   ←   QAA (a, b, q) JAA (a↑b, b↑c)
                                                                   AR (a↑c, r) F (q, r)

Attributes b of q and c of r have the same StrongType or EntityType. Therefore a join between q and r w.r. to b and c is possible. To express this, the renaming c ↑ b is performed in the newly constructed query q r (c↑b) on the left side.

QAA (a, d, q r (c↑b, d))   ←   QAA (a, b, q) JAA (a↑b, b↑c)
                                                           RAA (r, a↑c, b↑d) F (q, r)

Here *r has two attributes c and d, c is joinable with attribute b of q.*

Note: This might yield unnatural queries, if a = d or b = d, containing more joins than expected. We could eliminate these queries by adding further predicates, namely a ≠ d and b ≠ d to the right side of the rule. Practical experience with the rule system will have to show, which variant of the rule is preferable. This ease of presenting the structure and changing the behaviour of a rather complicated system is, of course, one of the main reasons for writing IQ as a rule system.

QAA (a, d, q r (c↑b, d))   ←   QAA (a, b, q) CE (a↑b, b↑c)
                                                          RAA (r, a↑c, b↑d) F (q, r)

Here *attributes b of q and c of r can be coerced to be of the same StrongType or EntityType, then the join can be performed.* The coercing is automatically performed by the renaming c↑b on the left side in r (c↑b, d). c↑b always coerces to the lower type, i.e. it coerces either c to b or b to c, depending on which of the two attributes has the lower type.

QAA (a, c, q r (c↕b))   ←   QAA (a, b, q) CE (a↑b, b↑c)
                                             AR (a↑c, r) F (q, r)

We now define further rules for queries involving not just two attributes, but types, entities and relations as well.

QAT (a : Attribute, t : StrongType, q : Query)
       q is a query involving a and t.

QAT (a, t, q)   ←   QAA (a, b, q) AT (a↑b, t)

If we have a query q for attributes a and b and b is of type t, then q is also a query for a and t.

QAT (a, t, r (a))   ←   AT (a, t) AR (a, r)

QAE (a : Attribute, e : EntityType, q : Query)
       q is a query involving a and e.

QAE (a, e, r (a))   ←   AE (a, e) AR (a, r)

QAE (a , e , q) ← QAA (a , b , q) AE (a↑b , e)

QAR (a : Attribute , r : Relation , q : Query)
q is a query involving attribute a and relation r.

QAR (a , r , q) ← QAA (a , b , q) AR (a ↑ b , r)

QAR (a , r , r (a)) ← AR (a , r)

QTT (t1 : StrongType , t2 : StrongType , q : Query)

QTT (t1 , t2 , q) ← QAA (a , b , q) AT (a , t ↑ t1)
AT (a ↑ b , t ↑ t2)

QTE (t : StrongType , e : EntityType , q : Query)

QTE (t , e , q) ← QAA (a , b , q) AT (a , t)
AE (a ↑ b , e)

QTR (t : StrongType , r : Relation , q : Query)

QTR (t , r , q) ← QAR (a , r , q) AT (a , t)

QEE (e1 : EntityType , e2 : EntityType , q : Query)

QEE (e1 , e2 , q) ← QAA (a , b , q) AE (a , e ↑ e1)
AE (a ↑ b , e ↑ e2)

QER (e : EntityType , r : Relation , q : Query)

QER (e , r , q) ← QAR (a , r , q) AE (a , e)

QRR (r1 : Relation , r2 : Relation , q : Query)

QRR (r1 , r2 , q) ← QAA (a , b , q) AR (a , r ↑ r1)
AR (a ↑ b , r ↑ r2)

4. Queries as Responses to Hints

We still have to settle the question, how IQ will actually respond to a user input like

$$? \ X \ , \ Y$$

In general an application user may not even know enough about the database schema to specify.

whether the general concepts X and Y he has in mind are Relations, Attributes, EntityTypes or StrongTypes. Indeed, X and Y may be an arbitrary combination of Relations, Attributes, EntityTypes or StrongTypes. We introduce a superentity HintEntity

$$\text{HintEntiity} = \text{Relation} + \text{Attribute} + \text{EntityType} + \text{StrongType}$$

Such an entity can be given as a general hint to IQ about the desired query, which IQ should formulate. IQ will respond to an input like ? X , Y using the rule system for:

DQHH (x : HintEntity , y : HintEntity , q : Query).
Desired queries, if two hints are given.

First, hints like X, Y cause the insertion of a single tuple into a special realation

HH (x : HintEntity , y : HintEntity)

This triggers the system to respond by deriving the desired queries making up the relation DQHH. We list the rules defining DQHH:

    DQHH (x , y , q)   ←   HH (x , y) A (a↑x) A (a↑y)
                                                         QAA (a↑x . b↑y , q)

Note: This is really a control rule saying that the inference algorithm should look for responses, i.e. completely formulated queries, in QAA only. The following rules must all be seen as such control rules:

Rules to look for responses in QAT:

    DQHH (x , y , q)   ←   HH (x , y) A (a↑x) T (t↑y)
                                                        QAT (a↑x , t↑y , q)

    DQHH (x , y , q)   ←   HH (x , y) A (a↑y) T (t↑x)
                                                        QAT (a↑y , t↑x , q)

Rules for responses in QAE:

    DQHH (x , y , q)   ←   HH (x , y) A (a↑x) E (e↑y)
                                                        QAE (a↑x , e↑y , q)

    DQHH (x , y , q)   ←   HH (x , y) A (a↑y) E (e↑x)
                                                        QAE (a↑y , e↑x , q)

Rules for responses in QAR:

    DQHH (x , y , q)   ←   HH (x , y) A (a↑x) R (r↑y)
                                                        QAR (a↑x , r↑y , q)

    DQHH (x , y , q)   ←   HH (x , y) A (a↑y) R (r↑x)
                                                        QAR (a↑y , r↑x , q)

Rule for responses in QTT:

$$DQHH(x,y,q) \leftarrow HH(x,y)\ T(t{\uparrow}x)\ T(t{\uparrow}y)\ QTT(t1{\uparrow}x, t2{\uparrow}y, q)$$

Rules for responses in QTE:

$$DQHH(x,y,q) \leftarrow HH(x,y)\ T(t{\uparrow}x)\ E(e{\uparrow}y)\ QTE(t{\uparrow}x, e{\uparrow}y, q)$$

$$DQHH(x,y,q) \leftarrow HH(x,y)\ T(t{\uparrow}y)\ E(e{\uparrow}x)\ QTE(t{\uparrow}y, e{\uparrow}x, q)$$

Rules for responses in QTR:

$$DQHH(x,y,q) \leftarrow HH(x,y)\ T(t{\uparrow}x)\ R(r{\uparrow}y)\ QTR(t{\uparrow}x, r{\uparrow}y, q)$$

$$DQHH(x,y,q) \leftarrow HH(x,y)\ T(t{\uparrow}y)\ R(r{\uparrow}x)\ QTR(t{\uparrow}y, r{\uparrow}x, q)$$

Rule for responses in QEE:

$$DQHH(x,y,q) \leftarrow HH(x,y)\ E(e{\uparrow}x)\ E(e{\uparrow}y)\ QEE(e1{\uparrow}x, e2{\uparrow}y, q)$$

Rules for responses in QER:

$$DQHH(x,y,q) \leftarrow HH(x,y)\ E(e{\uparrow}x)\ R(r{\uparrow}y)\ QER(e{\uparrow}x, r{\uparrow}y, q)$$

$$DQHH(x,y,q) \leftarrow HH(x,y)\ E(e{\uparrow}y)\ R(r{\uparrow}x)\ QER(e{\uparrow}y, r{\uparrow}x, q)$$

Rule for responses in QRR:

$$DQHH(x,y,q) \leftarrow HH(x,y)\ R(r{\uparrow}x)\ R(r{\uparrow}y)\ QRR(r1{\uparrow}x, r2{\uparrow}y, q)$$

Note: All the double rules above could be replaced by a single rule, if we introduce one additional rule expressing the symmetry of HH, namely:

$$HH(x,y) \leftarrow HH(x{\uparrow}y, y{\uparrow}x)$$

If we use the condition $F(q, r)$ as in this paper, then DQHH will have a finite fixpoint. See [4], [2], [8] for the important role, which fixpoints, in particular finite fixpoints of rule systems play for efficient techniques to evaluate these rule systems. If we drop the condition F from the rules, then queries containing joins of relations with themselves are generated leading to infinite fixpoints. Most of these self-joins are not interesting. It would certainly be useful, to have a rule system which generates only "interesting" self-joins.

With the addition of the superentity HintEntity and the Relation HH to our database schema some of the internal relations making up the schema must now be completed. We only list the tuples that must be inserted into these realations as given in section 1:

E:	HintEntity	AE:	x	, HintEntity
			y	, HintEntity
R:	HH			
A:	x	AR:	x	, HH
	y		y	, HH
ER:	HintEntity , HH			
SE:	Relation , HintEntity			
	Attribute , HintEntity			
	StrongType , HintEntity			
	EntityType , HintEntity			

This completes the definition of the database schema and of the rule system for intelligent database querying. The rule system as presented in this paper is experimental and has not been implemented yet at the time this paper is being written. It is planned to implement it at the Technical University Munich using a deductive relational database system capable of processing rule systems including recursive rules very efficiently. This deductive dataabase system is based on the theories -in particular for handling recursive rules- presented in [4], [2], and [8].

Many applications have large databases and complicated database schemata. In such systems application users can probably not be expected to be familiar with database theory and to remember all the structural datails of the database schema to be able to formulate database queries and general transactions easily and properly.

Therefore an IQ-system as presented in this paper to support the application user effectively in dealing with the database should be a very useful tool.

REFERENCES

1. Astrahan et al.: *System R: Relational Approach to Database Management.*
    ACM TODS, 1, 2, 97-137 (June 1976)

2. Bayer, R: *Query Evaluation and Recursion in Deductive Database Systems.*
    Technical report TUM-I8503, Technical University Munich, 1985

3. Bayer, R.: *Q: A Recursive Query Language for Databases.*
    Technical notes, XEROX PARC, 1985

4. Bayer, R.: *Database Technology for Expert Systems.* Invited paper,
    to appear in proceedings of the Internationaler GI-Kongress '85, Wissensbasierte Systeme (Knowledgebased Systems), Munich, October 1985

5. Cattell, R.G.G.: *Design and Implementation of a Relationship-Entity-Datum Data Model.* Technical report CSL-83-4, XEROX PARC, 1983

6. Gallaire, H., Minker, J.: *Logic and Databases.* New York: Plenum Press 1978

7. Gallaire, H., Minker, J., Nicolas, J.-M.: *Logic and Databases: A Deductive Approach.* ACM Computing Surveys, **16**, 2, 153-186 (June 1984)

8. Guentzer, U., Bayer, R.: *Control for Iterative Evaluation of Recursive Rules in Deductive Database Systems.* Technical report TUM-I8513, Technical University Munich, 1985

NIL
A Very High Level Programming Language and Environment

Rob Strom
Shaula Yemini
IBM T.J. Watson Research Center

Abstract

This paper is a summary of ongoing research activities related to the programming language NIL, a very high level language for large, long-lived software systems developed at IBM Yorktown.

We first present a short summary of the major features of NIL. These include the NIL system model, which is a dynamically evolving network of loosely coupled processes, communicating by message passing; the very high level NIL computation model; and typestate, which is a refinement of type systems which enables a compiler to assume an important part of program validation.

We then describe our current research directions which include 1) developing transformations which map NIL programs to efficient distributed and parallel implementations; 2) including programs as first-class typed NIL objects; and 3) research toward a semantic definition of NIL.

Introduction

NIL is a very high level programming language intended for the design and implementation of large, long-lived software systems. Such systems are expected to undergo continuing evolution during their lifetime as both their functionality and the hardware on which they are implemented evolve to meet new demands. With the current architectural trends, such systems may have to run on multi-machine configurations as well as on single machines.

The major novel concepts in NIL include

1. the NIL system model, consisting of a dynamically evolving network of loosely coupled processes, communicating solely by asynchronous message passing, with no data sharing. Processes are the single modular building block, supporting both concurrency and data abstraction. The process model enables a system to evolve in a truly modular fashion, allowing newly compiled modules to be loaded and linked into a running system.

2. typestate, a refinement of type systems, which allows a compiler to assume an important subset of program validation. This subset is sufficient for enabling the compiler to guarantee security, detect "erroneous programs" (programs whose behavior is undefined by the programming language), ensure that user-defined constraints hold at specified program points and guarantee finalization of all dynamically obtained resources held by a program when it terminates without the run-time overhead of garbage collection.

3. a simple, abstract procedural computation model which hides performance-related and implementation-dependent details, while providing the full power of a general purpose programming language. Performance related details, such as data structure layout, physical distribution and parallelism, are controlled not by changing the source program, but instead by selecting from a menu of alternative translation strategies supported by NIL compilers. This property allows NIL to serve as a very high level "specification" language during the architecture phase; the architecture can then be tested and debugged using default implementations of the language primitives; and then multiple production quality implementations can be derived by selecting for each different target environment, the most cost-effective implementations of language constructs used in the program.

Language Summary

The NIL System Model

system structure

A NIL system consists of a network of processes. Each process encapsulates private data, its *state*. The state consists of a "program counter", and a collection of typed objects, including input and output *ports*, *owned* by the process. Processes may affect one another only by message passing over queued *communication channels* connecting the ports. There is no form of shared data in a NIL system -- all data is distributed among the processes. Each data object is at any time owned by exactly one process, although objects may change ownership by being communicated in messages. Before a process terminates, the compiler ensures that it finalizes any object it owns.

NIL processes are not nested nor is there any other form of statically defined system configuration such as binding by matching port labels. A NIL system is configured by *dynamically* instantiating processes and linking them by means of dynamically created communication channels. A communication channel is created when a process A, the owner of an input port P, performs an operation to *connect* an output port Q (also owned by A) to P. Q then becomes a connected output port, also called a *capability*. If A then passes Q in a message received by process B, B will then be able to communicate over the channel from Q to P. Multiple output ports may be connected to a single input port.

A process always communicates by sending messages to one of its *own* output ports. A process cannot explicitly refer to an external process or port. Thus each process is unaware of the identity of the individual processes comprising the rest of the system. If a new port is created somewhere in the system, another process will not become aware of that port unless it receives a capability to that port by explicit communication. At any point in time, a process sees the rest of the system as a "cloud" whose output and input ports are connected to the given process's input and output ports.

communication

NIL communication is by *ownership transfer*, in contrast to being by value (copy) or by reference. During communication, the *ownership* of a message is transferred from the process owning an output port to the process owning the input port at the other end of the communications channel.

Both queued *synchronous* ("rendezvous" or "remote procedure call") and queued *asynchronous* ("send without wait") communication are supported. The two forms of communication differ in whether or not the sender of the message waits until the message is returned.

Queued asynchronous communication is initiated by the operation *send*, specifying a message object and an output port object. The ownership of the message is transferred from the process issuing *send* to the process owning the input port at the other end of the communication channel. The sending process no longer has access to the message. The message is queued at the input port until the owner of the input port dequeues it using a *receive* operation. It is possible for the queue to contain more than one message, since the sender may continue to send additional messages through the output port, and since there may be several output ports connected to the same input port. Successive messages sent over a single output port will be received in FIFO order, but no specific order other than a *fair merge* is guaranteed for messages sent over different output ports which arrive at the same input port. An asynchronous message may be of any NIL data type. Typically, a message will be a *record*, which is an aggregate consisting of one or more components (fields) which may be of any data type, including ports, or processes.

Queued synchronous communication is initiated by the operation *call*. This operation builds a *returnable* message called a *callrecord*. A callrecord, like any record, is composed of components (the parameters), which may be of any data type. A call transfers the message through the designated output port to the input port at the end of the channel. The caller then waits until the called process dequeues the callrecord, possibly operates on its components (parameters) and then issues a *return* operation on that callrecord. The return operation causes the callrecord to return to the caller.

A process may wait for messages on any of a set of input ports, by issuing a **select** statement whose alternatives correspond to guarded input ports (only input ports may be guarded). A process in a select statement waits until one of the input ports having a true guard has received a message, at which point it then executes the appropriate select alternative, which will usually dequeue the message. In both synchronous and asynchronous communication, a process owning an input port dequeues the message by issuing a *receive* operation. A typical process body consists of a loop containing one or more select statements. In both NIL's "synchronous" and "asynchronous" communication, there is never any concept of simultaneity of action in two distinct processes, since all communication is queued.

process creation and termination

Processes are instantiated dynamically by a **create** operation specifying the name of a *compiled* program which the process will execute, and a list of creation-time *initialization parameters*. The initialization parameters constitute a callrecord passed from the creating process to the created process. The initialization parameters will consist of: (1) "in" parameters, typically including a number of bound output ports (capabilities) supplied by the creator to give the created process access to some of the input ports in the system to which the creator has access to, and (2) "out" parameters, typically consisting of output ports bound to those input ports owned by the newly created process, on which the created process wishes to receive communication.

Since there is no globally visible environment, a newly created process can initially see only that part of the environment passed in the initial callrecord, and can only export capabilities to those ports of its own which it has returned in the initial callrecord. A process can acquire further capabilities through subsequent communication.

Since a process is itself an object, it has a single owner process at each point during its lifetime. Processes terminate upon reaching the end of their program, or may be finalized earlier either by explicit finalization performed by their owner, or as a result of their owner's own termination. A process sends an *obituary message* to an owner-determined port upon termination. The owner of a process can thus wait for the process to terminate by waiting for its obituary message. A process can live beyond the lifetime of its owner if its ownership is transferred (i.e., the process itself is sent in a message) to another process.

A typical NIL system will include some processes that do not terminate, i.e., their program includes an infinite repetitive loop. Other processes may represent long-lived data objects, e.g. files, which generally live longer than the programs which use them. Yet other processes may be short-lived.

Abstract Computation Model

NIL presents the programmer with a simple and abstract computation model, in which all implementation-dependent and performance-related details are transparent.

In more conventional programming languages, e.g., Pascal, C or Ada, features such as dynamically created objects, dynamically changing data structures, very large or long-lived objects, etc., are supported by exposing implementation details through language constructs such as pointers, files, and operating system calls.

In contrast, the NIL programmer is presented with a very simple and abstract set of primitives. For example, in the abstract model seen by the NIL programmer, all objects are created dynamically, arbitrary recursive data structures are supported, and records, variants, and *relations*, modeled after database relations ([COD 70]), unify all forms of dynamic aggregates such as trees, lists, stacks, queues, sets, and arrays. All communication, whether local or remote, is by message-passing. There is no distinction between "variables" and "file" storage.

This simplicity is not obtained at the expense of expressive power. Rather, the NIL compiler encapsulates both "unsafe" low-level constructs such as pointers, and implementation-dependent features in the *implementations* of these abstract language primitives. Thus, pointers, shared memory, the existence of different kinds of storage, multiple processors, different kind of communication media and processor failures, are all encapsulated in implementations of the NIL type families, allowing NIL compilers to exploit these features for achieving efficient implementations, while concealing them at the NIL language level.

At the *implementation* level some objects may be pre-allocated and some bindings may be performed statically. A relation may be implemented by contiguous storage, i.e., as an array, by a linked structure or a hash table, and if large, may be put on secondary storage. A large relation may even be split among the local memories of multiple processors, or replicated for recoverability. Similarly, multiple processes may be assigned to a single processor, or conversely a single process may be split among multiple processors, with different objects and/or parts of the program of a process (subactions) divided among the processors. Recovery from processor failures may be handled by a combination of compile-time and run-time actions.

In addition to a substantial simplification of the computation model, the abstractness of the language primitives gives NIL a tremendous portability advantage. It enables to incorporate into NIL compilers highly sophisticated mappings, which can exploit target-specific features to achieve highly efficient performance on very different hardware environments. This eliminates the need for modifying the code whenever it has to migrate to a different environment.

One of the major focuses of our current research is on designing highly efficient and reliable distributed implementations of NIL systems. We mention some specific results in this area in the last section of this paper.

Types

NIL is a strongly typed programming language. There is a fixed set of *type families* (types and type constructors) in NIL, each of which is associated with a particular set of operations. The type families are:

Scalars Simple integer and booleans are primitive types.

Enumerations The type definition defines a particular set of constants, representing the values of variables of this type.

Records Each record type defines an aggregate of one or more component *fields*, which may have any data type. Records allow groups of objects to be manipulated, e.g. sent and received, as a unit.

Relations Relations contain a variable number of *rows*. Rows are user-defined structured n-tuples. The relation type family was designed in accordance with the model of relational databases [COD 70].

Relation definitions may specify *functional dependencies* -- Any valid third normal form relational schema can be expressed as a relation type definition. Relations may be ordered or keyed. Operations are provided to locate, insert, or delete individual rows, or to iteratively operate on an entire relation. Thus relations provide a single type constructor for aggregating data of a uniform type, subsuming arrays, linked lists, sets, queues, stacks, etc.

Variants A variant is a structured object which may assume one of a number of types specified in the definition of the variant and distinguished by mutually exclusive *cases*. Variants are used when the type of an object cannot be determined at compile-time, and thus the compiler needs to generate a runtime type-check. The type and typestate rules of NIL guarantee that no field of a variant may be accessed except within a clause in which the case has been explicitly tested.

Ports Ports are either *input* or *output* ports. Port types determine the type of message which may be passed through the port. There is no type family corresponding to messages. *All* NIL object types may appear as messages, and thus be transferred between processes. Operations on ports include creating a communications channel by connecting an output port to an input port of matching type, and sending, calling, or receiving. Since only ports of matching type can be connected to form communications channels, the sender of a message is assured that the receiver is expecting a message of the same type.

callrecords A callrecord is a record which is formed as a result of a call specifying a parameter list. The callrecord contains the parameter objects. It is distinguished from other records by the following: (1) it contains information which enables the callrecord to return to the originator of the call, (2) some of the fields may not be fully initialized, since they may be **out** parameters, i.e., they are to be initialized by the called process.

Processes The operations on processes are: **create**, creating a new process, and **finalize** i.e., forcing the process to terminate.

All NIL object types are *first class*. If an object can be an operand, then it also has a type, and can appear in all context permitting objects, such as the value of a variable, or as an element in an aggregate or a message.

Control Constructs

NIL includes all the familiar go-to-less structured programming language control constructs: sequential composition, conditionals, loops (while and iterated), and case statements. In addition, there is a select-variant statement to select the case of a variant; a select-event for waiting to receive communication on a set of guarded input ports, selecting non-deterministically from among the input ports which contain a message and whose guard is true, (like guarded commands except that only input ports may appear); and exception handling.

Exceptions

NIL's type families do not specify how data is laid out in memory, or what the actual limit of supposedly unbounded objects such as integers and relations actually is. On any finite implementation, there is always a number, a table, a queue size, which is "too big". The specific cutoff point varies from implementation to implementation. The language therefore allows any operation which may consume machine resources to *raise an exception* thereby notifying the program that it did not yield the expected result. In addition to reaching implementation-dependent resource limits, exceptions can be raised by operations when the expected precondition to the operation is not satisfied, e.g. in an attempt to retrieve a non-existent item from a relation.

When an exception is raised, control transfers to an *exception handler* within the module.

As a result, every program which successfully compiles and loads has a defined result for the execution of every statement. This result may be a normal outcome, and may be an exception, but will never be "undefined and unpredictable".

Typestate

Typestate checking is a fully automated subset of program validation, enabling a NIL compiler to (1) detect and reject "erroneous programs", (2) ensure that user-defined constraints hold at specified points and (3) ensure clean termination of processes in all cases.

The *type* of an object determines the set of operations which may be applied to the object. However, not all operations of a type are applicable at all times. For example, an integer variable that has not been bound to a value cannot be an operand of an arithmetic operation. Similarly, a message that has not been received should not have its contained objects operated upon, an output port that has not been bound to an input port should not be used for a send or call, etc.. Such actions not only indicate logic errors that should be corrected; they may also cause the violation of security of other *correct* programs, since in typical implementations, they would result in dangling pointers and overwriting arbitrary storage.

Languages permitting erroneous programs to have arbitrary side-effects are *non-secure*, because no part of a system, no matter how simple, can be proven to meet any specification until *every* module in the system has been proven error free. Such languages inhibit modular verification. NIL security is guaranteed at compile-time by a mechanism called *typestate checking*. Because NIL is a secure language, even unverified programs are limited to returning incorrect results of correct type in the correct locations and nowhere else.

Typestate is a refinement of the notion of type, which captures the different degrees of initialization of objects. For primitive types, typestate captures whether the object has been created and whether it has been bound to a value; for composite objects, typestate captures to what degree their constituent objects have been initialized.

Each NIL type family has associated with it a set of typestates. Each operation of the family has a unique *typestate precondition* for each of its operands, which is the typestate the operand must be in for the operation to be applicable, and a unique *typestate postcondition* for each operand, resulting from the application of the operation. If the operation has both normal and exception outcomes, there may be a different postcondition for each possible outcome. In addition to the "built-in" typestates associated with each type family, NIL supports user-defined *constrained typestates*, or *constraints* for short. Constraints specify a predicate on the value of an object. Any valid NIL expression can serve as a constraint, allowing the specification of interrelations between components of structured objects such as the "A" field of some record type is greater than the "B" field. The **validate** operation checks that its operand satisfies the specified typestate constraint and raises an exception if the constraint does not hold.

The typestate information for each family can be represented by a *typestate graph* where the nodes correspond to the typestates of the type family, and the arcs correspond to operations of the type family. User-defined operations, which in NIL correspond to calls, must specify an *interface type* for the corresponding input and output ports. An interface type specifies in addition to the type of each field of the callrecord, the typestate transition resulting from the call. (Interfaces are described in more detail in the following section). The NIL compiler incorporates both the typestate graph and the interface type definitions in order to track the typestate of each variable declared by a process throughout its program.

Typestates of each type family are partially ordered and form a *lower semilattice*, that is, every pair of typestates has a unique *greatest lower bound*. Intuitively, a higher typestate is associated with a higher degree of initialization. NIL requires that for each variable, the typestate is an invariant at each statement in a program, independent of the control path leading to that statement. Thus, at control join points such as following a conditional or the end of a block, the typestate is defined to be the greatest lower bound of the resulting typestates of each of the joining paths. Since by tracking typestate, the compiler tracks the level of initialization of each object, it is able to insert *downhill coercions*, to force the typestate of an object to the greatest lower bound typestate at control join points.

Since typestate tracking enables the compiler to check the consistency of the typestate of each object with respect to the operations applied to it, it enables the compiler to detect and reject "erroneous programs", i.e., programs which in other languages would be compile-time correct, but whose behavior would be undefined, because operations were applied in invalid orders. It also enables the compiler to ensure that properties that cannot be verified before run-time are checked at the appropriate times and are handled by corresponding exception handlers if they do not hold. Finally, since the compiler tracks the level of initialization of each individual variable and inserts downhill coercions at join points, the compiler effectively performs all that is needed to ensure finalizing all objects owned by a process when it terminates. (The **end** of a program is a join point for which the greatest lower bound for all variables is *uninitialized*.)

Thus typestate tracking eliminates the need for expensive run-time garbage collection, as required by other languages which allow dynamic creation of objects.

Typestate-checking thus enables compile-time enforcement of a greater degree of reliability in a highly dynamic environment, than that enforced by static scope rules in a static environment. Typestate checking is more fully described in [STR 83a] and [STR 85b].

Interfaces

Ports are typed by *interface types*. Interface types define the type of message which can travel over the communications channel formed by connecting a matching set of input and output ports. Send interfaces need specify only the type of message sent. However call interfaces must specify not only the type of each parameter, but also the typestate prior to the call -- the *entry typestate*, and and the typestate change resulting from the call -- the *exit typestate* for each possible outcome (normal and each exception).

Interface types, along with all other type definitions, are separately compiled into *definition modules* that may be imported by any NIL program. A caller (sender) program and a receiver program must both be compiled against the *same* interface type definition for their corresponding ports in order for a binding to be possible. At the call site, the compiler checks that all parameters are in the entry typestate, and assumes all parameters after the call are in the exit typestate. Conversely at the site receiving the call, all parameters are assumed to have the entry typestate, and the compiler checks that they are set to the exit typestate prior to the return from the call.

Ongoing Research Activities

NIL's initial design took place between 1978 and 1981. A compiler for the full NIL language (minus real numbers), for an IBM 370 was completed early in 1982, and between 1982 and the present, experiments have been conducted in which significant subsystems were written in NIL and in which performance of the compiled code has been measured and enhanced by adding compiler implementation options without modifying the source. We have been very pleased by the results of these experiments, and a new compiler for targets running C/UNIX is currently being written.

Our current research activities are divided among 3 areas.

The first research area, and the one which has been the main focus of our recent work, is that of developing strategies for implementing NIL on various types of distributed system architectures. Developing practical distributed implementation strategies is essential if we are to continue to maintain the abstract level of the NIL computation model.

A second research area is that of including programs as first-class, typed, constrained NIL objects. This work will result in NIL being a fully integrated programming language and system, with programs

A third area of research is that of providing a semantic definition of NIL as a step towards the goal of formal verification of NIL programs. We expect the lack of shared variables, the existence of explicit narrow interface and the static information associated with them, and typestate checking and exception handling, which eliminate programs whose behavior might otherwise be undefinable, to make it easier to provide a full semantic definiton of NIL. On the other hand, we expect the dynamic nature of NIL systems to make our task harder.

Distributed and Parallel Implementation Strategies for NIL

Much of our more recent research is directed at developing sophisticated implementation strategies for NIL primitives. There is a trend in hardware technology to satisfy the requirement of very high performance and reliability of some environments by clusters of multiple processors. It is our conjecture that an effective way to design software for such systems is to first *specify* them as a network of simple serial NIL processes. Such systems would then be *implemented* by applying a set of *transformations*, reflecting different target-specific implementation strategies including concurrent and distributed realizations, to map the NIL "specifications" into efficient implementations *transparently*. This preserves the NIL goal of abstractness which requires that none of the features of the target environment be reflected in the source programs. The division into processes continues to be based on modularity, not on processor boundaries.

The goals of this area research are to identify transformations which are *independent* of any program to which they may be applied, prove that they preserve the intended semantics and quantify their performance properties. Such work will result in our having at our disposal a library of reusable program transformations that can be applied either individually or composed with other transformations, to obtain efficient distributed implementations.

A straightforward distribution of a NIL system requires only allocating processes to processors, and mapping communication that crosses physical processor boundaries into physical communication. However, if we wish to maintain our high level of abstraction, there are additional issues that must be handled transparently by implementations. One important issue is that of recovery from processor failures. If processes run on different processors which may fail independently, it is necessary to ensure that upon recovery, all processes states are *consistent*, i.e., that all processes have a consistent view of which communication between them has actually occurred and which has not.

We recently developed an efficient asynchronous algorithm for recovery in a distributed system called *Optimistic Recovery*([STR 85a]). Optimistic recovery can be fully embedded in a compiler and run-time environment for NIL.

Another important problem that may need to be supported in practical implementations is that of scheduling computations of *a single process* in parallel, to better exploit distribution and improve response time. In this context we have identified an interesting new and particularly useful *family* of transformations called *optimistic transformations*. Optimistic transformations allow logically serial computations such as C1; C2 to be executed in parallel whenever C1's effect on C2 can be guessed in advance with a high probability of guessing correctly. C2's computation is based on this guess. If the guess is wrong, C2 will have to be undone, but if the probability of a correct guess is sufficiently high, the losses due to undoing computations will be compensated by performance gains due to increased parallelism.

In [STR 84b] we show three examples of "guesses" which can lead to optimistic transformations of practical value: (a) the guess that multiple iterations of a loop will not conflict, (b) the guess that exceptional program conditions will not occur, and (c) the guess that machine failures will not occur. We demonstrate the practicality of synthesizing distributed implementations by composing transformations by presenting an improved distributed two-phase commit protocol derived systematically by composing the above three optimistic transformations.

We expect that many transformations will be fairly general, and therefore that transformations can be developed independently of individual programs, and made available as part of a compiler system. Some of the more obvious transformations include distributing the data among the "sites" of a distributed system, distributing computations among the sites, "paging" data to computations or computations to data, replicating data for high availability, checkpointing for recoverability, etc.. Applications of some of these transformations are already used in current distributed systems, though to our knowledge, these have never been considered as transformations of simpler serial algorithms but rather as special algorithms in and of themselves.

If our conjecture proves to be true, we will have developed a simpler software methodology for designing and implementing software for distributed systems by applying reusable transformations to serial, implementation-independent programs instead of designing a specialized version of each system for each individual target. Our initial results in this area have been encouraging, and we are continuing to pursue this direction.

Programs as Objects

In conventional programming languages, the domains of programs and data objects are mutually exclusive. A program can be represented by a data object which is a character string, but there are no operations which can take this character string, compile it and "instantiate" it as part of the executing program.

Languages of the LISP family are an exception to this rule, since both programs and data all have the same (typeless) representation: s-expressions. The ability to treat programs as data and to write programs which themselves write programs is one of the most attractive features of LISP. Unfortunately, LISP lacks features that are considered as supporting good software engineering such as strong typing, statically checkable scope and binding rules, constructs for abstraction and modularity, etc.

By supporting constrained typestates in NIL, it becomes possible to define *syntactically correct* programs as constrained structured objects. A program is simply a record containing a declarations relation and an (ordered) statements relation. The constraints on program objects have the following flavor: variable names appearing in statements must appear in the declarations relation, type names appearing in a declaration must appear in the definitions relations used by the program, etc. By treating programs as first-class objects, one can now write programs which write other programs, validate them against the constraints to ensure syntactic correctness, instantiate them via the **create** statement already supported in NIL, and link them into the system, making the whole process recursive. The result seems to provide a truly integrated language and system, which we expect to give rise to novel concepts in integrated programming environments.

A Formal Semantic Definition of NIL

NIL interface types characterize communication ports by both the type and the typestate transitions for each object in callrecords passed along the port. We would like to be able to augment this information with some form of behavioral specification. Since the only way a process can affect its environment is via communication, it would be desirable to be able to specify the externally observable behavior of a process by relationships among events on its ports.

We would like any semantic model that we apply to be modular, abstract and compositional. *Modularity* implies that the behavior of a process be specifiable independently of any other particular process that may be in the same system. *Abstractness* implies that the process' behavior is specifiable independently of any particular implementation of the process. *Compositional* implies that there are rules for composing multiple individual process' behavior specifications, given a particular connection configuration, to derive the behavior of a system consisting of those processes connected in the given configuration.

Finally, we would like the semantic model to be able to handle the dynamics of the system, though we suspect that this issue might be rather difficult to address.

We are currently studying several formal models of concurrency in the hope of providing a semantic theory for NIL. The most problematic issues that we have encountered in existing models of concurrency include the assumptions of global time and simultaneity of actions in different processes and the low-level "operational" flavor of some models. Some of these problems seem to be handled by [KEL 83] which we would like to try to apply to NIL.

References

[APT 80] Apt, Krzysztof, Nissim Francez and Willem P. de Roever "A Proof System for Communicating Sequential Processes" *ACM Transactions on Programming Languages and Systems* , July, 1980.

[COD 70] Codd, E. F. "A Relational Model of Data for Large Shared Data Banks", *Communications of the ACM*, vol 13, No 6., June 1970.

[HAI 82] Hailpern, B., "Verifying Concurrent Processes Using Temporal Logic", *Springer Verlag Lecture Notes in Computer Science*, no. 129, 1982.

[KEL 83] Keller, Robert M., and Prakash Panangaden, "An Archival Approach to the Semantics of Indeterminate Operators" Draft Report, University of Utah.

[LAM 83] Lamport, L., "Specifying Concurrent Program Modules" *ACM Transactions on Programming Languages and Systems* 5:2, April 1983 July 1978.

[MIL 80] Milner, Robin "A Calculus of Communicating Systems", *Springer Verlag Lecture Notes in Computer Science #92*, 1980.

[MIS 81] Misra, Jayadev, and K. Mani Chandy "Proofs of Networks of Processes" *IEEE Transactions on Software Engineering* July, 1981.

[STR 83a] Strom, R. E., "Mechanisms for Compile-Time Enforcement of Security", *Tenth Symposium on Principles of Programming Languages*, Austin, January 1983.

[STR 83b] Strom, R. E., and Yemini, S. "NIL: An Integrated Language and System for Distributed Programming", *Proc. ACM SIGPLAN Conference on Programming Language Issues in Systems*, June, 1983.

[STR 84a] Strom, R. E. and Yemini, S., "Synthesizing Distributed Protocols through Optimistic Transformations" *Proc. Fourth International Workshop on Protocol Specification, Testing and Validation*, Skytop Penn. North Holland, 1984.

[STR 84b] Strom, R. E. and Halim, N., "A New Programming Methodology for Long-Lived Software Systems" *IBM Journal of Research and Development*, January 1984.

[STR 85a] Strom, R. E., and Yemini, S. "Optimistic Recovery in Distributed Systems" to appear in *ACM Transactions on Computer Systems* August, 1985. also available as IBM Research Report RC 10353.

[STR 85b] Strom, R. E., and Yemini, S. "Typestate: A Programming Language Concept for Enhancing Software Reliability", to appear in *IEEE Transactions on Software Engineering*, Special Issue on Software Reliability.

DF Language
A Program Development Tool Applying a Concept of Data Flow

Hidenobu Ishida
Michio Ohnishi
Mitsui Engineering & Shipbuilding Co., Ltd.

ABSTRACT

The DF language (DF-COBOL and DF-PLI) developed by a new concept of data flow is a software for designing, developing, and maintaining business applications. It consists of the DF chart and DF language.

The DF chart is a powerful tool for describing briefly and appropriately the specifications of applications by using the technique of data flow and is the basis for DF language.

The DF language consists of elements which correspond to elements of the DF chart, so it is very easy to understand. It also generates automatically COBOL or PLI programs. Basic processing units prepared in the DF language are designed by using a complete modular concept, so the module units can be combined without any restriction, and the DF language has high generalities and low redundancies. The DF language can provide great improvements in application development productivity by using the technique of data flow.

INTRODUCTION

Growing performance and decreasing prices of computers have extended the range of their applications in business fields and have caused an increased demand for high-quality application programs. However, the software development still depends on the "craftsmen" who determine the quality of each program. The maintenance work has also increased year by year, and so the data-processing staff in each company need to consider how they can obtain higher productivity in programming.

The software has so far been improved by:

1) Standardizing the programs
   Complete utility programs and subroutine packages
2) Structuring the programs
   Modular programming, Warnier Diagrams, and structured programming
3) Forming teams for programming
   The chief programmer team.

These programming techniques and IBM's improved programming technologies (IPT), which combined the above techniques, have been used to improve programming. But these methods are basically production techniques used with current high-level languages; they do not contribute to drastically increasing the productivity of software.

This difficulty can be attributed to the nature of current high-level languages, which are procedure-oriented languages designed to express problem-solving procedures. Essentially, the procedures themselves are hard to understand and communicate. Consequently, it would seem to be very important for the improvement of software productivity that the programs should not be procedure-oriented or at least that they should have minimum numbers of procedure entries.

We made a study on the expression of data-processing contents for business applications with minimum description of procedure. As a result of this study, we found that the data flow theory, which has recently gathered public attention, could be effectively applied to a problem, and we developed the new programming languages DF-COBOL and DF-PLI on the basis of COBOL and PLI, which have been popularly used for business data processing.

The concept of the DF language, details of the development, and the result of DF language application are reported below.

DATA FLOW THEORY

1. Data Flow

The data flow theory drew public attention when it was introduced because it provides an excellent way of expressing data-processing contents (specifications) and it offers a theoretical basis to a parallel-operation computer with multiple built-in processors.

The data flow is believed to eliminate the drawbacks of conventional procedure flow. Procedure flow is procedure-oriented, whereas data flow is data-oriented. This is a major and fundamental difference between the two.

Procedure flow (Fig.1) can be expressed by a data flow method as in Fig.2. The two figures contrast markedly: Fig.1 shows a flow of procedure while Fig.2 shows a flow of data; Fig.1 shows three possible types of flow, while Fig.2 shows only one.

With the procedure flow, three different flows are available for one processing (as shown in Fig.1), and each flow indicates that one procedure starts after the previous procedure has finished. However, it is not desirable that one processing flow be expressed in different flows of procedure in different sequences when procedure sequence is required.

A data flow defines mutual relationships of data and can show processing contents clearer than a procedure flow.

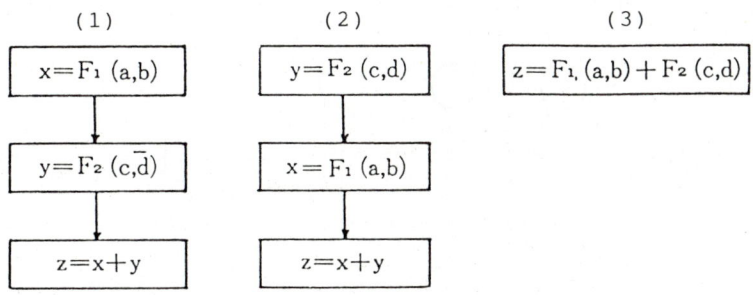

Fig.1. $Z = F_1(a,b) + F_2(c,d)$ by procedure flow

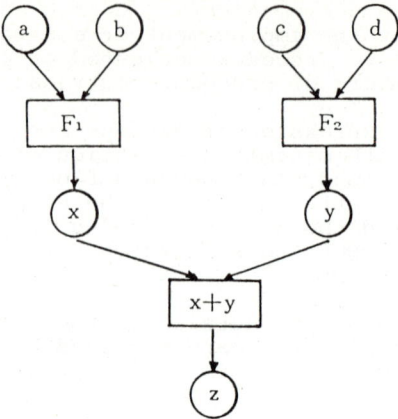

Fig.2. $Z = F_1(a,b) + F_2(c,d)$ by data flow

In addition, the data flow in Fig.2 does not fix the calculation sequence but specifies it, as opposed to the procedure flow in Fig.1, which fixes the calculation sequence. Either of the calculations F1 and F2 in Fig.2 can be done at any time.

A data flow indicates mutual relationships of the data. It does not fix the calculation sequence but specifies it. Consequently, a calculation with complete conditions and multiple calculations can be executed at any time (in parallel).

2. Data Flow Theory Application to the Software

This section describes how the concept of data flow is applied to the DF language.

1) Data-processing content expression by data flow

The data processing generates new data by means of combination, conversion, and integration of data. In this sense, data flow showing the mutual relationships of data can be the clearest expression of data-processing contents.

A chart for the flow of data (Fig.3) to express data-processing outline has been used conventionally for business data processing. This flow, however, only shows the input/output files with no mutual relationship of data being indicated, so it is difficult to understand the processing contents in detail.

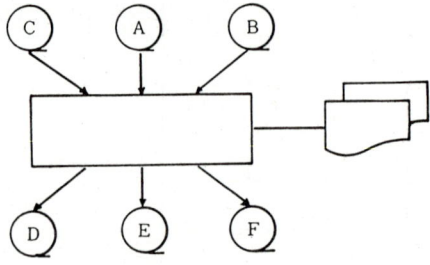

Fig.3. Conventional input/output flow

The conventional input/output flow is expressed by using the data flow as shown in Fig.4. Figure 4 shows the mutual relationship of the data clearly and also helps in understanding the processing contents.

The DF language uses the flow shown in Fig.4 supplemented by data areas (DA) (Fig.5) to describe processing contents. This flow is called the DF chart. A DA is located in the main memory and used to store data. The data are transmitted from a processing step to the subsequent step via the DA.

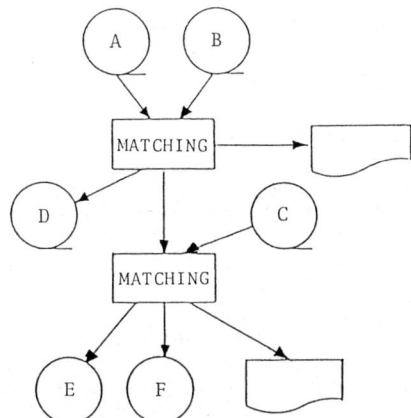

Fig.4. Data flow

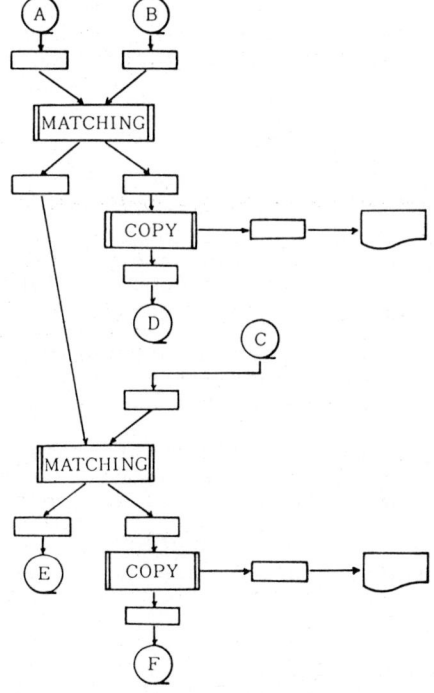

Fig.5. DF chart

2) Modular processing by data flow

The data flow is also suited for use with the modular concept. With a data flow, highly independent modules can be generated. In the DF languages, the modules are generated according to the following principles.

* Processing is divided until only the data directly related to the processing are input
* Even with a single input, a processing without identical input and output must be regarded as a single processing; example — summary, format conversion, data removal
* To prevent single data from being input for different processings at one time, data copying may be performed if required.

The modular units can be combined via DA without any restrictions.

3) Program structure by data flow

The data flow concept gives the program structure the two following characteristics. First, the logic of each functional unit is standardized because it is exclusively related to the corresponding DA. Second, the control logic for entire flow is simplified because it should only check each DA status. The outline of the control logic is as follows.

Each DA is assigned by one of the two possible types of status - "full" and "empty". One basic processing can be executed at any time when the input DA is "full" and output DA is "empty". When a basic processing is executed, required input data are set to "empty" (already used) and required output data are set to "full" to be output. The control logic checks each DA status and initiates processings ready to be executed, regardless of processing sequence.

In the DF language, a processing is classified into input, processing, and output, each of which is applied to a corresponding basic logic in the data flow described above. Figure 6 shows the activation condition for each module and DA status after the activation.

DATA FLOW LANGUAGE

The DF language, designed on the basis of the data flow concept, generates a source program of compiler language (COBOL or PLI) by using macro language. It allows the use of compiler language (COBOL or or PLI) as an auxiliary language to enter partial processings.

The DF language consists of DF-COBOL and DF-PLI. The DF-COBOL generates COBOL programs and uses COBOL as the auxiliary language; the DF-PLI generates PLI programs and uses PLI as the auxiliary language.

Although the DF language generates COBOL programs or PLI programs, a system developed with the DF language can be maintained by the DF language. Generated programs (COBOL or PLI) are only the intermediate results.

Major DF-language elements (Table 1) are as follows.

1. Module

The modules correspond to all elements of the DF chart. Use of the input/output modules, including ¥INPUT, ¥OUTPUT, ¥VSIN, and ¥DBIN, eliminates the entry of statements (such as OPEN, CLOSE, READ, WRITE,

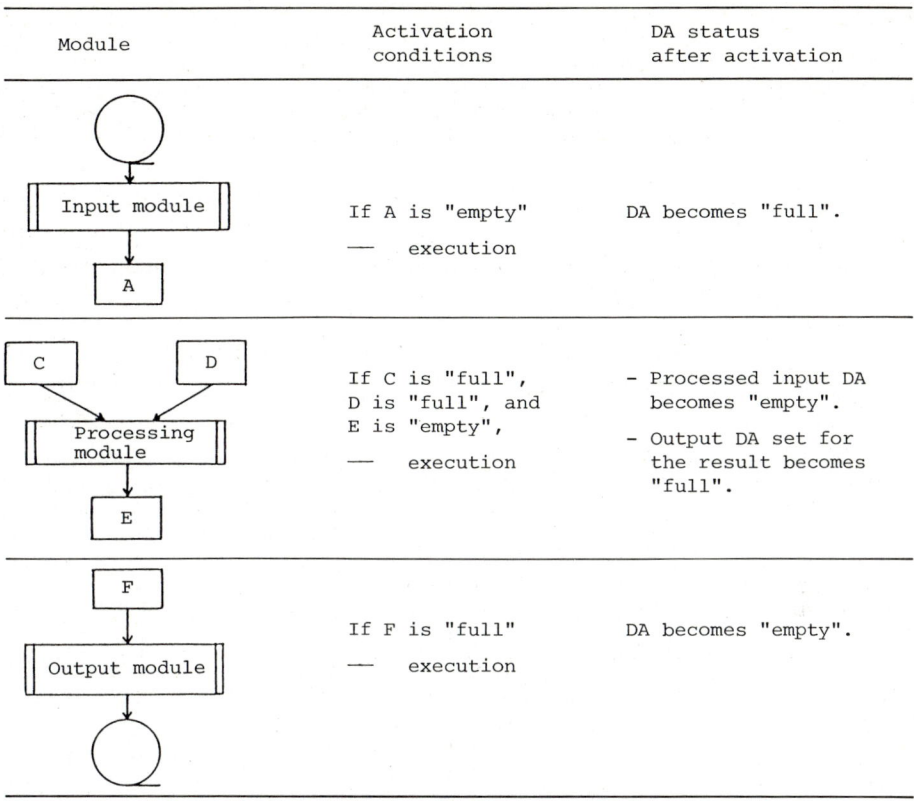

Fig.6. Activation conditions for modules

etc.) for the files during normal processing. These instructions are automatically generated by the DF language so that they are executed in proper timing.

Creating the DF charts for a variety of programs reveals that almost all programs are made by combining several basic elements, including the input/output modules described above and the processing modules — ¥SUMMARY, ¥MERGE, ¥MATCH, and others. Conventional programming of matching and merging processings in a procedure flow requires fine setting of processing timings.

However, the ¥MERGE module eliminates the timing setting, and the ¥MATCH module can be used correctly without the need for timing because the control of data flow replaces the timing setting for the module.

2. User Routine

The user routine is used for the coding of a partial processing defined in each module, such as a matching processing with ¥MATCH module by using COBOL or PLI statements and DF-language functions.

Main COBOL or PLI statements used for the coding consist of the statements of the four operations in arithmetic and move operations. There is almost no necessity for the entry of input/output statements such as READ and WRITE.

3. ¥DA

The ¥DA defines data areas (DA) in the main memory. Data are sent from one module to other modules via the DA.
The DA specifies the boundaries between modules. The user routine statements specified in a module only apply the variables in the DA and user area (UA).

4. Function

The functions are used to execute fundamental and simple processings, other than those done by modules, in combination with COBOL or PLI statements in user routine.

Table 1. Summary of DF-language elements

Element	DF-language statement	Outline
Indentification	¥ID	Description of program identification
Definition	¥DA	Definition of data area
	¥UA	Definition of user area
	¥TEMPFMT	Definition of temporary format
	¥TABLE	Definition and preparation of sequential look-up tables
	¥VSAM	Definition of VSAM files
	¥DB	Definition of IMS DB
Module	¥INPUT	Input of sequential files
	¥VSIN	Input of VSAM files
	¥DBIN	Input of IMS DB
	¥ISORT	Input of sequential files with sort function
	¥OUTPUT	Output of sequential files
	¥VSGEN	Creation of VSAM files
	¥SUMMARY	Summary process
	¥MERGE	Merging process
	¥MATCH	Matching process
	¥USER	User process
	¥COPY	Copy process
	¥LIST	Report process
	¥GROUP	User routine with time lag process
	¥CARD	Input of source data
Function	#LOOKUP	Table look-up process
	#GET	Random read for VSAM files or IMS DB
	#INSERT	Random insert for VSAM files or IMS DB
	#REPLACE	Random update for VSAM files or IMS DB
	#DELETE	Random deletion for VSAM files or IMS DB
	#SETDA	Function to make data area active
	#RESETDA	Function to make data area inactive
	#SETCD	Function to make condition area true
	#RESETCD	Function to make condition area false
	#SETNPRT	Non-print function used within ¥LIST module
	#BRK	Key break function used within ¥GROUP module
	#CERROR	Function to set error flags and messages, used within ¥CARD module
User routine		User routines which consist of COBOL, PLI statement, and DF-language functions are defined in and performed from each module

The functions #SETDA and #RESETDA are specific to the DF language. They are used to control data flow. There are two types of modules: one type controls data flow automatically and the other entrusts all or parts of the data flow control to the user. The ¥MERGE is of the former type and the ¥MATCH is of the latter, for which the user needs to control part of the data flow. The functions #SETDA and #RESETDA are required when the data flows become different from one another during partial processings — for example, with differences in data flows due to the condition of parts balance in updating the parts master file.

However, the concept of data flow control can be understood more easily than the timing control for the procedure flow.

5. System Development by DF Language

Figure 7 shows the system development approach by the DF language. Use of the DF language not only changes programming procedures but also updates the designing of data processing.

The DF-language coding steps correspond to each of the steps in the DF-chart (as shown in Fig.8); the coding is easily performed even by the people with little knowledge of programming.

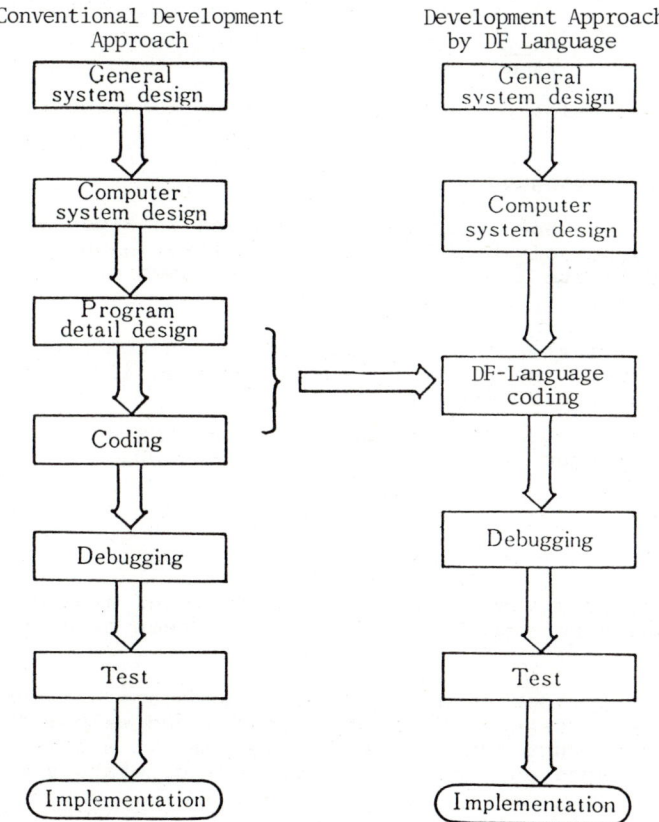

Fig.7. Difference between conventional development approaches and the development approach by DF language

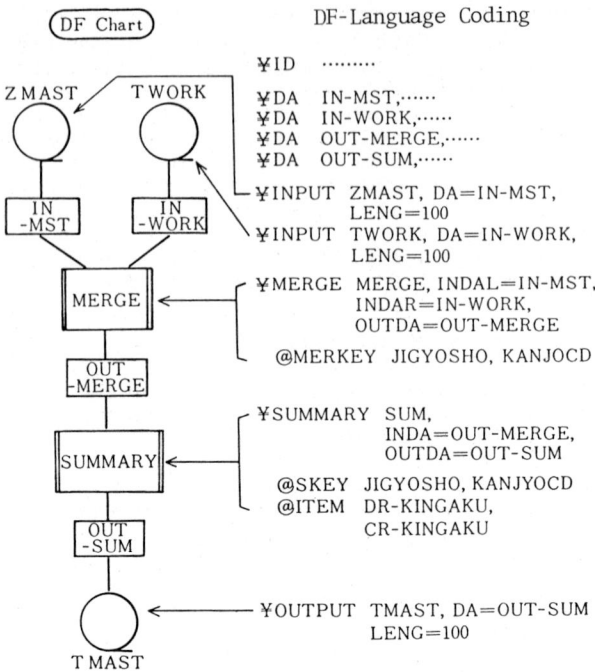

Fig.8. Relation of DF-chart with DF-language coding

6. Examples of DF-Language Application

Examples of DF-language application are given in Figs. 9 and 10. Figure 9 shows the coding only by DF-language modules, in which DF-COBOL and DF-PLI statements are identical. Figure 10 shows examples of DF-COBOL and DF-PLI coding.

When the number of steps of DF-COBOL coding and COBOL programming in Fig.9 are compared, the DF-COBOL coding requires 15 steps and the COBOL programming requires 90 steps.

Table 2 lists the pages of program specifications and number of statements for the programming with COBOL and DF-COBOL.

DISCUSSION

There are signs of great advances in the field of application program development where no remarkable progress has been made in the past two decades.

The new concept for designing application programs and systems is based on the idea that "changes are always present and welcome" for designing systems. To construct such "flexible" systems at a lower cost requires high-level languages and application program generators with a data base.

The DF language appears to be an effective tool for realizing new application program development. The DF-COBOL was first marketed in 1979 and has been used by 25 customer companies. The DF-PLI was first marketed in August, 1985 and used on a trial basis by pilot users. The DF language will be further improved by incorporating suggestions from the current users.

The DF language, applying the features of data flow, has a new capability to realize decentralized data processing, so that it can also be improved and developed for that purpose.

(1) DF Chart

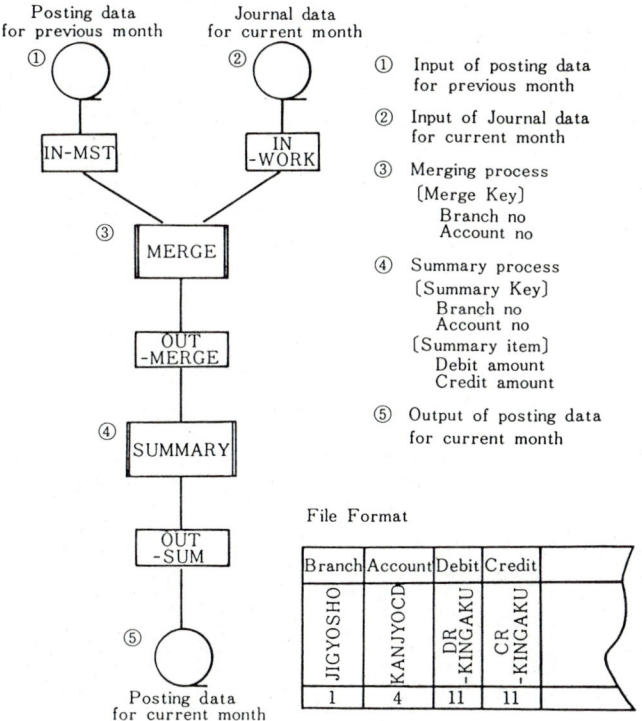

(2) DF-Language Coding

```
¥ID         PGM=SAMPLE, AUTHOR=T. YMADA,
            REMARKS="…SAMPLE…"
¥DA         IN-MST, FORMAT=KAIKEI
¥DA         IN-WORK, FORMAT=KAIKEI
¥DA         OUT-MERGE, FORMAT=KAIKEI
¥DA         OUT-SUM, FORMAT=KAIKEI
¥INPUT      ZMAST, DA=IN-MST, LENG=100
¥INPUT      TWORK, DA=IN-WORK, LENG=100
¥MERGE      MERGE, INDAL=IN-MST, INDAR=IN-WORK,
            OUTDA=OUT-MERGE
  @MERKEY   JIGYOSHO, KANJYOCD
¥SUMMARY    SUM, INDA=OUT-MERGE, OUTDA=OUT-SUM
  @SUMKEY   JIGYOSHO, KANJYOCD
  @SUMITEM  DR-KINGAKU, CR-KINGAKU
¥OUTPUT     TMAST, DA=OUT-SUM, LENG=100
```

Fig.9. Example of merging and summarizing process

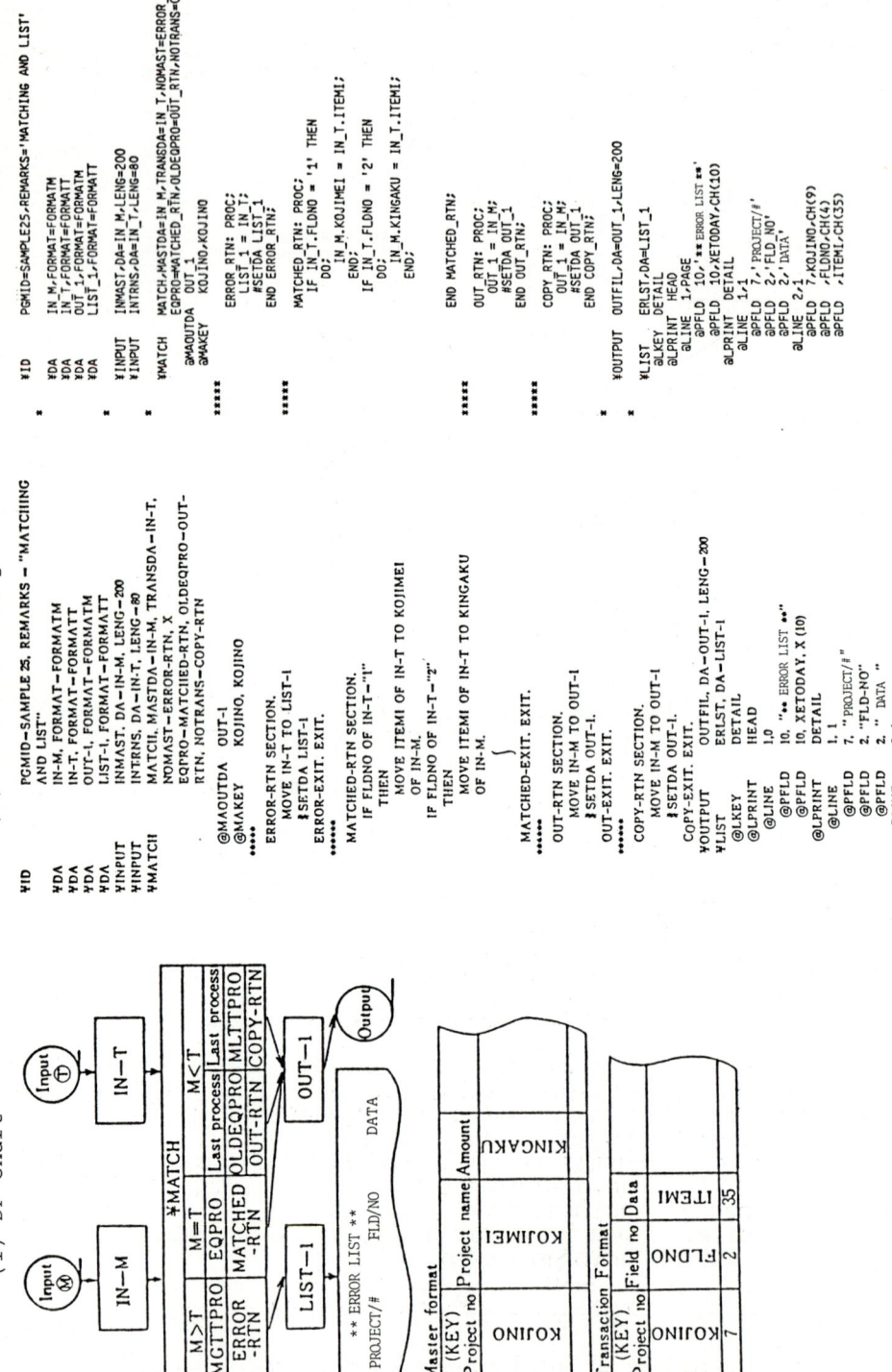

Fig.10. Example of reporting process

Table 2. Comparison of number of DF-COBOL statements with number of COBOL statements

Process	DF-COBOL		COBOL	
	Pages of program specifications	Number of statements	Pages of program specifications	Number of statements
Total check	4	183	6	550
Input check	6	392	10	1310
Data generation	6	295	10	495
Data correction	5	420	10	1001
Data conversion	5	325	12	1113
Merging process	4	204	10	550
Matching process	7	353	11	618
Matching process	6	200	8	377
Detail report	3	168	15	801
Detail report	3	166	18	774
Summary report	3	123	11	585
Total	52	2829	121	8174
Average	4.7	257	11	743

CONCLUSION

The advantages and disadvantages of the DF language are as follows:
1) Data flow expression of processing contents standardizes specifications and reduces the pages of specificaitons
2) Even an inexperienced programmer can easily learn the programming with the DF language
3) The number of coding steps is about one-third of that with conventional compiler languages
4) Program development productivity is two to three times higher than with conventional compiler languages
5) Modular programming simplifies and standardizes programming logic and makes the maintenance easier
6) Because the DF language is a precompiler, the period of time to generate the compiler language and compiling time for generated programs become longer than with conventional languages
7) The programs generated using the DF language generally have a longer structure than those coded by a compiler language; the execution time is increased by 10% - 20%.

ACKNOWLEDGMENTS

We express our gratitude to Prof. Tosiyasu L. Kunii of the Faculty of Science, The University of Tokyo, and Mr. Hideki Myoko for their suggestions in making this report.

REFERENCES

1. K.P. Gostelow and R.E. Thomas: "A View of Data Flow," AFIPS Conf. Proc., Vol.48, 1979 NCC, AFIPS Press, Reston, V.A., pp.626-636.

2. James Martin: "Application Development without Programmers," Prentice-Hall, Englewood Cliffs, N.J., 1978.

3. Tadao Ezaki:"A Macro Language," Technical Report of Mitsui Enginerring & Shipbuilding, No.8, Tokyo, 1976.

4. Hideki Myoko, Hidenobu Ishida and Michio Ohnishi: "DF-COBOL," IBM Review, No.79, Tokyo, 1980.

A Document Management System

Yahiko Kambayashi
Kyushu University

In this paper, a document management system under development is discussed. Besides conventional relational database functions and word processing capabilities (a spell checker, a word index generator), the following functions are required.

(1) History of documents, including merging and splitting of documents, must be handled.
(2) The correspondences between two consecutive documents can be detected, so that the influence of changes on one document to the other can be calculated.
(3) Since there may be a lot of similar documents, data compression facility is required.
(4) Powerful retrieval functions as well as a user friendly interface are required. Especially, handling of time and versions is important.
(5) Besides retrieval functions, we need functions to identify required documents among a set of candidate documents.
(6) In order to handle related documents, a new authorization mechanism must be developed.

The major controbution of this paper is to introduce the following concepts.
(a) A graph structure to represent a document history is used for the purpose of handling (1) and (4). There are database systems proposed so far which can handle time, but they cannot handle merging and splitting of attributes.
(b) For (2) and (3), an editor is proposed which has a capability to detect correspondences between documents as well as a data compression capability.
(c) For (5) an overview function of documents is introduced.
(d) For (6) a new authorization mechanism is developed to handle a set of documents including closely related ones, since conventional authorization mechanisms handle only independent data items.

INTRODUCTION

Systems handling large amount of data are usually classified into three categories, information retrieval systems, database management systems and transaction processing systems. One of the major objectives of database management systems is to realize efficient sharing of a large amount of time-varying business data by many users. Due to the recent expansion of database application areas, various kinds of new database management systems are proposed, such as personalized database systems, multi-media database systems, CAD database systems etc. In this paper, we will discuss functions of document management systems together with implementation techniques. Such a system can be also used for version management of programs as well as their manuals.

For the purpose of managing versions, we need to handle time intervals showing the effective periods of documents. There are papers on historical databases [GINST8408, GINST8505, KLOP8310, LUM8406, LUM8505, SNOD8404]. As time can be expressed by a one-dimensional line from the past to the present (see Fig. 1 (a)), history handled by these papers is represented by a combination of a value and the time interval when the value is effective (or the starting time only, when the value is currently effective). Even if there is a schema change, there is a correspondences of attributes before and after the change. Many kinds of history can be expressed by such models, like job history, address history. There are, however, history which cannot be expressed by a one-dimensional line. For example, for history of a company, we must consider the cases of splitting and merging companies. Such history must be expressed by a graph shown in Fig. 1 (b). Besides each person's job history, we must store company history.

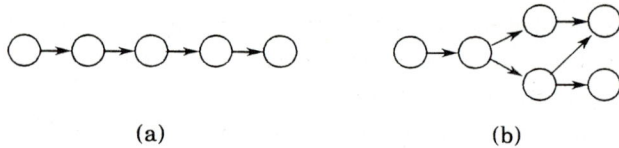

(a)   (b)

Fig. 1 — Representation of history

Such a graph structure is necessary to manage versions in CAD systems, programs and various kinds of documents. Thus a historical database system which can handle such a graph is very important, but known systems only handle structures shown in Fig. 1 (a). For simplicity we will discuss a document management system in this paper, as a special case of a historical database system handling graph-type history. Development of formal data models to handle such structures is a future research topic. In such a document management system, we need to handle one history graph for documents in each family. In general, if a document consists of relations, figures etc., besides sentences, history of attributes, relation names, figures should be handled separately. Furthermore, each chapter or each section of a document can have independent history. The history of the whole document consists of these histories, which becomes very much complicated. By extending techniques similar to ones developed in this paper, such complicated cases can be also handled .

In conventional database systems, data items are considered to be independent. To handle mutually related data items, the following problems must be handled.

(1) Correction of one document may cause correction of succeeding documents. We need to identify the places required to be corrected in the succeeding documents.

(2) Since there are similarity among documents, a data compression method without much sacrificing retrieving speed is required.

(3) Identification aids of a required document, especially among documents in the same

family (i.e., in the same history graph) need to be developed.

(4) An authorization mechanism considering mutually related data items is necessary.

In the next section an outline of the system is discussed. An editor for the system, user interface and an authorization mechanism are discussed in the succeeding sections.

OUTLINE OF A DOCUMENT MANAGEMENT SYSTEM

Fig. 2 shows the organization of the proposed document management system. Information of documents which is used to identify required documents is stored in a relational database system. As the size of a document is variable, a separate file to store documents is prepared, which is called a document file. In order to handle documents with identical subcontents efficiently, data compression techniques must be used. Data compression information is also stored in the document file. By the reason discussed later, data compression is realized by the document editor while editing documents. In order to handle data history shown in Fig. 1 (b), we need to develop a powerful query language which is an extension of a conventional relational database language.

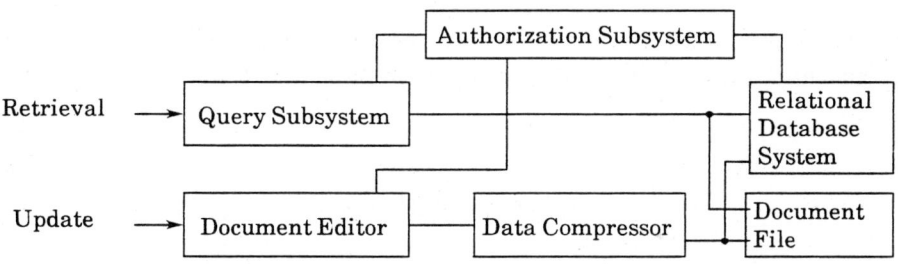

Fig. 2 — Organization of a document management system

The following functions supported by conventional powerful editors should be supported by the editor of a document management system.

(a) UNDO, REDO functions for editor instructions.
(b) Screen editor.
(c) Utilities such as a spell checker, a word index generator (see, for example, writer's work bench[MACDF8201]).

The following additional functions are considered in this paper.

(d) Correspondences between the document before editing and the one after editing are obtained, so that the effect of corrections of the former document to the latter one can be calculated.
(e) Similarity information between the two documents are obtained in order to compress data.

(f) Overview functions to be discussed later can be implemented easily using the data structure of the editor.

There are the following two typical methods to compute the similarity between the two documents.

[Method 1] Comparison of the two documents to find identical subsequences (see for example, [KAMBN8105]).
[Method 2] Usage of editor instruction sequences, which show logical correpondences.

Method 1 uses a program which will find identical subsequences in the two documents. Method 2 calculates the difference from the sequence of editor instructions. We selected Method 2 by the following reasons.

(1) Method 1 requires computation time after editing, but Method 2 can generate information during the editing process.
(2) Logical correspondences and correspondences determined by identical subsequences are usually different. If the original document has more than one identical subsequence, the correspondence cannot be determined uniquely by Method 1.
(3) Logical correspondences are very important to compress data when splitting and merging of documents are permitted.

For the purpose of explaining (3), consider the case shown in Fig. 3. From document D_1, two documents D_2 and D_3 are generated, and from these documents, D_4 is generated. A part

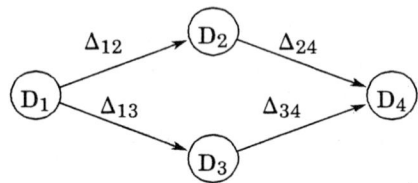

Fig. 3 — Splitting and merging of documents

of D_4 is generated from D_2 and the other part of D_4 is generated from D_3.

Method 2 has advantages over Method 1 for case (3), in the following three points.

[Computation of Δ_{24} and Δ_{34}] By Method 1, correspondence between D_2 and D_4, and correspondence between D_3 and D_4, are both separately computed. In order to obtain the best correspondence (i.e., the amount of information to store differences is minimized), we need to solve a minimum cover problem which is known to be very hard to solve (NP-complete), to determine whether each part of D_4 must be covered by a part of D_2 or a part of D_3. Thus Method 2 is better than Method 1.

[Computation of composition] Consider the the case when D_2 and D_3 are not necessary any more although D_1 and D_4 are still necessary. We must compute the differences between

D_1 and D_4. Using logical correspondences between D_1 and D_2 and these between D_1 and D_3, together with these between D_2 and D_4 and these between D_3 and D_4, the logical correspondence between D_1 and D_4 can be computed efficiently by the second method. If the first method is used, we need to compute the difference between D_1 and D_4 from the beginning.

[Computation of influence of a change] Required changes on D_2, D_3 and D_4 caused by changes on D_1 can be computed, if logical correspondences obtained by Method 2 is used.

The query subsystem must have a facility to identify required documents among candidate documents obtained by the relational database system. It must include facility to handle a history graph and overview functions.

The authorization subsystem must handle, for example, the problem to determine an appropriate user set for a document newly generated by two documents with different user sets.

DOCUMENT EDITOR AND ITS DATA STRUCTURE

Let a set of documents be $\{D_1, D_2, \ldots, D_n\}$. First we assume that the document history is expressed without merging (this structure is shown in Fig. 1 (a)). The difference information Δ_{ij} is used to generate D_j from D_i.

$$D_j = f(D_i, \Delta_{ij}) \qquad (i)$$

Note that the expression for Δ_{ij} is not unique. Thus if there are Δ_{ij} and Δ'_{ij} satisfying

$$f(D_i, \Delta_{ij}) = f(D_i, \Delta'_{ij}) \qquad (ii)$$

Δ_{ij} and Δ'_{ij} are equivalent with respect to D_i, and it is simply denoted by (D_i can be omitted)

$$\Delta_{ij} = \Delta'_{ij} \qquad (iii)$$

We can define a composition of difference information.

$$D_j = f(D_i, \Delta_{ij}) \qquad (i)$$
$$D_k = f(D_j, \Delta_{jk}) \qquad (iv)$$
$$D_k = f(D_j, \Delta_{jk}) = f(f(D_i, \Delta_{ij}), \Delta_{jk}) \equiv f(D_i, \Delta_{ij} * \Delta_{jk}) \qquad (v)$$

Thus we define a composition * of Δ's as follows.

$$\Delta_{ik} = \Delta_{ij} * \Delta_{jk} \qquad (vi)$$

Let k and j be identical in (vi), we get

$$\Delta_{ii} = \Delta_{ij} * \Delta_{ji}.$$

Since Δ_{ii} is empty we use the following notation.

$$\Delta_{ij} = \Delta_{ji}^{-1} \qquad (vii)$$

When merging of documents is involved, formal expressions become complicated. For example, in Fig. 3, only a part of D_4 is generated from D_2 and Δ_{24}. Besides D_2, Δ_{24}, D_3 and Δ_{34}, we need information how to combine $f(D_2, \Delta_{24})$ and $f(D_3, \Delta_{34})$. Thus D_4 can be expressed by

$$D_4 = g(f(D_2, \Delta_{24}), f(D_3, \Delta_{34}), \Delta_4) \qquad (viii)$$

where Δ_4 is such additional information.

The editor generates Δ_{ij} when D_j is obtained from D_i. Δ_{ij} must have the following property.

(1) As discussed in the previous selection, logical correspondence between D_i and D_j must be shown.

(2) Instead of storing both D_i and D_j, we can store either D_i and Δ_{ij}, or D_j and Δ_{ji}, for the purpose of reducing storage space. If D_j is accessed more frequently than D_i, it is better to store D_j and Δ_{ji}, in order to shorten the time to get D_j. Thus $\Delta_{ij}^{-1} = \Delta_{ji}$ must be obtained easily, since the system generates Δ_{ij}.

(3) Intermediate version may be erased. For example, if D_j is not required when (i) and (iv) are satisfied, we need to compute (vi). Thus composition of difference information must be calculated easily. Furthermore, if merging is involved as shown in Fig. 3, we need to compute Δ_{14} from Δ_{12}, Δ_{13}, Δ_{24}, Δ_{34} and Δ_4 in $(viii)$.

(4) There should be a method to reduce redundancy, since Δ_{ij}^{-1} and $\Delta_{ij}*\Delta_{jk}$ may have some redundancy.

If Δ_{ij} is obtained from editor's instruction sequences, it may also have redundancy in the following cases.

(a) A modified subsequence is modified again.
(b) An inserted subsequence is deleted or modified.
(c) A modified subsequence is deleted.

By the above discussion we need to realize the following functions.

(1) Computation of Δ_{ij}.
(2) Simplification of Δ_{ij}.
(3) Computation of Δ_{ij}^{-1}, when Δ_{ij} is given.
(4) Computation of $\Delta_{ij}*\Delta_{jk}$, when both Δ_{ij} and Δ_{jk} are given.
(5) Expression capability of document merging.

Fig. 4 shows the proposed data structure for the editor. It is a line-oriented editor and each line of D_i and D_j has an identification number. A word-oriented editor can be implemented similarly. D_j is generated by removing line 2 (bbbb) and inserting a new line eeee. Lines not in D_i (generated by modification or insertion) are stored in file Δ_{ij} by the order of generation. In order to distinguish the line number of D_i, we use notation $-h$ (negative integer). Δ_{ij} consist of two parts. One part stores sentences newly generated by modification and insertion. The second part shows the contents of D_j by a sequence of line numbers. By a similar method we can generate Δ_{ji} using line numbers of D_j. Instead of computing Δ_{ji} from Δ_{ij}, the editor generates the both during the editing process. Unnecessary one is deleted later.

Using the above data structure, redundancy caused by editing process can be detected. Composition of Δ's is also straightforward. To handle document merging, we use a vector

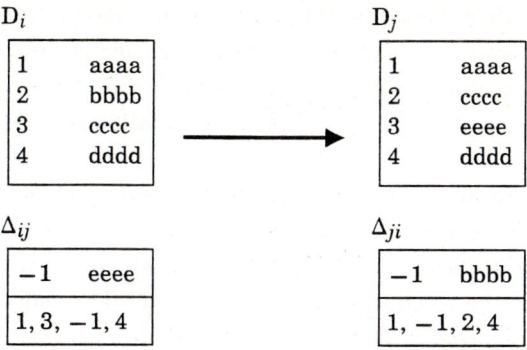

Fig. 4 — Documents and difference information

which shows origin of each line. If the i-th value of Δ_4 is j, it shows that i-th line of D_4 is obtained from D_j. By this way, computation of $\Delta_{12}*\Delta_{24}$ and that of $\Delta_{13}*\Delta_{34}$ can be done independently. Data compression of the vector is possible using run-length coding etc.

USER INTERFACE

The user interface consists of the following functions.

(1) Registration of a new version, and deletion from registration.
(2) Modification is handled in three different ways.
   (2-a) Modification of a currently registered version is done and only the version after the modification is required to be registered.
   (2-b) Generation of a new version.
   (2-c) The corresponding part of the successor documents must be modified.
(3) Besides retrieval function using metadata of documents stored in the relational databases, we need functions to retrieve documents using history graphs.
(4) As there may be similar documents, retrieval by comparisons (overview functions) is also important.

Methods for (1) and (2-b) are discussed in the previous section. The situation of (2-a) is shown in Fig. 5. If D_2' which is obtained from D_2 is required to be registered instead of D_2, we

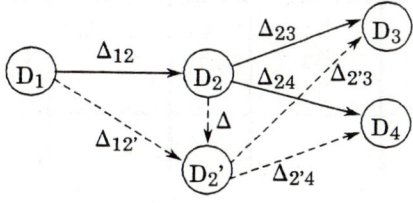

Fig. 5 — Modification of one version

need to calculate difference information shown by dotted lines. It can be done by computing Δ^{-1} and compositions of Δ's. For (2-c), consider the example shown in Fig. 4. If line 1 of D_i is modified, line 1 of D_j is automatically modified without changing Δ_{ij}. In this case we need a facility to check the effect of modification in order to avoid unnecessary modification.

In one connected component of a history graph, we need to store at least one document in the original form. For selecting such documents, we have to handle a trade-off problem between time and storage space. For example, consider the documents D_1, D_2, D_3 and D_4 in Fig. 5. If D_3 is stored in the original form, retrieving D_4 takes long time since computation using D_3, Δ_{23}, Δ_{24}, is required. If both D_3 and D_4 are accessed frequently, storing the both documents may be appropriate if storage space is not a problem. If we cannot store two documents in original form, to store D_2 in the original form may be the best way, since both D_3 and D_4 can be computed easily.

If retrieval frequency f_i for each document D_i and computation time t_{ij} for $f(D_i, \Delta_{ij})$ are given, we can compute average time for retrieval as well as total storage space (computed from the sizes of D_i's and Δ_{ij}'s), when documents to be stored in the original form are selected. We can implement an automatic tuning function which automatically determines which documents should be stored in the original form. This method requires large amount of computation time, thus for simplicity documents without successors can be selected, since such documents are the newest versions and considered to be retrieved more frequently than others. If the number of the newest versions with the same predecessor exceeds some predetermined threshold value, the predecessor is selected instead of these versions. In this case, computation of difference information during the computation of new version is rather easy. If more than one document is stored in the original form in one connected graph, we need a path selection method to determine the path of the shortest computation time.

A document is retrieved by the following steps.

(1) Using metadata and history graphs candidates are selected.

(2) By comparing candidate documents, proper documents are selected.

An example of metadata for documents is shown in Fig. 6. For each document (in this case document ID=4) various information is stored. Sets of direct predecessors and direct

<table>
<tr><th>ID</th><th colspan="3">Direct Predecessors</th><th colspan="2">Direct Successors</th><th>Keywords</th><th>Gen. Date</th><th>Effective Periods</th><th>Prepared by</th><th>Right to Modify</th><th>Right to Read</th></tr>
<tr><td rowspan="2">4</td><td>2</td><td>Δ_{24}</td><td rowspan="2">Δ_4</td><td>5</td><td>Δ_{45}</td><td rowspan="2">K_1
K_2
K_3</td><td rowspan="2">84.12</td><td>85.1-85.3</td><td rowspan="2">Yamada</td><td>Yamada</td><td rowspan="2">All except Okamoto</td></tr>
<tr><td>3</td><td>Δ_{34}</td><td>6</td><td>Δ_{46}</td><td>85.8-</td><td>Yoshida</td></tr>
<tr><td></td><td></td><td></td><td></td><td></td><td></td><td></td><td></td><td></td><td></td><td></td><td></td></tr>
</table>

Fig. 6 – An example of a metadata relation for documents

(*viii*) of the previous section if there is more than one predecessor we need to store merging information (in this example, Δ_4). A set of keywords, the generation date and the effective periods are shown. Authorization information is also stored. A person who has a right to modify the document is also considered to have a right to modify metadata for the document.

For users, unnormalized form (relations with set values) is preferred as shown in Fig. 6 [KITAK8111, KITAK8201, SHUL8209, KAMBT83]. There are the following two methods to get unnormalized relations using conventional relational database systems.

(1) First get a normalized relation from the system and convert it into an unnormalized relation.
(2) Decompose the given query into subqueries and obtain normalized relations for these subqueries from the system. Compute unnormalized relations using these relations.

Although the second method is rather complicated, we can reduce the computation time if there are no update operations during the execution of these subqueries. Using the method discussed in [KAMBT83], we have the following method.

(1) From the given query and dependencies satisfied by each base relations, compute dependencies satisfied by the result of the query.
(2) For a join dependency, decompose the query into subqueries, since a normalized relation satisfying a join dependency has redundancy. We will get the result in decomposed form using the join dependency.
(3) If each decomposed relation satisfies one or more more functional dependency, apply appropriate group-by operations, each of which produces a horizontally decomposed relation. Here, each subrelation has identical values for the set of attributes specified by the group-by operation.
(4) These results are merged to obtain a relation in an unnormalized form.

In order to retrieve proper documents, we need to develop a query language which can handle history shown by a graph (Fig. 1 (b)). A query language for a simple case (corresponding to Fig. 1 (a)) was developed by [SNOD8404]. Since a history graph corresponds to a binary relation, a language to handle binary relations shown in [KAMBT7710] is also related to the language design. Together with the query capability, we need a facility to cope with the graph structure in an multi-window environment. Detailed discussion on the query language is omitted here.

The following two functions can be used to compare documents.

(1) By a multi-window function two documents are shown and by moving a cursor for one document, a cursor for the other document moves automatically to show the corresponding places.
(2) Comparison of more than one document using an overview function.

If a document appears serially, it is not easy to compare. We need to show documents as much as possible for efficient comparison.

An overview function is realized by showing following information.

(1) Section organization.
(2) Users can specify words and subsentences, sections containing them are shown with these words and subsentences.
(3) By specifying a predecessor (predecessors, a successor or successors), the parts containing specified words or subsentences are shown.

Fig.7 shows an application of an overview function to this paper by specifying section names and figure names. These functions can be easily implemented by the data structure used in the document editor together with a partial matching functions.

AUTHORIZATION MECHANISM

Among various kinds of authorization mechanisms, the mechanism used by System R is considered to be the most advanced one, since it is decentralized (no security officer is required) and dynamic (assignment of access rights can be altered). As the contents of a database change, it is not appropriate to use fixed (assignments of access rights cannot be changed) mechanism. This authorization mechanism was proposed by Griffiths and Wade [GRIFW7609] and Fagin showed a correct algorithm for it [FAGI7809]. Thus we call the mechanism a GWF-mechanism. For a document management system we need to develop a new authorization mechanism to handle mutually related data items, where conventional authorization mechanisms handle only mutually independent data items.
The key ideas of the GWF mechanism are as follows.

(1) Any user may be authorized to create a new relation.

(2) If the user wishes to share his relation with other users, he may use the GRANT command to give various privileges on that relation to various users. A GRANT command is shown by

    GRANT (operations) **on**   (relation)   **to** (user's names)

Examples of these privileges are READ, INSERT, DELETE, UPDATE, INDEX, EXPAND, ALL PRIVILEGES etc.

(3) The user may grant a set of privileges with the grant option, which permits the grantee to further grant his acquired rights to other users.

    GRANT (operations) **on**   (relation)   **to** (user's names) **with GRANT OPTION**

(4) Any user who has granted a privilege may subsequently withdraw it by issuing the REVOKE command,

    REVOKE (operations) **on**   (relation)   **from** (user's names)

In order to make algorithm work correctly in the GWF mechanism, the same grantor has to issue GRANT command to the same recipient repeatedly[FAGI7809], which makes the graph to analyze the effect of a REVOKE command complicated. To avoid this problem we assume that the meaning of GRANT command is "the recipient has the privilege whenever the grantor has it" (algorithms and other generalization, see [KAMB8010]).

INTRODUCTION

        Fig. 1   Representation of history

OUTLINE OF A DOCUMENT MANAGEMENT SYSTEM

        Fig. 2   Organization of a document management system

        Fig. 3   Splitting and merging of documents

DOCUMENT EDITOR AND ITS DATA STRUCTURE

        Fig. 4   Documents and difference information

USER INTERFACE

        Fig. 5   Modification of one version

        Fig. 6   An example of a metadata relation for documents

        Fig. 7   Application of an overview function

AUTHORIZATION MECHANISM

CONCLUDING REMARKS

        Fig. 7   —   Application of an overview function

For document management system we need to modify the GWF mechanism, since there are two different data items, metadata and documents, and there are related data items (documents in the same connected history graph).

(1) Privileges for a document and its metadata should be related.

    A user who has a right to modify metadata should have the right to modify the corresponding document.

    Reading of metadata is necessary to identify a document, so a user who has the right to read a document should have the right to read the corresponding metadata.

    Some user can read metadata for searching, but the corresponding documents are not permitted to read.

(2) When a new document is prepared using existing documents by a user, the user must have the right to read and modify the documents and their corresponding metadata. The privileges for the new document can be determined from privileges for these documents.

    If a new document is generated by modifying one document, the privileges for the original document should be transferred.

    If new documents are generated by splitting one document, the privileges for the original document should be transferred.

    If more than one document is merged, the use set for each privilege for the new document is determined by an intersection of the user sets of original documents.

(3) When an error of document is corrected, the effect of the correction will propagate to succeeding documents. The user who corrected the error may not have the right to modify some of these documents. In such a case, the correction is not done automatically. It is performed when one of the users who have the right to modify the document permits the correction.

CONCLUDING REMARKS

In this paper we discussed functions and some implementation techniques of a document management system. Such a system can be considered to be a generalization of a writer's workbench and various kinds of editors. Document management systems are different from conventional database management systems by the following points.

(1) History of documents are handled.
(2) Similar data are stored, so that we need to develop functions to retrieve using similarity, data compression techniques and new authorization mechanisms.
(3) Since each data item consists of more than one word, we need facilities to see inside structure, for such a purpose string matching functions are not enough.

As there are many data having these characteristics, we believe that this kind of systems will be used widely. Development of a historical database system which is a generalization of the document management system as well as specification of a data model handling graph-structure history are future important research topics.

Acknowledgment

The author would like to thank to Miss Hai-Yan Xu and Mr. Hidetoshi Yoshimura (now at Kyushu Matsushita Elec. Co.) who are working to implement a document management system discussed in this paper. The author also express his appreciation to Mr. Keizou Saisho for helping the preparation of the final version of this paper. The work is partially supported by a grant of the Ministry of Science and Culture of Japan.

REFERENCES

[CLIFW83]   J. Clifford and D. Warren, "Formal Semantics for Time in Databases," ACM TODS, Vol. 8, No. 2, pp. 214–254, June 1983.

[FAGI7809] R. Fagin, "On an Authorization Mechanism," ACM TODS, Vol 3, No. 3, pp. 310–319, Sep. 1978.

[GINST8408] S. Ginsburg and K. Tanaka, "Interval Queries on Object Histories : Extended Abstract," Proc. 10th International Conference on VLDB, pp. 208–217, Aug. 1984.

[GINST8505]   S. Ginsburg and K. Tanaka, "Computation-tuple Sequence and Object Histories : Extended Abstract," Proc. International Conference on Foundations of Data Organization, Kyoto, Japan, May 1985.

[GRIFW7609]   P. P. Griffiths and B. W. Wade, "An Authorization Mechanism for a Relational Database System," ACM TODS, Vol. 1, No. 3, pp. 242–255, Sep. 1976.

[KAMBT7710]   Y. Kambayashi, K. Tanaka and S.Yajima, "A Relational Data Language with Simplified Binary Relation Handling Capability," Proc. 3rd VLDB, pp. 338-350, Oct. 1977.

[KAMB8010]   Y. Kambayashi, "Generalized Dynamic Authorization Mechanisms," Operating Systems Engineering (Lecture Notes in Computer Science 143), pp. 63–77, Springer-Verlag, Oct. 1980.

[KAMBN8105]   Y. Kambayashi, N. Nakatsu and S. Yajima, "Data Compression Procedures Utilizing the Similarity of Data," AFIPS National Computer Conference, Vol. 50, pp. 555–562, May 1981.

[KAMBT83]   Y. Kambayashi, K. Tanaka and K. Takeda, "Synthesis of Unnormalized Relations Incorporating More Meaning," Information Science, Vol. 29, pp. 201–247, 1983.

[KITAK8111]   H.Kitagawa, T.L.Kunii and Y.Ishii, "Design and Implementation of a Form Management System APAD Using ADABAS/INQ DBMS", IEEE COMPSAC, pp.324-334, Nov. 1981.

[KITAK8201]   H.Kitagawa and T.L.Kunii, "Form Transformer - Formal Aspect of Table Nests Manipulation", Proc. Hawaii Conf., pp.132-141, Jan. 1982.

[KLOPL8310]   M. R. Klopprogge and P. G. Lockemann, "Modeling Information Preserving Databases : Consequences of the Concept of Time," Proc. 9th International Conference on VLDB, pp.399–417, Oct. 1983.

[LUM 8405]   V. Lum, P.Dadam, R.Erbe, J.Guenauer, P.Pistor, G.Walch, H.Werner and J. Woodfill, "Designing DBMS Support for the Temporal Dimension," Proc. of SIGMOD International Conference on Management of Data, pp. 115–130, June 1984.

[LUM 8505]   V.Lum et. al., "Design of an Intergrated DBMS to Support Advanced Applications," Proc. Foundation of Data Organization, pp.21-31, May 1985.

[MACDF8201]   N.H.MacDonald, L.T.Frase, P.S.Gingrich and S.A.Keenan,"The writer's workbench : Computer Aids for Text Analysis," IEEE Trans. Vol. COM-30, No.1, pp.105-110, Jan. 1982.

[SHUL8209]   N.C.Shu, V.Y.Lum, F.C.Tung and C.L.Chang, "Specification of Forms Processing and Business Procedures for Office Automation," IEEE Trans. Vol.SE-8, No.5, pp.499-512, Sept. 1982.

[SNOD8404]   R. Snodgrass, "The Temporal Query Language TQuel," Proc. ACM SIGACT-SIGMOD International Symposium on PODS, pp. 204–213, Apr. 1984.

An Overview of CrossoverNet
A Local Area Network for
Integrating Distributed Audio Visual and Computer Systems

Tosiyasu L. Kunii
Yukari Shirota
The University of Tokyo

Abstract

CrossoverNet is a new type of local area network for integrated management of audio visual and computer facilities. It has been already over two years since CrossoverNet was installed on a university campus to aid actual class room activities. Both for scheduled and interactive use of the facility, CrossoverNet has been proven to be effective and user-friendly. Especially, audio visual devices, which are designed mainly for use independent of one another, have been integrated into computer workstations and their network to increase their controllability remarkably. CrossoverNet has further enhanced its controllability by an interactive and user-friendly iconic menu program through which a user issues commands to control the devices. On the other hand, these menus also have burdened programmers with extra time and labor because they were all custom-made. To solve this problem, we have designed and implemented a prototyping system called InteractiveProto that aids an end-user to develop a menu himself/herself without programmer's help. Another technological key point of InteractiveProto is that it helps a user to develop a prototype menu using a visual or graphical programming language instead of a traditional textual programming language. The manner InteractiveProto adopts can be applied to wide areas beyond the control of audio visual devices.

1. Introduction

The use of audio visual (AV) devices as new media tools increases the effectiveness of communication in an office, a home, a plant and an educational environment. For example, in a university campus a number of AV devices are usually distributed over several classrooms and auditoriums. But most of the devices are used independently of one another. To use them more effectively, an integrated and improved AV control system is required.

After extensive analysis of the problems of current AV control systems, we find that they can be summarized as the following nine items:

   1. AV devices located in a remote site cannot be controlled;
   2. The status of the remote devices such as whether a switch is on or off, what the level of a volume dial is, and which channel is selected cannot be sensed;
   3. A user must vary hard-wired connections among devices according to his or her desired tasks to be performed; rewiring work bothers the user, and what is still worse, a number of AV cables and lines are a mess behind the devices;
   4. It is difficult to integrate and coordinate multiple AV devices;

5. Complicated scheduling of AV device control is impossible;
6. Control programs are device-dependent and cannot be generalized as software packages;
7. A beginner cannot simply decide what to do to accomplish his or her desired task, which puts a lot of stress on the user (so-called an allergy to computers);
8. In each division of devices, different experts in handling those devices are required; therefore, personnel expenses are mounting;
9. A user sometimes makes a mistake absent-mindedly while he or she controls the system.

To overcome these problems, we have designed and developed a new type of network operating system for a local area network (LAN) system to control AV devices effectively. The system is named CrossoverNet [KUN85a, KUN85b and KUN86e]. To develop the CrossoverNet system, we utilized the following technology:
* a broadband LAN,
* computer generation of icons and menus,
* device-independent software design/development, and
# a database management system.

The first and second problems are solved by connecting AV devices by a LAN. Even the devices located in the different buildings can be controlled through the LAN, and the user can monitor the device status via the network. Especially, by using a broadband LAN, only one coaxial cable may accomplish several different types of communication services (voice, video and data) simultaneously, which decreases the number of AV cables.

The third through fifth problems are solved by computer controlling. When the user sends information through coaxial cables and optical fibers, channel selection and communication path selection are done by computer programs, not by manual hard-wiring. CrossoverNet handles channels and communication paths simply as kinds of hardware resources. CrossoverNet, therefore, dynamically changes them depending on his or her needs; rewiring work is unnecessary. CrossoverNet solves the fourth problem by the job oriented control (See section 2). As for control scheduling, control information of all the devices connected to the LAN can be programmed, stored and executed to drive devices automatically.

The sixth problem is solved by device-independent design of control software. To do this, a LogicalLayer and a PhysicalLayer are introduced to the software development. Because the CrossoverNet system holds the diversity of its devices, the database is necessary for effective management of these devices. The database of CrossoverNet holds the data of each device model such as
* physical tasks that the device has to perform (we call those PhysicalDeviceTasks.),
* a procedure by which a LogicalDeviceTask is mapped to PhysicalDeviceTask(s).
Such data as
* an inventory list of all the devices in the CrossoverNet system,
* a detailed inventory of all the repairs to be done,
* logging data on replacements of consumption goods (e.g., a light bulk and an electric battery for backup),
are also stored in the database.

The seventh through ninth problems are on user interface. One of the most important factors in judging whether a device control system is easy to handle or not, is the time required to accomplish a user's desired job. In CrossoverNet, to reduce the time, we do not let an end-user use the same interface program, but a different program of which contents match the individual application requirement specifications. To realize it, we provided the users a computer-aided prototyping system that helps them develop, by themselves, menu programs to their liking.

When an administrator handles a large-scaled system like CrossoverNet, an accident might happen because of human errors. For example, he or she may misjudge the situation because of the diversity of its circumstantial data, or may make a mistake absent-mindedly. To prevent such human errors, the system is needed to have pilot facilities by which the system can direct, advise, or guide the user through difficulties. Reducing his or her labor burden by using pilot facilities, leads to great reliability of the system management.

The pilot facilities have the following features and the following databases become required:

1. All the actions a user has to take are prompted on a computer screen or directed by means of a computer-synthesized voice.
   ======> a database of standard job procedures;
2. The system shows an important point or predicts a point on which a user tends to make a mistake. Then, CrossoverNet calls his or her attentions to this place or let the user double-check.
   ======> a database of the past operation errors and accidents,
   ======> a database of a special knowledge on the management of the system;
3. After circumstantial analysis, the system shows a user the data in the form so that he or she can easily understand and judge the situation (e.g., a chart and a short statement).
   ======> a database of analytical materials,
   ======> a database of algorithms/rules for data analysis;
4. The system predicts what will happen if a user executes a certain operation.
   ======> a database of algorithms/rules for prediction;
5. The system advises a user which operation he or she should select among several ones under certain circumstances.
   ======> a database of a special knowledge on the management of the system.

When an error or an accident happens, the above-mentioned third through fifth facilities help the administrator to judge the situation aptly, and let him or her become calm down, not in a panic.

2. Features of CrossoverNet

CrossoverNet has the following features.

(f1) Job Oriented Control

Most of AV devices are designed to be used independently of one another, and how to control them depends on their models as well as their types such as video cassette recorders, video discs, video projectors and record players. Instead, in CrossoverNet end-users can consider these devices at a higher and more abstract level according to their intended use or job. To complete the job, a computer called a manager integrates and coordinates several AV and digital devices on the network. The job is translated by a manager computer from a high level source description to a sequence of concrete device commands at a low level (See section 4.3.2.).

(f2) Digital and Analog Information Control

CrossoverNet handles both digital and analog information. Handling analog information without digitizing improves the efficiency of transfer, storage and sharing of analog images and voice. AV devices that input and output analog information are controlled with digital control information. Table 1 shows information types managed by CrossoverNet.

(f3) Remote Control

In CrossoverNet all the devices are connected by a LAN. Therefore an end-user can control even devices located remotely, for example, in another building (Figure 1). The following key points must be considered to ensure remote control:

1) If a command is sent to a device, the device must respond to it;
2) A manager can read out the status information of the devices under its control.

CrossoverNet has facilities to offer such capabilities. The status of the remote devices, such as
    whether the switch is on or off,
    which video channel is selected,
    what the level of the TV display sound_volume is, and
    how the zoom ratio of the video camera is,
can be directly controlled and monitored through the LAN. Consequently, an end-user can manipulate them without going all the way to where the device is located just to push its buttons by his or her finger. Moreover, because all users can share a device connected by the LAN, the network resource may be used effectively.

Table 1. Information types on CrossoverNet.

control information	information contents
digital * a message for control sent among managers * a message for control sent between a manager and a device(e.g., a status report message from a device to a manager)	a digital data file containing coded characters and digital images and voice
	an analog data file containing uncoded characters, video and voice

Figure 1. In <u>CrossoverNet</u> an end-user can control even devices located remotely, for example, in another building, through a LAN which connects these devices.

(f4) <u>Device-Independent Design</u>

One of the problems of the current AV device control systems is that control programs are device-dependent and cannot be generalized as software packages. In <u>CrossoverNet</u>, this problem is solved by device-independent design of control software. The software consists of the following two layers:

1) <u>LogicalLayer</u>
   The software in the LogicalLayer is designed for simple and logical definition of the functions a user wishes the devices to perform.
2) <u>PhysicalLayer</u>
   The software in the PhysicalLayer maps the above functions onto their corresponding low level descriptions to control the actual or physical devices.

A device in the LogicalLayer/PhysicalLayer is called a LogicalDevice/PhysicalDevice.

Let us look at some examples of the layered architecture to keep the system's architecture independent of the development of physical AV devices. Suppose we have a logical TV display. Physically, this is a usual CRT-based color TV or a modern liquid crystal display TV. It can also be a high quality digital TV that is going to be announced in a market in a few years. Independently of which physical TV display we use, by defining the LogicalTasks of a logical TV in software, these LogicalTasks can be converted to those of actual PhysicalDevices (Figure 2). The status of the remote devices aforementioned is monitored by a manager in the form of LogicalDevice status information. For example, the status of a logical TV display is represented by the combination of the following five items:

      (1) source = { broadcasting, video, RGB };
      (2) channel = { 1, 2, ... , n };

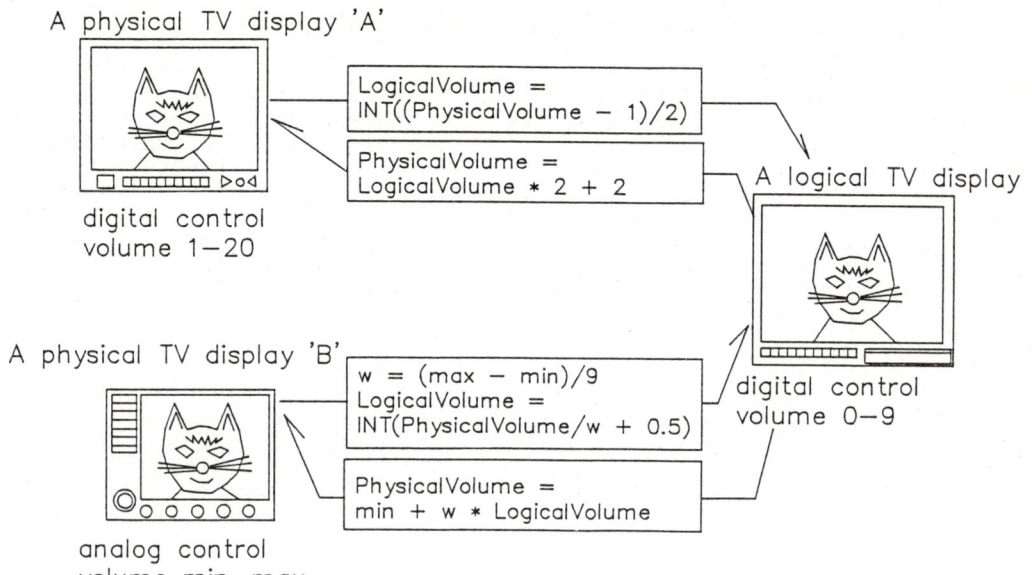

Figure 2. LogicalLayer to PhysicalLayer mapping information for TV displays. Independently of which physical TV display we use, by defining the LogicalTasks of a logical TV display such as the control for volume in software, these LogicalTasks can be converted to those of actual PhysicalDevices.

       n is twelve in the Japanese ordinary broadcasting system.
(3) sound_mode = { stereo, sub, mono };
    In case of a bilingual broadcasting program, the <u>sub</u> sound_mode switches into English from Japanese.
(4) sound_volume = { 0, 1, ... , m };
(5) power = { on, off }.

By incorporating a database management system (DBMS) into the <u>CrossoverNet</u> system, we can add more extensibility to the system by allowing users to easily create new software packages, especially user interfaces. <u>CrossoverNet</u> holds such a DBMS that is capable of storing and handling these LogicalTasks and PhysicalTasks. The LogicalDevice models, containing some information on the LogicalDevices' structures, functions and so on, are readily available in the DBMS of <u>CrossoverNet</u>. Therefore, by using the LogicalDevice to PhysicalDevice mapping information, a user can utilize his or her programs to drive even recently purchased new types of AV devices by changing only portions of them in the PhysicalLayer.

The device-independent design, furthermore, prolongs lifetime of the system beyond those of the individual devices connected to the network. In [COU85] Coutaz describes that abstractions of devices are necessary in designing device-independent and portable application programs.

(f5) User-Friendly Man-Machine Interface Program

CrossoverNet can enhance its controllability by an interactive and user-friendly interface program through which a user issues commands to control devices. In CrossoverNet it is a menu-oriented program called MacroMenu [KUN85c and KUN86d]. A remarkable feature of MacroMenu is that the user can control AV devices without issuing a lengthly command sequence. All that he or she needs to do is to touch an icon on the touch panel equipped on a manager in charge of his or her desired job. MacroMenu takes care of the rest, sending commands to several devices and automatically prompting the next menu. MacroMenu has been very popular among all the levels of users including beginners and experts.

(f6) Pilot Facilities

In advance, we collect the past error/accident data, the device model data, the special knowledge on the management of the system, and so forth, and store those data in the database of CrossoverNet. Then, by using the database, CrossoverNet navigates the administrators and the users through their jobs to avoid any possible trouble.

3. Hierarchical Structure of CrossoverNet

We describe the structure of CrossoverNet in this section. Only the terms and functions at the LogicalLayer are illustrated here, but the same discussions are applicable to those at the PhysicalLayer. A prefix "Logical" ("Physical") is placed at the beginning of a term defined in the LogicalLayer (PhysicalLayer). A symbol used at the LogicalLayer (PhysicalLayer) has a superscription 'L' ('P') in the upper right hand corner.

LogicalObject

CrossoverNet includes a number of devices that communicate through a LAN transmission medium. In CrossoverNet, any device or a group of devices as a functional unit at the LogicalLayer is called a LogicalObject.

CrossoverNet is a tool by which a human organization (e.g., a company and a school) attains its organization wide goals. In general, the human organization consists of a family of interacting, hierarchically arranged, decision-making units[MES70]. We introduce the use of this organization hierarchy to our CrossoverNet so as to accomplish the goals fast and reliably. The LogicalObjects on CrossoverNet are arranged hierarchically as illustrated in Figure 3. There is a strong resemblance between the structures of the two. The CrossoverNet LogicalObjects are grouped into three levels, namely:

* a NetworkLevel;
* a WorkstationLevel;
* a DeviceLevel.

Each level may be further divided into two sub-levels. One sub-level consists of LogicalObjects to perform management functions such as control of the other LogicalObjects and bookkeeping of control information. The other sub-level consists of LogicalObjects to be

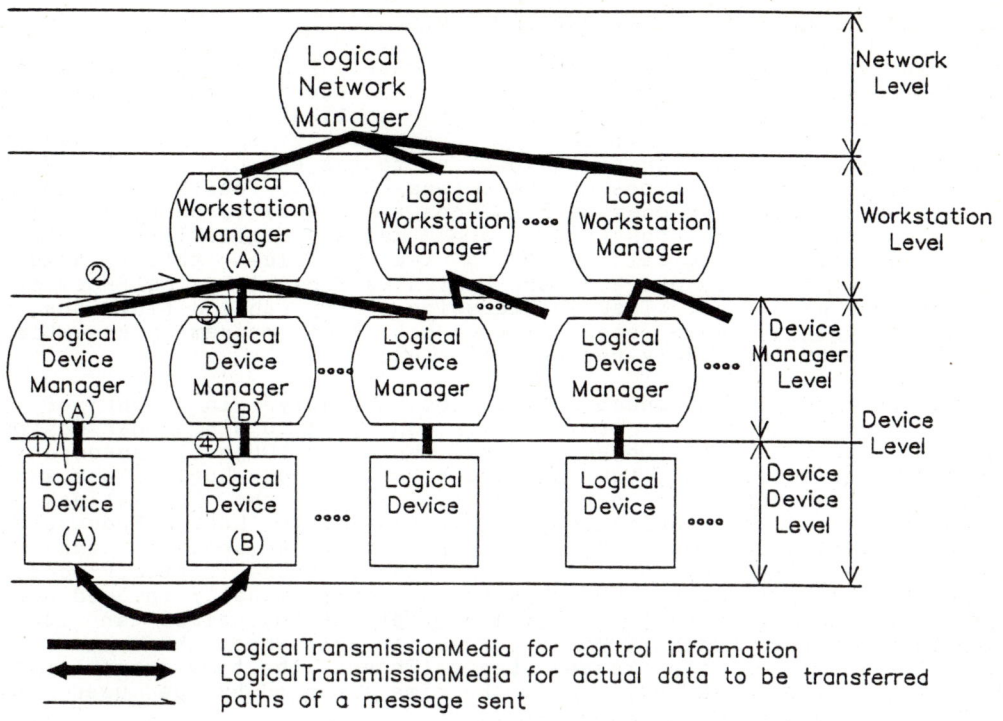

Figure 3. The hierarchy of <u>CrossoverNet</u> LogicalObjects. The set of LogicalObjects consists of LogicalManagers and LogicalDevices.

managed. So we may well say, in fact, that the <u>CrossoverNet</u> LogicalObjects exhibits a six-level structure, instead of a three-level structure.

LogicalObjects are connected by bidirectional logical transmission media. These logical transmission media connect LogicalObjects at the different levels, and construct a LogicalObject hierarchy. Suppose, in Figure 3, the LogicalDevice(A) wishes to send some control information to the LogicalDevice(B). The control information from LogicalDevice(A) must reach LogicalDevice(B) via LogicalDeviceManager(A), LogicalWorkstationManager(A) and LogicalDeviceManager(B). In <u>CrossoverNet</u>, a parent LogicalObject and its child LogicalObject communicate directly. A message from a LogicalObject to another LogicalObject must be sent up to the closest common ancestor LogicalObject first, and then down to the destination LogicalObject. When the message is actually sent, its routing is decided dynamically at the PhysicalLayer level, different from the logical routing.

LogicalJob, LogicalTask and LogicalProcess

To begin with, we define the terms, LogicalJob, LogicalTask and LogicalProcess. An end-user utilizes the LogicalObjects in CrossoverNet so as to carry out his or her application. We call this application to be accomplished a LogicalJob. A LogicalObject is equipped with a lot of functions to offer to the end-users. They may be, depending on the types of the LogicalObjects, facilities inherent in the LogicalObject or command sequences (programs) to perform certain LogicalJobs. These functions are called LogicalTasks of the LogicalObject. A LogicalTask of a LogicalDevice implies a unit operation when a user operates the LogicalDevice. A LogicalProcess is here defined as a LogicalTask in execution. Many other definitions of the term process have been given, but "the program in execution" concept is most frequently referenced[DEI83].

First, an end-user (conceptually, not actually) in front of a LogicalObject, inputs a command corresponding to his or her desired LogicalJob to the LogicalObject. Then, the LogicalObject, accompanied with the execution of the LogicalJob, selects one LogicalTask among many of its own and then executes it. Third, the LogicalTask in execution becomes the LogicalProcess. Moreover, the LogicalProcess may invoke execution of other LogicalProcesses one after another, and may construct a hierarchy of LogicalProcesses; those LogicalProcesses belong to the LogicalJob. Though a LogicalJob is usually invoked by a user, as we have seen, in CrossoverNet the LogicalJob can also be invoked by another LogicalProcess already in existence; the user sets up in advance on the target LogicalObject, the time-scheduling of LogicalJobs to be executed, and at the time arranged by the user, the LogicalJob will be invoked automatically.

A LogicalJob may be, if necessary, divided into groups of sub-LogicalJobs. It is a user that determines the sub-divisions of the LogicalJob; in advance, he or she divides a LogicalJob into sub-LogicalJobs (sub-divisions may be nested,) and allocates them to the LogicalManagers, according to his or her own application. As a result, the LogicalJob on CrossoverNet constructs a hierarchy of LogicalJobs, and the LogicalJobs are grouped into three levels, namely, a network-wide LogicalJob, a workstation-wide LogicalJob and a device-wide LogicalJob.

Mapping from LogicalLayer to PhysicalLayer

Although we have described only the concepts related to the LogicalLayer, the same discussions are applied to the PhysicalLayer; in the PhysicalLayer, the prefix "Logical" and the superscription 'L' are replaced by "Physical" and 'P' respectively. So far, the hierarchies in the LogicalLayer and the hierarchies in the PhysicalLayer have been defined independently.

Here we explain the procedure to map the LogicalLayer into the PhysicalLayer. A LogicalDeviceTask (an operation unit of a LogicalDevice) may be constructed by collecting several PhysicalTasks of several PhysicalObjects in the real world; in general, a LogicalObject may be mapped to several PhysicalObjects. First, we select one LogicalDeviceTask, and assign to the LogicalObject that owns the LogicalDeviceTask, the set of the PhysicalObjects that own the corresponding PhysicalDeviceTasks. Then, each LogicalObject can be mapped to the corresponding PhysicalObjects at any level, from the bottom level (the DeviceDeviceLevel) towards the top level (the

```
    Logical Layer                               Physical Layer
                        Network Level
   LogicalNetworkManager                     PhysicalNetworkManager(s)
              ⇅                                        ⇅
                       Workstation Level
   LogicalWorkstationManager                 PhysicalWorkstationManager(s)
              ⇅                                        ⇅
                      Device Manager Level
   LogicalDeviceManager                      PhysicalDeviceManager(s)
              ⇅                                        ⇅
                       Device Device Level
   LogicalDevice                             PhysicalDevice(s)
              ⇅                                        ⇅
                              map
   LogicalDeviceTask  ─────────────────→     PhysicalDeviceTask(s)
```

Figure 4. The relationship of mapping from LogicalObjects to PhysicalObjects.

NetworkLevel), one level at a time, because there is the hierarchy of the objects in each layer, i.e., the LogicalLayer or the PhysicalLayer (Figure 4). In the same manner, the LogicalProcess hierarchy and the LogicalJob hierarchy are mapped to the related PhysicalProcess and PhysicalJob hierarchies, respectively.

4. Menu Generator

In this section, we show a user-friendly menu program to control AV devices called MacroMenu that is defined as a tree structured menu-oriented system, and its prototyping system InteractiveProto for MacroMenu [KUN85c and KUN86d].

4.1. Requirements

In CrossoverNet, user-friendly menu system called MacroMenu runs on the WorkstationManager. A remarkable feature of MacroMenu is that the user can control AV devices without issuing a lengthy command sequence. Just by touching a graphical cell icon on a touch panel of the WorkstationManager, an end-user can accomplish his or her desired job; MacroMenu executes the desired macro command that consists of a set of lower level DeviceTasks resistered in the Manager's database. The internal structure of MacroMenu is a tree of menus.

One thing that we have learned from our experience in programming a number of <u>MacroMenus</u> is that the end-user's requirement specification for one <u>MacroMenu</u> differs from that for another. Possible factors causing these differences are:

(1) Screen layouts:
Colors, arrangements, sizes and figures of icons on the screen. In an interactive graphical program such as a menu, every user has his or her own particular idea.

(2) Device configuration:
The number and the types of the devices installed at a given location.

(3) Purpose of using the <u>CrossoverNet</u> system:
Depending on the application being considered, the <u>MacroMenu</u> contents are changed.

This type of individual menu development involves extensive programming, and a number of prototypes must be developed before the end-user is satisfied with the menu; what is still worse, the user rarely has his or her requirements explicitly specified at the beginning. One solution to overcome this serious problem is to make a prototyping system for <u>MacroMenu</u>. We call this the <u>InteractiveProto</u> prototyping system.

4.2. <u>Visual Approach</u>

Using <u>InteractiveProto</u>, end-users themselves can specify menu requirements easily, and input this source specification into a menu generator that automatically generates the target <u>MacroMenu</u>. Another feature of the <u>InteractiveProto</u> system is that it is equipped with a visual or graphical programming language instead of a traditional textual programming language, so that even non-programmers can operate the system easily; the user programs through interaction with icons displayed on the screen without writing and reading text lines. Visual programming enables the users to program directly from images in their mind. Therefore the number of bugs decreases, and the users no longer need to waste a great deal of time in programming textually. Recent visual programming researches are surveyed in [GLI84], [GRA85] and [RAE85]. E. P. Glinert and S. L. Tanimoto indicate the following points about visual programming[GLI84]:

* A user needs to handle so many objects at a time that they cannot be displayed simultaneously on a single screen. Even if a user organizes the objects hierarchically by separating them into several graphical modules, there would be many problems to overcome in transforming the object images in their mind into the object images on the screen.

* It is difficult to visualize abstract data objects such as arrays, linked lists and trees, and to identify the objects with icons.

Compared with other visual programming systems such as <u>Programming by Rehearsal</u> [FIN84], <u>Pict</u> [GLI84] and <u>ThingLab</u> [BOR81], <u>InteractiveProto</u> offers more desirable solutions to the above two problems, by imposing the following restrictions:

* The workstation area is limited so that all the devices in one area may be displayed on a screen;

* Objects accessed, that is, the AV devices and other devices such as electronic black-boards and electrically controlled curtains, must be easily represented by using icons on a screen, and these icons must offer users the object status information. Figure 5 shows an example of the icons corresponding to the objects in the classroom.

InteractiveProto needs no host language such as C or Smalltalk-80[†].

In the field of prototyping systems, it is often seen that programmers define their specification by using a state transition diagram (STD) language and that from this specification is generated an interactive program. For example, there are State Transition Language [JAC85] and RAPID/USE [WAS83] prototyping systems. A STD is suitable for describing interactive interface program specifications. A few of them including State Transition Language [JAC85] use visual programming paradigms when defining the specification. InteractiveProto for AV controlling is one of these prototyping systems with visual programming. Because InteractiveProto is a special-purpose system for AV control, the user needs not write any text lines after the system initialization. The MacroMenu can also be simulated on the screen for verification.

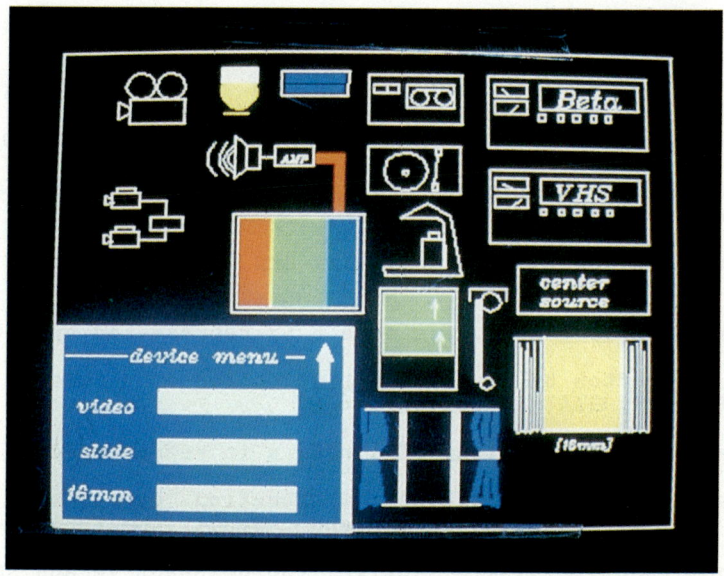

Figure 5. An example of the InteractiveProto screen layout. The big rectangle in the down left area shows an WorkstationManager screen.

[†] Smalltalk-80 is a trademark of Xerox Corp.

4.3. MacroMenu

In this section, a _menu transition graph_ and _MacroMenu_ are defined. While the _menu transition graph_ is a general definition for menu-oriented (menu-driven) systems, _MacroMenu_ is restricted to be the _menu transition graph_ in the form of a tree of menus.

4.3.1. Menu Transition Graph

A _menu transition graph_ is defined as a directed graph by the eleven-tuple

$$G = (N, T, tf, s, E, M, LB, NA, menu, button, act)$$

where
(1) N: a set of nodes;
(2) T: a set of transition arcs;
(3) tf: a transition arc function $tf: T \Rightarrow N \times N$;
(4) $s (\in N)$: the start node;
(5) $E (\subset N)$: a set of end nodes, having no menus;
(6) M: a set of menu images/frames displayed on the screen;
(7) LB: a set of logical buttons;
(8) NA: a set of actions;
(9) _menu_: a menu function $menu: N-E \Rightarrow M$;
(10) _button_: a button function $button: T \Rightarrow LB$;
(11) _act_: an action function $act: T \Rightarrow NA$.

This _menu transition graph_ G has to satisfy the following conditions to maintain the consistency of the system:

(c1) $\forall n \in N, \exists path(s, ..., n)$;
(c2) $\forall e \in E, \forall n \in N, (e, n) \notin tf(T)$;
(c3) $\forall n \in N-E, \exists n' \in N$ such that $n \neq n', (n, n') \in tf(T)$;
(c4) $\forall n \in N-E, \forall t, t' \in T, \exists 1, m \in N, tf(t) = (n, 1)$
   and $tf(t') = (n, m)$ and $t \neq t' \Rightarrow button(t) \neq button(t')$.

Here we introduce an _active arc_ and an _active button_ of a node to simplify the description.

A transition arc $t \in T$ is an _active arc_ of a node n.
$\longleftrightarrow \exists n' \in N, tf(t) = (n, n')$.

The set of _active arcs_ of a node n is represented by $AA(n)$.

A logical button $b \in LB$ is an _active button_ of a node n.
$\longleftrightarrow \exists t \in AA(n), button(t) = b$.

The set of _active buttons_ of a node n is represented by $AB(n)$.

For given sets $AA(n)$ and $AB(n)$, a function _button_ is injective. Moreover, for given sets $AA(n)$ and $AB(n)$, the function _button_ is surjective because its image is the whole codomain $AB(n)$. As the result, the function is bijective (both injective and surjective). This restriction of the function _button_ and its inverse are now represented as follows:

$bt_n : AA(n) \Rightarrow AB(n)$;
$bt_n^{-1} : AB(n) \Rightarrow AA(n)$.

We will explain the rule by which a menu transition graph changes the state. Initially the system is in the initial state that is depicted by the start node s. Suppose that the system is in the state represented by a node n. (Here we can interchangeably say "the system is in a node n.") Then the menu(n) is displayed on the screen. When an end-user chooses one of the active buttons, the state of the system changes. The action associated with the transition arc is carried out, and the new state is decided as the destination node of the arc. Then the image of the new node is displayed. If the destination node is an end node, the system will halt after the action of the arc is executed. We introduce a function dest that assigns to a transition arc the destination node where the arc terminates. The nth node of a menu transition graph, node(n), is described with this dest as follows:

(1) $node(1) = s$;
(2) Suppose the system is in the $node(n-1)$ and $b \in AB(node(n-1))$ ($n \geq 2$):
 (2i) If $dest(bt_{n-1}^{-1}(b)) \notin E$, then $node(n) = dest(bt_{n-1}^{-1}(b))$;
 (2ii) If $dest(bt_{n-1}^{-1}(b)) \in E$, then the system terminates and $node(n)$ is undefined.

4.3.2. MacroMenu

MacroMenu is defined as a restricted menu transition graph. In MacroMenu, the logical buttons displayed on the screen are called cell icons. A set LB and a function button are therefore replaced with a set C of cell icons and a function cell $T \rightarrow C$ respectively. A set M and a function menu are also changed into a set B of background menu images and a background menu function $bgm \ N-E \rightarrow B$, to emphasize the distinction between the background and the cell icons that are the foreground. The whole screen image is called a menu frame, and the menu frame of a node n is represented by $bgm(n)$ and $cell(AA(n))$. Figure 6 shows an example of a menu frame.

We define MacroMenu imposing the following five restrictions on a menu transition graph.

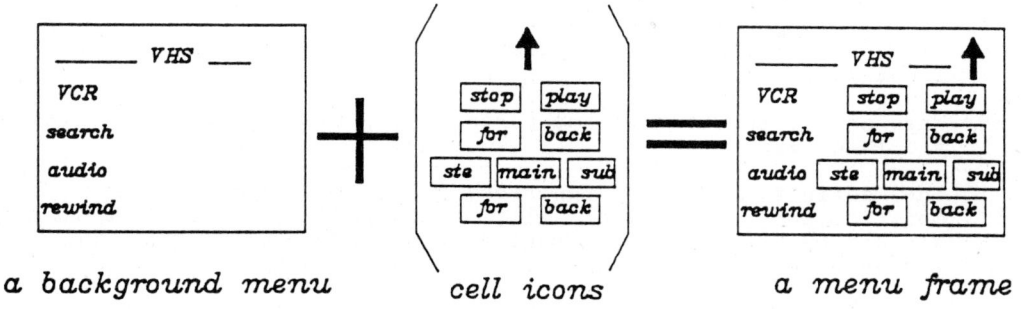

a background menu   cell icons   a menu frame

Figure 6. An example of a menu frame.

(r1) A function $\underline{tf}$ is injective, i.e.

$$\forall \underline{t}, \underline{t}' \in T, \underline{t} \neq \underline{t}' \rightarrow \underline{tf}(\underline{t}) \neq \underline{tf}(\underline{t}').$$

This implies that there are not such two arcs that have the same ordered pair of nodes but carry the different actions. An arc is, therefore, identified with an ordered pair of nodes.

(r2) No loop arc is included, i.e.

$$\forall \underline{n} \in N, (\underline{n}, \underline{n}) \notin \underline{tf}(T).$$

(r3) For any transition arc, there exists the corresponding reverse arc, i.e.

$$\forall \underline{n}, \underline{n}' \in N, (\underline{n}, \underline{n}') \in \underline{tf}(T) \rightarrow (\underline{n}', \underline{n}) \in \underline{tf}(T).$$

Then a pair of the directed arcs can be treated as one undirected arc, with a pair of actions executed depending on the direction of transition.

(r4) For any node except the start node $\underline{s}$, there exists exactly one simple path from the start node to the node. A simple path cannot include the same node twice. Then MacroMenu constructs a tree structure the root of which is the start node $\underline{s}$. Each arc has two directions up and down. Up is defined to be the direction to the start node, and in other words, from a child node to the parent node. To take the full advantage of the tree structure, we introduce the following four functions that are substituted for the two functions $\underline{act}$ and $\underline{cell}$.

- $\underline{ucell}$: an up cell function mapping a non-root node to an icon cell that is associated with a trigger for the transition to its parent node;
- $\underline{dcell}$: a down cell function mapping a non-root node to an icon cell that is associated with a trigger for the transition from its parent node;
- $\underline{uact}$: an up action function mapping a non-leaf node to an action to be taken in the transition from its child node;
- $\underline{dact}$: a down action function mapping a non-root node to an action to be taken in the transition from its parent node.

Figure 7 presents the relationship between the domains and codomains.

(r5) When arcs point to the same parent node, their associated actions are identical.

The definition of a function $\underline{ucell}$ now requires a slight modification. First of all, an equivalence relation E on T is defined by: (A function $\underline{parent}$ assigns to a transition arc the parent node of the ordered pair of nodes.)

$$\forall \underline{t}, \underline{t}' \in T, \underline{t} \; E \; \underline{t}' \longleftrightarrow \underline{parent}(\underline{t}) = \underline{parent}(\underline{t}').$$

Then the quotient set T/E is defined. We introduce a function g N-L $\Rightarrow$ T/E, which assigns to a non-leaf nodes, an equivalence class the representative of which is an arc $\underline{t}$ including a node $\underline{n}$ as the parent node. Applying this function g and an action function $\overline{\underline{act}}$ T/E $\Rightarrow$ NA in succession, we obtain a new version of function $\underline{uact}$

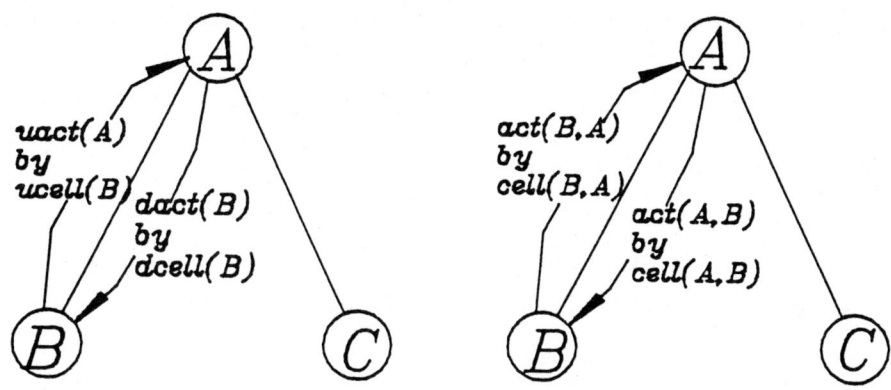

Figure 7. The relations of (cell and act) and (ucell, dcell, uact and dact). A is the parent node of B and C.

$$uact = \widetilde{act} \circ g.$$

It can be easily shown that uact is well-defined.

Before the definition of MacroMenu, an action leaf and an end leaf deserve some explanation.

An action leaf is a node that does not have a menu frame. When a transition arc, the destination node of which is an action leaf, is chosen by means of an icon, the action is executed and the system returns to the parent node of the action leaf. An end leaf is included in the set of action leaves, and after the action executed the system terminates. A node that is not an action leaf is called a menu node.

MacroMenu is a tree of menus and it is defined by the thirteen-tuple

M = (N, T, A, r, E, B, C, NA, bgm, ucell, dcell, uact, dact)
where
(1) N: a set of nodes;
(2) T: a set of unordered pairs of nodes, called transition arcs;
(3) A($\subset$ N): a set of action leaves;
(4) r($\in$ N-A): the root of the tree (start node);
(5) E($\subset$ A): a set of end leaves;
(6) B: a set of background menu images displayed on the screen;
(7) C: a set of cell icons (areas for interaction on the screen);
(8) NA: a set of actions, which is defined below;
(9) bgm: a background menu function bgm: N-A $\Longrightarrow$ B;
(10) ucell: an up cell function ucell: N-{r} $\Longrightarrow$ C;
(11) dcell: a down cell function dcell: function N-{r} $\Longrightarrow$ C;
(12) uact: an up action function uact: N-L $\dashrightarrow$ NA;
where
L: a set of leaf nodes
(13) dact: a down action function dact: N-{r} $\dashrightarrow$ NA.

Conceptually, the transition rule of _MacroMenu_ is identical with that of a menu transition graph.

An action is defined using the BNF as follows:

```
<action>::={<command>}.
<command>::=<statement>|<macro command>.
<macro command>::=<name>{<argument>}<action>.
<statement>::=<set statement>
             |<if statement>
             |<delay statement>.
```

The three types of statements are, in more detail:

    set(device_name, task_name)
        Execute the task on the specified device.

    _if_(condition, action)
        If the condition is TRUE, then execute the action.

    delay(time)
        Delay _MacroMenu_ motion.

In _MacroMenu_, a user can define a new macro command by grouping statements and predefined macro commands, i.e., neither a recursive call to itself nor the use of a coroutine is allowed. The _twin_ below is an example of a macro command:

```
twin(cell1, cell2){
    set(cell1, on);
    if (cell2 == on, set(cell2, off));
}
```

The above definition is given by a textual programming language like C, but in _InteractiveProto_ a user does it visually, not textually. When twin cells, a pair of flip-flap commands such as _stop_ and _play_ of a video cassette recorder, are displayed on the _MacroMenu_ screen, touching one cell switches its status to "on," and the other's status automatically changes to "off"; externally, the cell background color is switched between white("on") and black("off").

4.3.3. Example of _MacroMenu_

Here we illustrate the use of _MacroMenu_ through an example (Figure 8). Three screen images are snapshots of menu frames. By a dashed line, a node is related to the screen image. In the figure, the down cell of the video node is a rectangle icon on the right-hand side of the string video on the device menu node screen. The up cell, in this example, is an arrow-shaped icon.

When a transition between two menu nodes takes place, an up action or a down action is executed and the current position is transferred to the destination menu node. When the down cell that points to an action leaf is picked by the user, only the down action corresponding to the action leaf is executed; the system returns to the last menu node without executing any further action.

For example, when the play icon on the VHS menu node is selected, _dact_ ("play") is executed and the menu frame, in effect, never changes. The contents of _dact_("play") are a command sequence to start the VHS cassette recorder and to change the colors of the play and stop icons.

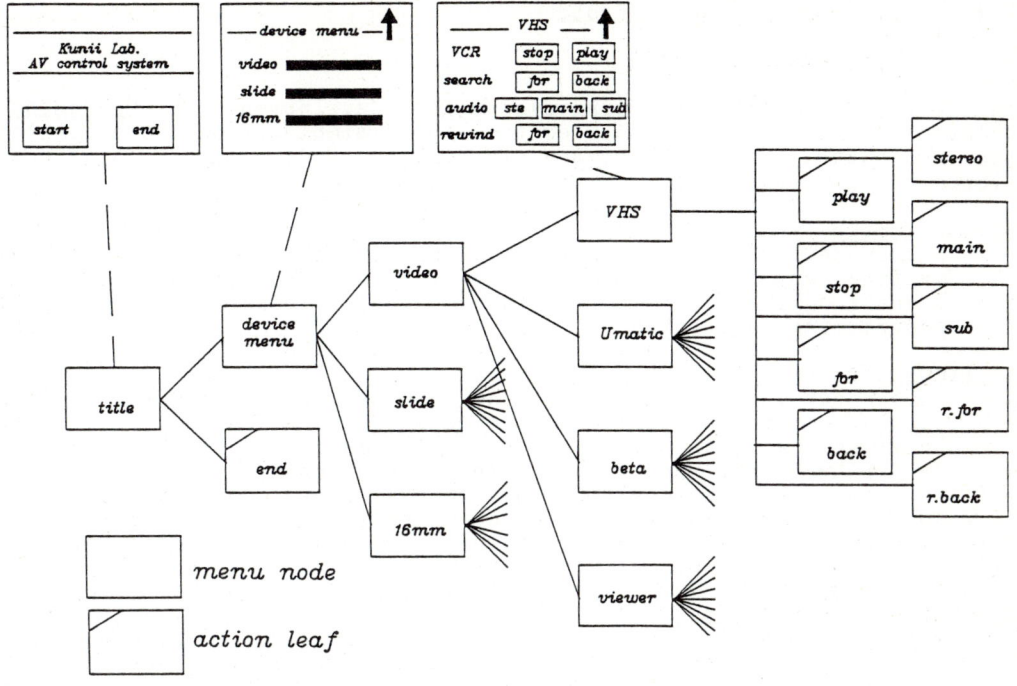

Figure 8. An example of MacroMenu.

4.3.4. Advantages of MacroMenu

Though by using a menu transition graph, most general-purpose menu-oriented systems can be described, among them, the menu-oriented systems the form of which is a tree of menus is used very frequently. Then we define MacroMenu by imposing some restrictions on the general menu transition graph. Compared with a usual menu-oriented system (or a menu-oriented program), MacroMenu has the following advantages:

(a1) Once a cell is selected, a transition and an action are uniquely determined, independently of other conditions. Therefore, a user does not have to, unlike usual, consider a more complicated branching condition.

(a2) What you see is what you can select; visible cells are all selectable as triggers for users to activate desired actions.

A procedure to invoke a state transition is, in most cases, more complicated.

The structure of MacroMenu is simple and straightforward; the ease of understanding its structure leads to better utilization of the system. As a direct consequence of these advantages, a user can easily specify desired MacroMenu through InteractiveProto.

4.4. Design of InteractiveProto

4.4.1. System Architecture

InteractiveProto, a prototyping system for development of MacroMenu, consists chiefly of a visual editor, an AV simulator and an automated menu generator. Figure 9 shows a global view of the InteractiveProto system architecture.

Visual Editor
   A user defines the menu specification with a visual editor without writing any text lines. Once the system is invoked, every operation is done in an interactive, graphically oriented programming environment.

AV Simulator
   A user can verify and analyze the menu specification using an AV simulator on the same computer as with the visual editor. Using only the graphical images, a user can simulate AV device actions as if he or she were actually in the same room as the AV devices. This saves a great deal of time, because a user does not need to go over the remotely located room. Editing and simulation cycle is repeated until the user obtains the menu specification that satisfies his or her requirements.

Automated Menu Generator
   A user inputs the graphical source specification of MacroMenu into an automated menu generator, which automatically generates the target MacroMenu. This MacroMenu code is written in the target computer's machine language.

There is another vital element in InteractiveProto, that is, a database management system (DBMS). It is not listed above because this DBMS is not resident in InteractiveProto itself but in the CrossoverNet as a

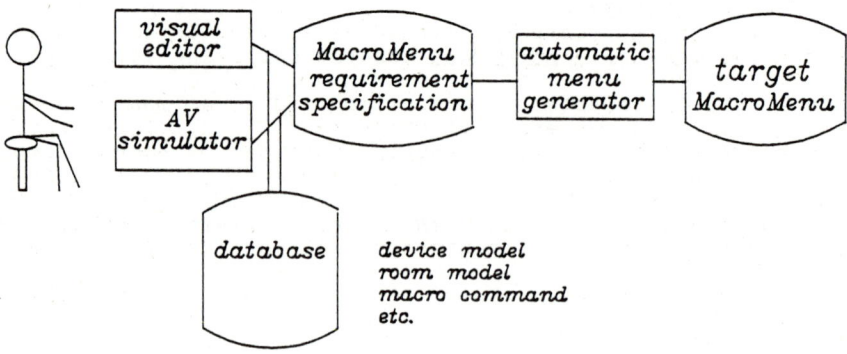

Figure 9. A global view of the InteractiveProto system architecture.

whole. It stores all of the graphical and nongraphical information characterizing the devices and the AV room as device models and a room model.

* Device Model *

At first, it is necessary to construct a device model in the database. The device model contains some information on the structures, functions and other characteristics of the device, and can be used readily in the CrossoverNet DBMS. InteractiveProto makes use of the LogicalDevice model thus provided and holds the status readout from the PhysicalDevice in the form of the LogicalDevice status. In order for the user to recognize internal or external status changes of the devices visually through InteractiveProto, the icons must be designed explicitly to reflect each status change.

* Room Model *

The database also holds the models of AV rooms where the AV devices are actually used. The AV room model includes the following items:

* names of the LogicalDevices in the room,
* names of PhysicalDevices that implement the LogicalDevices,
* the number of each type of LogicalDevice/PhysicalDevice,
* device icon layout on the screens (This layout information is used only by InteractiveProto and has no relation to the real PhysicalDevice layout in the physical AV room.).

In describing geometrical or graphical models, hierarchical structures are very common [FOL82]; and the majority of CAD/CAM data models adopt hierarchies of graphical and nongraphical data. The DBMS described here is intended to manage not only the hierarchies of two dimensional and solid model information but also the hierarchies of other modeling information; based on the latter information the CrossoverNet system can answer such inquiries as Which components are assembled into a LogicalDevice?, What PhysicalDevice implements this LogicalDevice?, When and how was a PhysicalDevice repaired recently? and What PhysicalDevices are portable?

4.4.2. State Transition Diagram of InteractiveProto

In this section, we explain how a user defines the specification of MacroMenu using InteractiveProto. The main flow of this definition process consists of the following four processes.

(Process1) cell icon definition

Frequently used cell icons, such as an arrow, a rectangle and a school badge (for a title screen), are predefined and stored in the database. When a user wants to use a new cell icon on the MacroMenu, he or she must define the cell icon.

(Process2) room model definition

Next, the way in which devices in the AV room (e.g., a classroom, an auditorium or an office) are interconnected, is defined. The already defined device icons are called from the database and placed at proper positions on the screen. This screen represents the current configuration of the devices in the AV room.

(Process3) <u>MacroMenu</u> structure definition

Third, a user must draw a <u>MacroMenu</u> structure on the screen by using a two-dimensional (2D) CAD to tell <u>InteractiveProto</u> the structure of <u>MacroMenu</u> being defined. These graphical data for one screen is called a <u>MacroMenu</u> structure chart. Figure 10 shows examples of these <u>MacroMenu</u> structure chart screens available to the user while <u>InteractiveProto</u> is running. When the structure is too large or too complicated to be displayed on a single screen, a <u>MacroMenu</u> tree structure is separated into several sub-trees and each sub-tree, when chosen, is drawn on a separate screen; these graphical data are stored in the database hierarchically. In Figure 10, slide, video and 16mm are sub-trees. A number there shows the depth of charts, not the depth of nodes. <u>InteractiveProto</u> aids the user to define the <u>MacroMenu</u> specification at the next process using this <u>MacroMenu</u> structure information, in the same way as a structured editor knows a target programming language grammar and helps the user to edit a source program file using that knowledge.

These three steps are to set up <u>InteractiveProto</u> for <u>MacroMenu</u> generation. Through these processes the user is aided by a 2D CAD tool that functions as a 2D graphical database query language.

(Process4) <u>MacroMenu</u> action definition

Finally, a user assigns an up action or a down action to each transition. Because this process is more complicated than the first three, we shall describe how to operate <u>InteractiveProto</u> to define the actions using a state transition diagram (STD).

Generally speaking, in describing an interactive man-machine interface program, an STD helps the user to understand the program behavior [JAC85]. An STD is composed of the following three elements:
<u>NODE</u>
In a node, a system displays a prompt and a menu window, and waits for a user's operation input and a screen display. A sub-node is a node that calls another STD for its details.

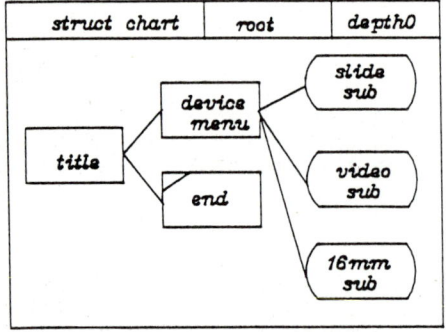 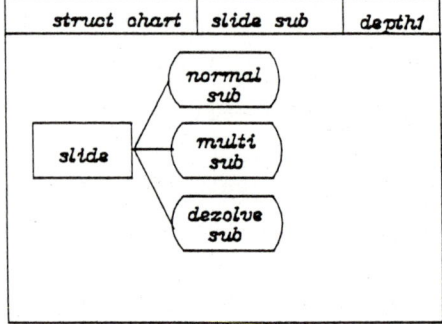

Figure 10. Examples of <u>MacroMenu</u> structure charts.

ARC
An arc is labeled with a character string. When an operation is input, the arc with the same label as the input character string is selected and the transition from the current node to its destination node happens depending on the arc.
ACTION
An action is started at the same time with the initiation of a transfer from one node to another.

An STD starts its motion at the predefined initial node, and ends at an end node. A transition from a sub-node depends on the string returned by the end node of the STD called by the sub-node. The above-mentioned elements are symbolized as shown in Figure 11.

There are three ways for a user to select: (1) select an operation from the menu window using a mouse (denoted as MW in the STD); (2) pick a device icon or a cell icon on the screen using a mouse (PICK in the STD); and (3) type in through a keyboard (KEY in the STD).

The STDs in Figures 12 through 19 (except 16 and 18) illustrate the user behavior while defining the specification of MacroMenu.

MacroMenu action definition (Figure 12)

* In the STD node called MacroMenu structure chart, the current MacroMenu structure chart is displayed. A transfer between charts may be made by selecting a sub-node name when moving downward to a child or by selecting return on the menu window when moving upward to a parent.

* The background menu image is defined using a 2D CAD tool.

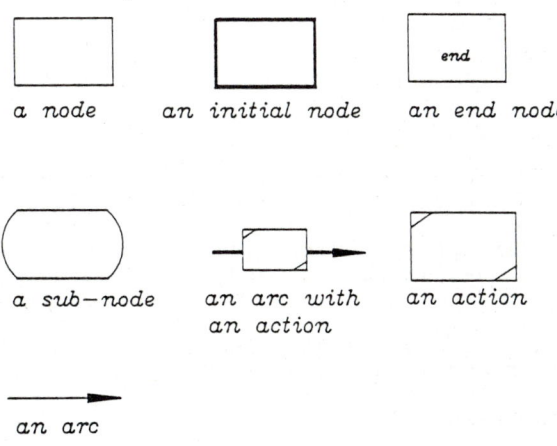

Figure 11. Symbols of STD.

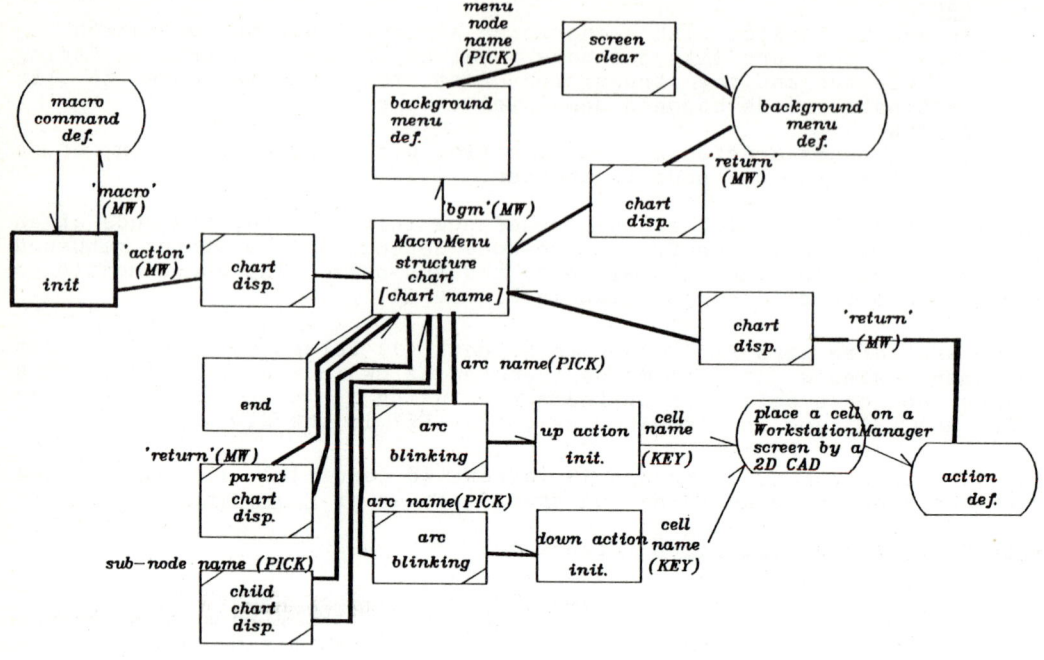

Figure 12. MacroMenu action definition STD.

* An up action/down action is defined as follows:
  At first the user picks a transition arc between nodes on the MacroMenu structure chart by using a mouse. Next, he or she selects an icon to be used to invoke the transition associated with the transition arc and places the icon on the WorkstationManager screen by a 2D CAD tool. Finally he or she defines an action to be taken when the transition occurs.

macro command definition (Figure 13)

* After a user indicates argument icons necessary for a macro command, blinking color designates the available icons.

* Actions are defined by using icons as explained above. Each argument of the macro command may be parameterized as a set of icons.

* When using a cell icon as an argument, it is placed inside the WorkstationManager by using a 2D CAD tool.

action definition (Figure 14)

* A command sequence is defined to achieve the desired action. In this definition process, three primitive editing tools are provided: insert, delete and list (forward and backward).

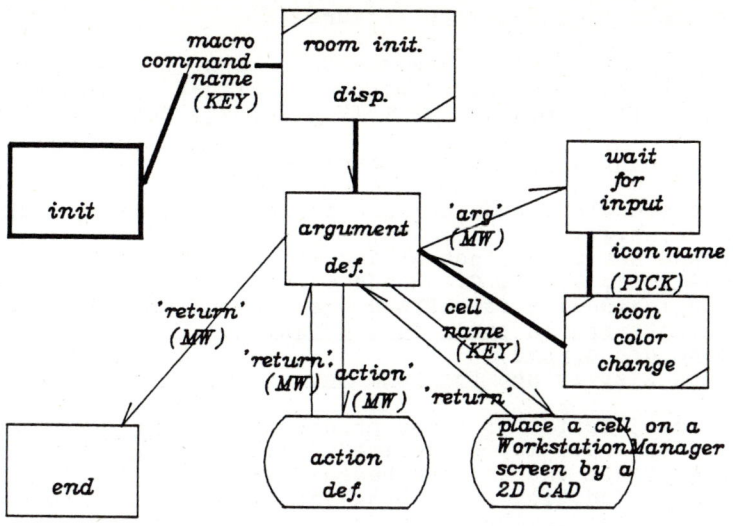

Figure 13. Macro command definition STD.

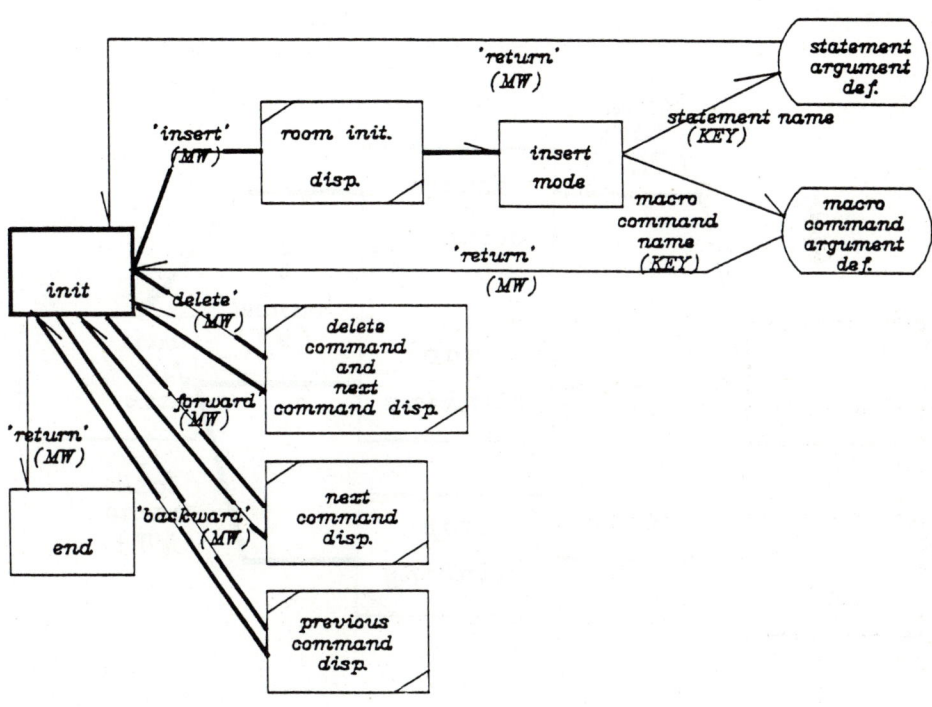

Figure 14. Action definition STD.

set statement argument definition (Figure 15)

* After a device name is picked, the task to be performed is specified through a menu window. Figure 16 shows the screen immediately after defining set("TV display", "on").

if statement argument definition (Figure 17)

* The if statement's condition part is described using a disjunctive normal form; e.g.,
(A and B and not C) or (not A' and B' and C') or (A" and C")
where A, A', A", B, B', C, C' and C" are predicates.
On each LogicalDevice is defined one or more predicate as follows:
down lights are in their full brightness,
curtains are opened,
a screen size is set for 16mm movie projection.

Forming a conjunction of these predicates or their negations, the status of the room may be defined. Note that we need not specify the status of all the devices because we care about only some of them. The conjunction describing the status of the room is called the room condition. Then we may use a disjunction of the room conditions as the condition in an if statement. Figure 18 shows the condition such as "(curtains are opened) and (a screen is closed) and not (down lights are bright fully (100V))."

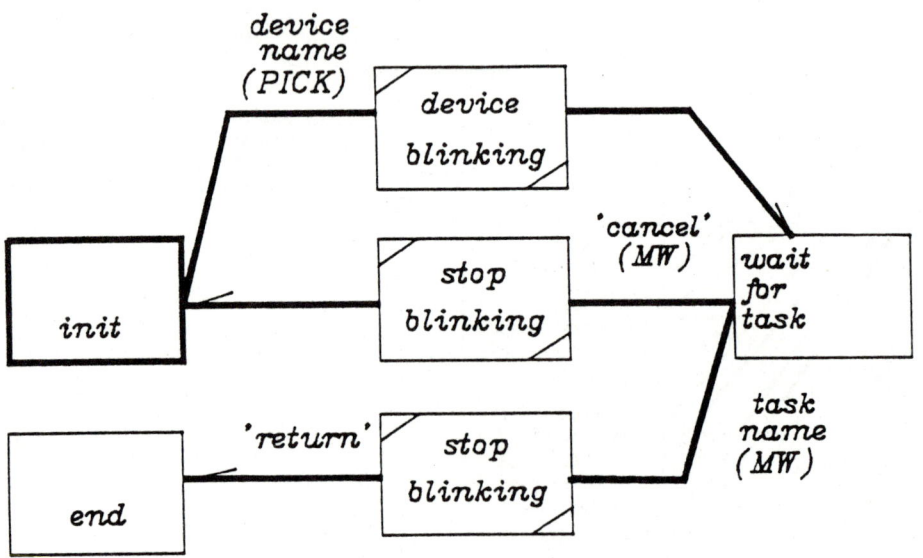

Figure 15. Set statement argument definition STD.

Figure 16. The screen immediately after definition set ("TV display", "on").

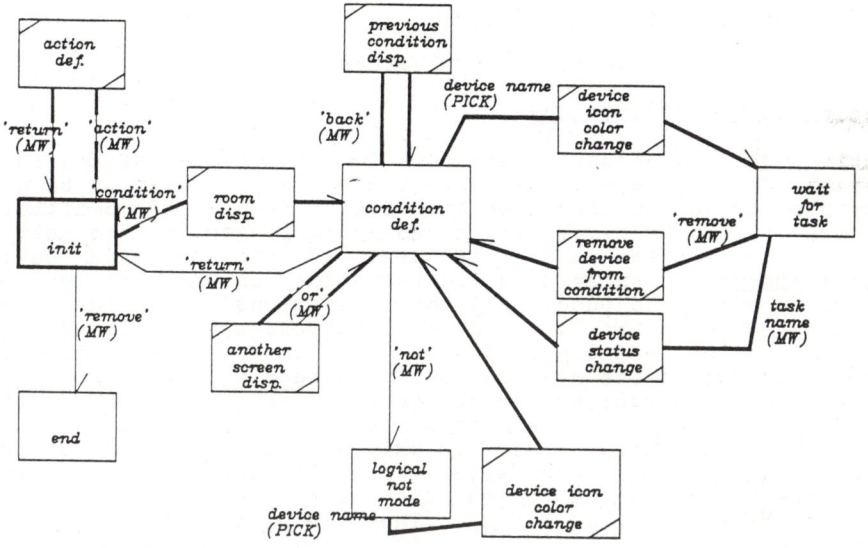

Figure 17. If statement argument definition STD.

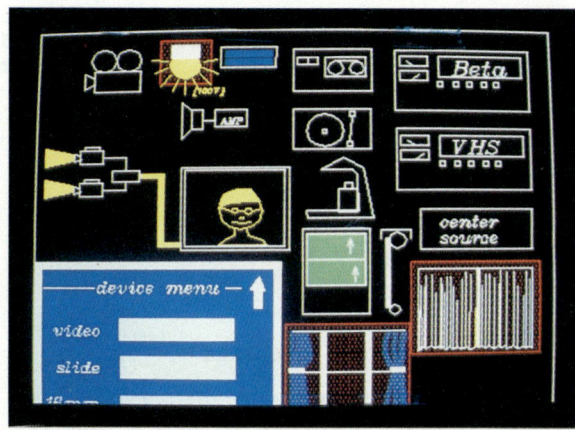

Figure 18. A condition of an if statement. The screen means "(curtains are opened) and (a projector screen is closed) and not (down lights are bright fully (100V))."

simulation (Figure 19)

* During the simulation, the following items may be registered as static parameters:

(1) break point(menu node name)
   Establish a break point just before the specified menu node;
(2) speed(n)
   Reduce the speed of simulation to $1/n$ when the actual MacroMenu action is too fast to follow;
(3) step(n)
   Select the mode of simulation. When $n > 0$, n statements are simulated step by step. When $n = 0$, simulation proceeds continuously.

4.5. Features of InteractiveProto

InteractiveProto has the following features.

(f1) Uniformity of Operations

To simplify user's operations and to reduce the number of the rules for the user to follow, we have designed the operation rules to be uniform. In many places, we found an operation input method of Smalltalk-80's pop-up menu [GOL80] very useful, because few finger and arm motions are required. A one-button mouse is used as the pointing device in InteractiveProto in order to avoid the situation where the user cannot decide which button to push. All the operations that the user can input are shown on the STD. These operations are displayed in a menu window (Figure 20), and the user moves the mouse (cursor) and selects an operation within the menu window. The menu window appears when the user presses the button; at the cursor position, when the button is pressed, is placed at the upper left corner position of the menu window. The menu window is visible only when the button is pressed. The user moves the mouse with the button pressed. When he or she releases the button, which causes the operation pointed to by the cursor is executed immediately. Then the window disappears. If the button is released when the cursor is outside the menu window, no operation is executed.

If there are too many possible operations (e.g., to decide a new macro command name) and the user wishes to input a text line from a keyboard, then a prompt message appears in a text input area at the bottom of the display. At any state of InteractiveProto, the same method is employed for operation input. There is a single regulation that governs the operation input process.

Operation input method
<OperationInput>::= <MenuWindowInput>
                  | <TextInput>
                  | <PickObject>.
<MenuWindowInput>::=
     <Positioning> <OpenMenuWindow> <SelectOperation> <Execution>.
<TextInput>::= (type a string and a return).
<PickObject>::= <Positioning> <SelectObject>.
<Positioning>::= (move a mouse).
<OpenMenuWindow>::= (press a button and continue pressing).
<SelectOperation>::= (move a mouse).
<Execution>::= (release a button).
<SelectObject>::= (press and release a button without moving a mouse).

(f2) Help Function

Generally, when a user interacts with a man-machine interface program, the following three types of information are required [JAC85]:

* What can I do next?
* How can I do that?
* Where am I?

InteractiveProto, having a powerful help function, meets such demands as follow. The user has only to choose one operation out of the operations listed in the menu window, which saves the user from being perplexed with what to do and how to do it. To understand the current state position in the InteractiveProto STD illustrated in 4.4.2., he or she can look up and display the STD chart, in which the current state is colored red. By taking a glance into this chart, instead of reading a manual, the user of InteractiveProto can get the aforementioned three type information. These charts have been already stored in the database, and when the user presses a help key on the key-board, InteractiveProto selects the most suitable chart and displays it.

(f3) Introduction of MacroMenu Structure Charts

To tell InteractiveProto the MacroMenu structure, the user lays out MacroMenu structure charts and stores them in the database. While running, InteractiveProto often refers to these charts, and so can the user. The charts are helpful for the user in acquiring the following types of information:

* What is the entire structure of MacroMenu?
* What cell icons are included in a menu?
* Where are break points located for an AV simulator?
* Where is the user when the control is returned from simulation to editing? (Figure 21)

(f4) MacroMenu formal definition

In this chapter, a menu transition graph is defined to describe a general-purpose menu-oriented system, and then, based on the menu transition graph, MacroMenu is defined to realize a menu-oriented system with a tree structure menu organization. InteractiveProto relies on the advantages of MacroMenu. When a user has the interaction with MacroMenu, the destination menu node can be selected simply by pointing out a cell icon. In general, interactive programs are complicated by these branching conditions. The user needs not specify complicated conditions; go to and call/return statements become unnecessary in defining a specification of MacroMenu. And also a user describing MacroMenu specifications needs not indicate the time when the transition occurs; the transition will occur when the cell is activated.

(f5) Device Icon

The device icons have been designed to illustrate the AV room in a realistic manner. As for a device whose changes proceed slowly and visibly (e.g., blackboards and curtains), the state transition is displayed on the screen through an animation so as to give users the feeling of being at a live performance.

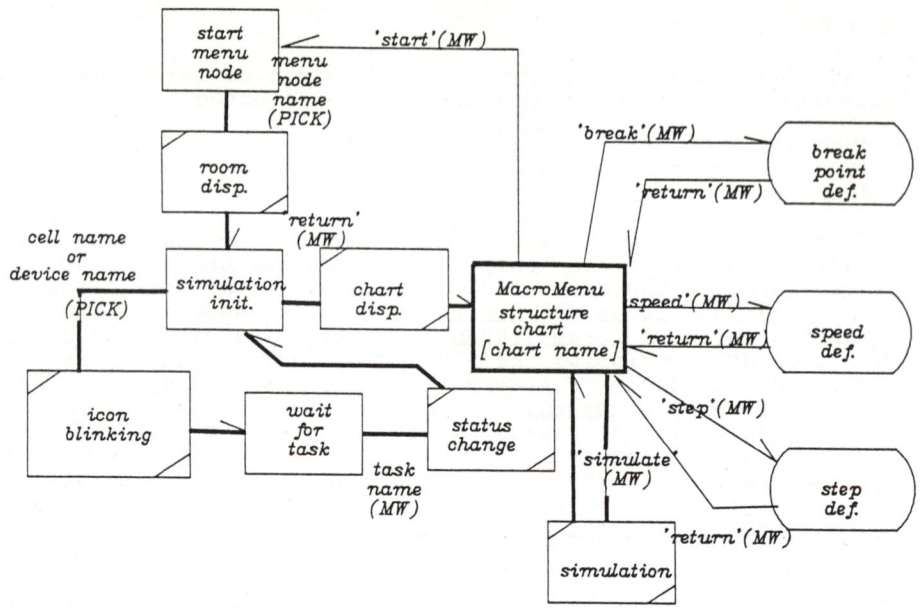

Figure 19. Simulation STD.

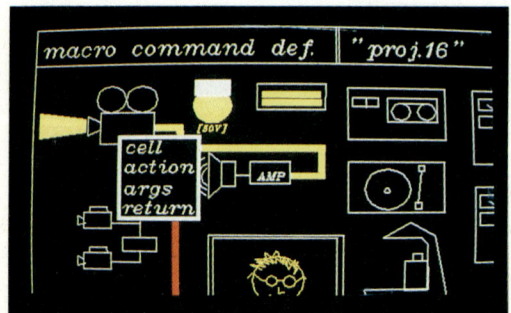

Figure 20. A menu window.

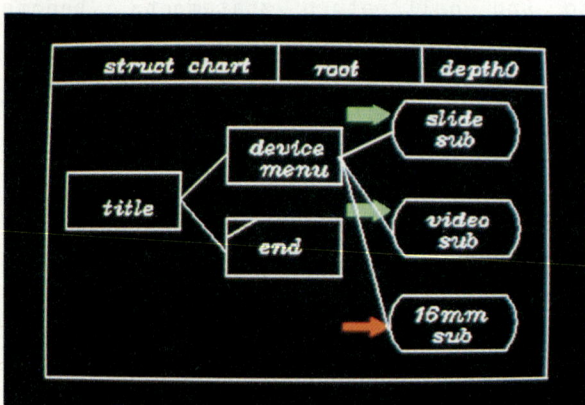

Figure 21. An InteractiveProto screen immediately after simulation. Green arrows show break points established by the user and a red arrow shows the menu node where the control stopped by an error or at a break point.

(f6) **Visual Programming**

Because the user defines the requirement specification with visual programming, *InteractiveProto* allows even a non-programmer to make a *MacroMenu* program by himself or herself.

(f7) **Integration of an Editor and a Simulator**

It is inconvenient for a user to switch between a visual editor and an AV simulator again and again just to repeat the job cycle of editing and simulation. Such inefficiency can be avoided by uniting the editor and the simulator into one system. As seen in Figure 19, the user can transfer from editing to simulation smoothly and naturally.

(f8) **Shared Macro Commands**

A macro command can be stored in the database and shared by other users. The macro command defined by an experienced user contains the knowhow of the AV control, such as the operation timing and sequence to get the desired results. As an example of their knowhow, let us look at an AV classroom and a teacher there. The knowhow says that, taking ergonomic[†] factors into consideration, the teacher should proceed as follows:

- * Dim down room lights gradually at the beginning of projection until the eyes of the students are used to seeing in the dark.
  (It would smoothly turn the students' attention to the screen.)

- * Light up the room gradually after the projection.
  (Lighting abruptly disturbs the students.)

- * Bring the video projector into focus. Depending on the sight of students/audiences, their optometric muscles, too, change with time. The focal distances vary between younger and elder people.

By sharing the macro commands, every user can control AV devices in the manner of an experienced user. It also saves the user from defining the duplicate commands.

4.6. Future Perspective of User Interface Programs

In order that an end-user can easily control AV devices, the interface between the user and the AV control system, and how the relationship and communication between the two can be improved are the significant problems. One solution to these problems is the development of interactive and visual interface programs and tools. In *CrossoverNet*, to make users familiar with the system, we have developed *MacroMenu* and *InteractiveProto*.

[†] Ergonomics is the study of the conditions in which people work most effectively with machines [Longman Dictionary of Contemporary English 1978].

In future a number of application programs will be necessary for the individual application field of CrossoverNet (e.g., TV conferencing and security monitoring); the special-purpose tool that is suitable for the application should be developed. Among these tools the following functions are essential and shared.

(1) a remote device control function by which an end-user can issue commands to devices located remotely and reads out a device status through a network system;

(2) a monitoring function by which an end-user can sense what's happening in remote buildings and rooms
(Figure 22 shows the application of monitoring for intrusion to buildings.);

(3) a monitoring function by which a supervisor can look up the performance and the activity status of the CrossoverNet system.

As we have seen, under the visual programming environment, the user no longer wastes a great deal of time and care in programming textually. Therefore we have several plans for introducing visual programming paradigm into the interactive man-machine interface tools. For example,

* a query language for the database of CrossoverNet holds both graphical and non-graphical data
(e.g., Query-by-Example by M. M. Zloof [ZL081], Form

Figure 22. The application of monitoring building intruders.

Transformer by Kitagawa and Kunii [KIT82] and FORMAL by N. C. Shu [SHU85] have been already implemented as two-dimensional database query languages.);

* a schedule definition tool allows device control schedules to be programmed, stored and executed automatically, so as to drive AV devices and to switch communication paths and channels.

The menu generator such as InteractiveProto can be applied to various fields. Application areas include office and factory automation. A menu can easily be tailored to suit the users' requirements without professional programmer's help.

5. Conclusion

Over two years of experience in using CrossoverNet has revealed its versatile applications beyond AV device control. We are seeing ever increasing demand of the core part of CrossoverNet which is a local area network operation system that can handle both digital and analog information uniformly. The applications include:
* Home automation;
* Home security and safety monitoring;
* Information transfer among distributed offices for office automation;
* Factory automation;
* Monitoring buildings for intrusion;
* TV conferencing;
* Integrated CAD/CAM system;
* Campus broadcasting system.

Recent potential applications go beyond these rather individualized application fields to cover further global areas such as intelligent buildings and intelligent cities.

Acknowledgement

We are grateful to Dr. Hideko S. Kunii of Ricoh Software Research Center for continuous encouragement and financial support. We are also grateful to Mr. Yoshiki Matsuoka of Meiji Gakuin University for his advices while implementing the system. We were benefited by discussions with members of Kunii Labolatory of Information Science the University of Tokyo, especially Mr. Tsukasa Noma, Mr. Issei Fujishiro and Mr. Yasuto Shirai. Managerial help from Mr. Hideo Takahashi of JVC deserves our special thanks.

References

[BOR81] A. Borning, "The Programming Language Aspects of ThingLab, A Constrain-Oriented Simulation Laboratory," ACM Trans. Programming Languages and Systems, Vol. 3, No. 4, Oct. 1981, pp.353-387.

[COU85] J. Coutaz, "Abstractions for user interface design," IEEE Computer, Vol. 18, No. 9, Sep. 1985, pp. 21-34.

[DEI83] H. M. Deitel, An Introduction to Operating Systems, Addison-Wesley, Reading, Massachusetts, 1983, pp. 53-73.

[FIN84] W. Finzer and L. Gould, "Programming by Rehearsal," Byte, Vol. 9, No. 6, June 1984, pp.187-210.

[FOL82] J. D. Foley and A. van Dam, Fundamentals of Interactive Computer Graphics, Addison-Wesley, Reading, Mass., 1982, pp. 217-243.

[GLI84] E. P. Glinert and S. T. Tanimoto, "Pict: An Interactive Graphical Programming Environment," Computer, Vol. 17, No. 11, Nov. 1984, pp.7-25.

[GOL84] A. Goldberg, Smalltalk-80, The Interactive Programming Environment, Addison-Wesley, Reading, Mass., 1984.

[GRA85] R. B. Grafton and T. Ichikawa, "Visual Programming," Computer, Vol. 18, No. 8, Aug. 1985, pp.6-9.

[JAC85] R. J. K. Jacob, "A State Transition Language for Visual Programming," Computer, Vol. 18, No. 8, Aug. 1985, pp.51-59.

[KIT82] H. Kitagawa and T. L. Kunii, "Form Transformer: A Formalism for Office Form Manipulation," in M. Maekawa and L. A. Belady (ed) Operating Systems Engineering, Proc. the 14th IBM Computer Science Symposium Amagi, Japan, Springer, Berlin and New York, 1982, pp. 382-406.

[KUN85a] T. L. Kunii and Y. Shirota, "CrossoverNet: A Computer Graphics/Video Crossover LAN System," Proc. Computer Graphics Tokyo (CG Tokyo), Apr. 1985.

[KUN85b] T. L. Kunii and Y. Shirota, "CrossoverNet: A Computer Graphics/Video Crossover LAN System," in T. L. Kunii (ed) Computer Graphics, Visual Technology and Art, Springer, Berlin and New York, 1985, pp.189-200.

[KUN85c] T. L. Kunii, Y. Shirota and T. Noma, "A Menu Generator for Audio Visual Networks," Proc. the 19th IBM Computer Science Symposium, Hakone, Japan, Oct. 1985.

[KUN86d] T. L. Kunii, Y. Shirota and T. Noma, "A Menu Generator for Audio Visual Networks," The Visual Computer, Vol. 2, No. 1, 1986, pp.15-30.

[KUN86e] T. L. Kunii and Y. Shirota, "CrossoverNet: A Computer Graphics/Video Crossover LAN System - Architecture, Design and Implementation -," The Visual Computer, Vol. 2, No. 2, 1986, pp.78-89.

[MES70] M. D. Mesarović, D. Macko, and Y. Takahara, Theory of Hierarchical, Multilevel, Systems, Academic Press, New York and London, 1970.

[RAE85] G. Raeder, "A Survey of Current Graphical Programming Techniques," *Computer*, Vol. 18, No. 8, Aug. 1985, pp.11-25.

[SHU85] N. C. Shu, "FORMAL: A Forms-Oriented, Visual-Directed Application Development System," *Computer*, Vol. 18, No. 8, Aug. 1985, pp.38-49.

[WAS83] A. I. Wasserman and D. I. Showmake, "A RAPID/USE Tutorial," Medical Information Science, Univ. of California, San Francisco, Nov. 1983.

[ZLO81] M. M. Zloof, "QBE/OBE: A Language for Office and Business Automation," *Computer*, Vol. 14, No. 5, May 1981, pp.13-22.

Human Interface Development Systems

Issues in the Design of Human-Computer Interfaces

Jurg Nievergelt
University of North Carolina

Abstract

Interactive use of computers in many different applications by users with a wide variety of backgrounds is ever increasing. This proliferation has increased the importance of designing systematic human-computer interfaces to be operated by casual users. Novel styles of interfaces that exploit bitmap graphics have emerged after a long history of interactive command languages based on alphanumeric terminals. This development is repeating, with a 20-year delay, the history of programming languages: Large collections of unrelated operations are being replaced by systematically structured modes that satisfy general principles of consistency. This tutorial illustrates some *general* issues that designers of human-computer interfaces encounter, by means of examples chosen from one *particular* research project. We do not attempt to give a comprehensive or balanced survey of the many approaches to human-computer interface design.

A survey and classification of design errors common in today's command languages leads to concepts and design principles for avoiding such errors. But the very notion of *command language*, which emphasizes the input, is of doubtful utility for understanding the principles that govern the design of good human-computer interfaces. Observation of users shows that the most common difficulties experienced are expressed well by such questions as: *Where am I? What can I do here? How did I get here? Where else can I go and how do I get there?* Such questions reveal the fact that the fundamental problem of human-computer communication is *how to present to the user what he cannot see directly, namely the state of the system, in such a manner that he can comprehend it at a glance.* If the *output language* of a system, what the user sees or can get onto the screen easily, is rich and well structured, the command language proper, or *input language*, can be simple - ideally the minimum set of *selecting the active data, selecting the active operation, and of saying "do it"*, all by just pointing at the proper places on the screen. Instead of emphasizing the *command part* of the interface, designers should focus their attention on the *display part*.

The questions above also provide hints about how the display of an interactive system should be structured. *Where am I?* and *What can I do here?* refer to the state of the system. A good design answers them by displaying to the user on demand his current data environment *(site)* and his current command environment *(mode)*. The questions *How did I get here? Where else can I go?* refer to the past and future dialog *(trail)*. A good design presents to the user as much of the past dialog as can be stored: to be undone (in case of error), edited and reinvoked. And it gives advice about possible extrapolations of the dialog into the future *(help)*. This *Sites, Modes, and Trails* model of interaction has been implemented in two experimental interactive system. They presents the state- and trail information by means of *universal commands* that are active at all times, in addition to the commands of an interactive application. This gives the user the impression that "all applications on this system talk the same language".

With the current spread of personal computers to be operated by casual users for short periods of time, the quest for systematic, standard man-machine interfaces must proceed beyond the design of "user-friendly" operating systems. It will be necessary to identify the most important dialog control commands, and to standardize these in a minimal piece of hardware that the user can carry with him and plug into any personal computer. We describe an experiment in this direction, which involves a *"mighty mouse"*.

Programming methodology has emphasized different criteria for judging the quality of software during the three decades of its existence. In the early days of scarce computing resources the most important aspect of a program was its *functionality* - what it can do, and how efficiently it works. In the second phase of *structured programming* the realization that programs have a long life and need permanent adaptation to changing demands led to the conclusion that functionality is not enough - a *program must be understandable*, so that its continued utility is not tied to its "inventor". Today, in the age of interactive computer usage and the growing number of casual users, we begin to realize that *functionality and good structure are not enough - good behavior towards the user is just as important.* The behavior of interactive programs will improve when the current generation of commercial software has outgrown its usefulness, and the next generation of programmers is free to make a fresh start.

1. The computer-driven screen as a communications medium

A versatile new mass-communications medium has come into existence: *The computer-driven screen.* Considering the computer to be a communications medium, in competition with television or overhead projector, is perhaps an unusual idea to the users of traditional "data processing"; but it is a useful point of view for the designer and user of an information system.

The computer-driven screen is the only *two-way mass communications medium* we have. It allows *two-way communication between user and author of the dialog.* You don't talk back to newspapers, radio or television - or if you do, at least you don't expect them to react to your outburst - there is no feedback. But you *can* control a well-designed computer-generated dialog: *"Show me that picture again", "go slower", "skip these explanations for the time being", "what else do you have on this topic", "I want more detail on this figure", "remind me of this fact again tomorrow"* - these are some of the commands a human-computer dialog *could* accept - but usually doesn't.

A dialog between man and machine is not symmetric. On the contrary, it should exploit the different capabilities of the two participants: The humans's prowess at trial-and error exploration based on pictorial pattern recognition, and the machine's ability to call upon large collections of data and long computations extremely accurately, and to display results rapidly in graphic form. There is no point in trying to mold the human-computer interface on the model of human-human communication. In fact, *implementing the "commands" above requires no breakthrough in artificial intelligence - it simply requires a good systems design.* Designing the interface so as to exploit the computer's speed and accuracy, and the human's abilities in pattern recognition and intuitive selection, may well be the problem of greatest economic impact computer scientist face in this decade.

2. Survey and classification of errors

Until recently only a small minority of computer users had the opportunity to use interactive systems - mostly professionals who use the computer daily. They are concerned with *the inherent power of the computer for their application* and get so used to its idiosyncrasies that a mysterious or even illogical interface doesn't bother them any more. If they are system designers, they are also concerned with the *internal structure of their software*, but have often treated the behavior of their system towards the user as a superficial problem, addressed in the style *"if you don't like it, change it"*. An example of this situation is provided by UNIX, widely acclaimed as a model of a well-designed operationg system. [No 81] reaches the conclusion "the system design is elegant but the user interface is not". The programmers' traditional attitude explains why today's interactive software is full of blemishes. Let us illustrate typical errors of dialog design, and try to explain them through a neglect of fundamental concepts of human-computer interaction. Two of the most fundamental concepts are the **state of the system** and the **dialog history**. Interface design is the art of presenting the state and the dialog history in a form the user can understand and manipulate.

The *state* is concerned with everything that influences the system's reaction to user inputs. The *interface* is concerned with interaction: what the user sees on the screen, what he hears, how he inputs commands on the key board, mouse or joy stick. Today's fad is to engineer fancy interfaces: fast animation in color, sound output and voice input. This is desirable for some applications, unnecessary for others, but in any case it does not attack the main problem of human-computer communication.

The fundamental problem of human-computer communication is how to present to the user what he cannot see directly, namely the state of the system, in such a manner that he can comprehend it at a glance.

So let us start with common errors of dialog design which hide vital state information from the user.

2.1 Insufficient state information

Imagine that you leave your terminal in the middle of an editing session because of an urgent phone call. When you return ten minutes later, the screen looks exactly as you left it. Even if the system state is unchanged, you may well be unable to resume work where you left off:

D: *"What file was I editing?"*, *"Is there anything useful in the text buffer?"*;

C: *"Am I now in search mode?"*, *"What is the syntax of the FIND command?"*;

T: *"Has this file been updated on disk?"*, *"Has it been compiled?"*.

Such questions indicate that part of the state information necessary to operate this system has to be kept in the user's short term memory. When the latter is erased by a minor distraction, the user needs to query the system to determine its state. Hardly any system lets him do this systematically, at all times.

Today's systems provide state information sporadically, whenever the programmer happened to think about it. Examples abound. The file directory can be seen in the file server mode, but not in the editor; in order to inspect it you must exit from the editor, an operation that may have irreversible consequences. In order to see the content of the text buffer you may have to insert it into the main text, thus polluting your text. To find out whether you are in search mode you may have to press a few keys and observe the system's response - not always a harmless experiment.

A designer who observes the following principle avoids the problems above:

> The user must be able AT ALL TIMES to conveniently determine
> the entire state of the system, WITHOUT CHANGING THIS STATE.

The notion of "entire state" is meant with respect to the user's model of the system, not at the bit or byte level of the implementation. Thus the details of the system state presented to the user will differ from system to system, but two of its components are mandatory. In any interactive system, the user operates on *data* (perhaps he only looks at it) by entering *commands* (perhaps he only points at menu items or answers questions). Thus the state of the system must include the following two major components: the *current data environment* (what data is affected by commands entered at this moment), and the *current command environment* (what commands are active). The questions above labelled D and C refer to data and command environments, respectively. The questions labelled T refer to the user's *trail*, i. e. the past and future dialog.

A particularly dangerous version of "insufficient state information" is the deceitful presence on the screen of outdated information. The programmer is aware of the moment he has to write some information on the screen, but he forgets to erase it when it no longer describes the state of the system. The user sees 'current file is TEMP' , when in fact his current data environment is another file. The bad habit of leaving junk lie around the screen is a relic from the days of the teletype, when messages written could not be erased. It should not persist on today's displays.

2.2 Ignoring the user's trail

"We have seen the enemy, and he is us!" The user is by far the most dangerous component of an interactive system - at least 90% of all mishaps that occur during operation are traceable to faulty user actions. Designers of interactive systems must accept the fact that high interactivity encourages trial-and-error behavior. Exhortations to discipline are worse than useless - they are counterproductive. A designer using a CAD system, a writer using a text processor, a programmer debugging his programs - they **must** concentrate on their creative task, and cannot allow a fraction of their conscious attention to be sidetracked, continuously double-checking clerical details. A system should be foolproof enough to absorb most inaccuracies and render them harmless.

It is surprising that virtually all research on reliability and security is directed against either hardware and software failure or deliberate attack by third parties. But the majority of users of interactive systems have no one but themselves to blame when their data is suspect or has been damaged. In cleaning up one's files the most recent version of a document is thrown away and an old one kept instead. You hit **D** instead of **C** on the keyboard, so something gets deleted instead of created. You forget to label a file permanent and so it vanishes. The catalog of plausible errors is different on each system; they all have in common that, as soon as we become aware of the blunder, we gasp: "How could I possibly have done that!". We can and do, at the rate of many oversights a day.

The safeguards built into today's systems to protect the user from his own mistakes are primitive. Are you sure? is the favorite question, followed by a sporadic request to press some unusual combination of keys if the action is really serious. This double-checking is effective only against accidental key pressing, not against an erroneous state of mind. When I delete a file I'm sure that's what I want to do - though I may regret it later.

The most effective protection of the user against his own mistakes is to store as much of the past dialog as is feasible. At the very least a universal command undo must always be active that cancels the most recent command executed. The memory overhead of keeping two consecutive system states is negligible, since these two states differ little - typically by at most one file. The error-prone and time-consuming Are you sure? is unnecessary when the user can always undo the last step. Ideally undo works all the way back to the beginning of the session, but that may be too costly: in order to replay the user's trail *backwards* the system must in general store *states*, not just the *commands* entered, as many operations have no inverses. A practical alternative is for the system to keep a log of commands entered and to allow the user to save the current state as a check point for future use, from which he can replay the session.

3. An interactive system as seen by the user

3.1 The Sites, Modes, and Trails model of interaction

Observation of casual users provides valuable insight into the fundamental design question of how a machine should present itself. Most of the recurring difficulties they encounter are characterized well by the following questions:

- **where am I?** (when the screen looks different from what he expected)
- **what can I do here?** (when he is unsure about what commands are active)
- **how did I get here?** (when he suspects having pressed some wrong keys)
- **Where else can I go and how do I get there?** (when he wants to explore the system's capabilities).

We are beginning to learn that the logical design of an interactive system must allow the user to obtain a convenient answer to the questions above **at all times**. In other words, the man-machine interface must include queries about the state of the system (without changing this state), about the history of the dialog, and about possible futures. This principle is much more important for today's computerized machines, *black boxes* that show the user only as much about their inner working as the programmer decided to show, than it is for mechanical machines of the previous generation, which by visible parts, motion and noise continuously reveal a lot of state information.

In order to answer the user's basic questions in a systematic way, the designer of an interactive system must also design a simple **user's model of the system**: a structure that explains the main concepts that the user needs to understand, and relates them to each other. The major concepts will certainly include many from the following list: the *types of objects* that the user has to deal with, such as files, records, pictures, lines, characters; *referencing mechanisms*, such as naming, pointing; *organizational structures* used to relate objects to each other, such as sets, sequences, hierarchies, networks; *operations* available on various objects; *views* defined on various types of objects, such as formatted or unformatted text; *commands* used to invoke operations; the *mapping of logical commands onto physical I/O devices*; the *past dialog*, how to store, edit, and replay it; the *future dialog*, or help facility.

The list is long and the design of the user's model of the system is an arduous task; but it should be possible to explain the overall structure of the model in half an hour. If the system's behavior then constantly reinforces the user's understanding of this structure, it will quickly become second nature to him. In contrast, if the system keeps surprising the user, it interferes with his memorization process and slows it down. Because questions will always arise during the learning phase, and because the right manual is rarely at the right place at the right time, an interactive system should be self-explanatory: it must explain the user's model of the system to the user, at least in the form of an on-line manual, preferably in the form of an integrated help facility that gives information about the user's current data and command environments at the press of a single key.

The user's model of the system is a *state-machine*: it has an *internal state* and an *input-output behavior*. Components of the state must include:

- the user's **data environment** (data currently accessible)
- the user's **command environment** (commands currently active).

The questions *Where am I?* and *What can I do here?* are then answered by displaying the current data and command environments, respectively [NW 80]. The user must be able to invoke this system display at all times, regardless of which applications program he is currently in, and its presentation must not change the state of the system in any way.

The system must have *universal commands*, which are active at all times. At least the state inquiry commands postulated above must be universal. In a highly integrated system many more commands can be made universal. General dialog commands that are needed in every interactive utility are omitted from the text editor, the diagram editor, the data base query language, and incorporated as universal commands into the system. The consequence on the user's view of the system is that all these utilities "talk the same language", and that in order to become proficient at using a new editor he only has to learn a small number of new commands: the data-specific commands are new, the data-independent commands remain the same.

3.2 Shortcomings of conventional operating systems

Developing good interactive software on existing systems is difficult. Today's programming languages and operating systems often fail to provide the building blocks necessary for processing user inputs and creating graphic output under the stringent real time conditions expected for highly interactive usage.

Today's commercial operating systems are answers to the demands of the computer center, to process a load dominated by batch jobs, on top of which slow-response time-sharing services with textual I/O were added later. Such an operating system provides both too much and too little for the interactive workstation, the computers on which most interactive work is done today. Too much by way of fancy resource management policies designed to accomodate many simultaneous users with different demands; the resulting overhead is wasted on the single-user computer. Too little by way of supporting a dialog with the user. For example, once a program is started, the operating system is typically unavailable to the interactive user, except for an *abort* command. However, in addition to "Control C" there are lots of utilities an operating system contains that should be made available to the end-user at all times, while he interacts with an application program. Directory inspection and manipulation commands, for example, are useful operations when the user wishes to enter a browsing phase, to search his files for useful data. Instead of forcing on the user the irreversible operation of exiting from an editor to inspect his files, the system should make its commands available at all times in a dedicated *system's window*.

3.3 Experimental interactive systems

The concepts described in the preceding sections have been realized in two prototype system XS-1 [Be 82] and XS-2 [St 84], intended as case studies of integrated interactive system for personal computers. Our present interest lies in the systems' kernel, rather than in the application programs that run on it, as an experiment in designing operating systems that support human-computer interaction in the form of the *sites, modes, and trails* model.

The functions and structure of the kernel

In conventional interactive systems the application program is responsible for defining, controlling and displaying its data and its dialog. The operating system does not communicate with the user except for starting or aborting the application program. By contrast, the kernel of XS-1 provides a structured space in which all the data and all commands are embedded, and provides operations to explore this space and manipulate its elements. The function of the kernel is to allow application programs to structure their data in the common space and to leave the dialog treatment to the system. As a consequence, the user interface is mostly determined by the kernel, rather than by the application program, and the user gets the impression that all programs in the entire system *talk the same language*.

In order to achieve this goal, the kernel of XS-1 has four major components:

EXPLORE makes sites, modes and trails visible to the user. It defines and controls all motion on these spaces and handles user requests for screen layout changes by means of universal commands. Explore allows the user to inspect the current system state without changing it.

TREE EDITOR provides universal commands for structural manipulations that are common to most editors regardless of the type of data objects they manage. It is syntax-directed in order to reflect the relationships between different data types.

CENTRAL DIALOG CONTROL is a front-end dialog processor that handles user input at all levels, from key press to command, checks it for syntactic correctness and records it. It directs information to Explore, the tree editor, and to modes.

TREE FILE SYSTEM realizes the data access and manipulation requirements imposed by Explore and the tree editor. It handles data of any size, from a single character to a large file, in a uniform way.

The kernel also contains a screen management package that handles an arbitrary number of windows. It can be considered to be a virtual terminal.

Interface between the kernel and modes

Conventional interactive application programs devote much code and time to screen layout and to dialog handling. In XS-1 these activities are centralized in the kernel. A mode presents its data on a virtual screen and lets the screen handling package map the virtual screen onto the physical screen in a window of arbitrary size and position. The kernel handles the dialog and provides universal commands for visualizing

```
|              U S E R   M O D E S                |
|                                                 |
|←←←←←←←←←←←←←←←←←←←←←←←←←←←←←←←←←←←←←←←←←←←←←←←←←|    ↑
|              Kernel-Mode Interface              |    K
|                                                 |
|←←←←←←←←←←←←←←←←←←←←←←←←←←←|←←←←←←←←←←←←←←←←←←←←←|    E
|        TREE EDITOR        |        EXPLORE      |
|                           |                     |    R
|←←←←←←←←←←←←←←←←←|←←←←←←←←←←←←←←←←←|←←←←←←←←←←←←←|
|    TREE FILE    |     CENTRAL     |    SCREEN   |    N
|     SYSTEM      |  DIALOG CONTROL |  MANAGEMENT |
|                 |                 |             |    E
|←←←←←←←←←←←←←←←←←←←←←←←←←←←←←←←←←←←←←←←←←←←←←←←←←|
|              Kernel-Host Interface              |    L
|                                                 |    ∨
|←←←←←←←←←←←←←←←←←←←←←←←←←←←←←←←←←←←←←←←←←←←←←←←←←|
|                    H O S T                      |
|           Hardware and System Software          |
```

Figure: Components of the XS-1 kernel

the state of the system and for performing structural changes to the site and mode spaces. A sizable portion of the code of typical application programs resides in the kernel; only the mode-specific operations remain to be written by the applications programmer.

The interface between the kernel and modes is syntax-directed with respect to data and to commands: A mode defines the syntax of the site space it works on, and the syntax of its commands. The former drives the tree editor, the latter drives central dialog control. Any collection of data, independently of their size or type, may be attached to a node in the site tree, and any collection of commands may be attached to a node in the mode tree. Data attached to the current site is affected by commands; commands attached to the current mode are active.

Interface between the kernel and host

This interface maps the requirements of the tree file system, of central dialog control, and the screen management package onto corresponding utilities of a conventional operating system, such as schedulers and drivers.

XS-1 is written in Modula-2 and runs on a Lilith workstation. The more recent version XS-2 has also been ported to a VAX equipped with a Bitgraph terminal and to several 68000-based personal computers.

3.4 The controversy about modes

The notion of a *"modeless system"* is one of today's fads. Amusing stories purport to document the disadvantage of systems that behave differently in different modes. For example, that of the user who reaches the conclusion that the computer is down because the terminal echos his key presses without executing any of the commands he enters. The unintentional *insert command* that got the system into the editor's insertion mode has long since been scrolled off the screen, so the user is unaware that he is interacting with a different subsystem than the one he thinks he is.

The major design error behind this story, however, is **not** that the system behaves differently in distinct modes, **but that the insertion mode is hidden.** This type of error stems from the days of printing terminals, on which a periodically appearing line *"now in insertion mode"* is indeed impractical. But on a display terminal a dedicated system status line is conventional. And if the application programmer does not wish to lose a single line on the screen, the system state can be displayed on demand, by way of a universal command which is **not** an insertable character!

How would a "trully modeless system" look? Fortunately we know that, although such systems are disappearing. Early Kanji typewriters for inputting thousands of Chinese characters had thousands of keys. "Trully modeless" means that no physical input device ever has its meaning affected by the system state. As a consequence, every logical command must have its own physical device. This may be practical for special purpose machines such as watches, pocket calulators, or games, but is impossible for general purpose work stations, on which an open-ended spectrum of applications will run.

Interactive systems have a lot of commands! They are not easily counted, as a complex command with many parameters can always be broken into simple commands without parameters, and vice versa. If we count the different decisions that the user may be called upon to make, i.e. we weigh each command with the number of its parameters, we easily reach hundreds of items of information that the user should know (think of a typical BASIC system on a hobby computer). After weeks of daily use, the professional has memorized these hundreds of items in his long-term memory. But the casual user, who must rely on his short-term memory of *"the magical number seven, plus or minus two"* chunks of information [Mi 56], finds himself constantly looking up the manual.

The *"sites, modes, and trails"* model implemented in XS-1 and XS-2 attempts to solve the problem of the "7 + or - 2 chunks in short-term memory" as follows. **The dozen universal commands define a modeless dialog machine** - the user must memorize these application-independent commands once and for all. An interactive application activates its own commands in addition. These are visible in the mode window, so the user need not memorize them. We tend to structure application packages into several modes, typically with less than 10 comands each, whose meanings are related.

4. Is a standard man-machine interface possible?

4.1 The user interface problem in a network of home computers

Once we have mastered the user interface of our first automobile, it only takes minutes to operate the steering wheel, pedals, and buttons of another car. Occasionally we must search for a light switch, but at least the car-driver interface has been standardized to such an extent that we need no manual in order to drive. Contrast this with today's interactive systems, where expertise with one text editor does not guarantee that you can use another without a manual.

The issue of standard man-machine interfaces will become acute with the coming integration of personal computers and telecommunications. I would like to illustrate this point with a study performed at Nippon Telegraph and Telephone. NTT's Information Network System INS [Ki 82] will link home communication terminals to a national fiber optics transmission network of sufficient bandwidth to carry animated pictures. The planned services include electronic mail with sound and pictures, access to libraries and information systems, financial services - whatever comes to mind in the domains of entertainment, education and business at home.

In the rapidly changing market of personal computers, the home terminal cannot be specified ahead of time - most of the needed 30 million terminals would be obsolete before their construction. One will have to accept as home terminals many of the personal computers that will come on the market over the next decades, with properties as yet unknown. The danger of a Babylonian chaos lurks in this scenario. The user who has mastered the terminals in his home and in his office will encounter another machine in the "telephone cabin". Will he be able to operate it without having to study a manual? Whereas the long-lived telephone set made standardized operation possible through essentially identical hardware, the short-lived personal computer will have to rely on a logical standardization as well.

4.2 The scope of existing standards

There is much standardization activity relevant to the problem of the network of home communication terminals, but standard dialog control is has barely been touched. Consider the different system components to be standardized by way of an example: The user wishes to interact with a teaching program obtained from a national library without any loss of quality.

First, some file(s) will have to be transmitted from the library into his home computer. Many standards govern network protocols for data communication. Their multitude is a problem, but a national carrier protected by a monopoly can choose its own, so file transfer is no problem.

When the files have arrived, they must be looked at in several ways. Text poses few problems. The easiest way to standardize text file formats is to limit the opaerating systems to be used to a few that are widely used on personal computers, such as MSDOS, CP/M, or Unix. Graphics is harder, as today's operating systems don't support the concept of a graphics file. Standards are available or underway from the computer field (ACM Core Graphics, GKS) and from telecommunications (Videotex, facsimile). Within the near future, they will be sufficiently stable to serve the purpose of viewing pictures on a home computer.

Text and pictures are only components of an interactive program that the user wants to execute under his control. What standards do we have to govern execution of an interactive program? A superficial answer is: Any programming language standardizes execution, so all we have to do is agree on a few widely used languages such as Basic, Fortran, or Pascal. This will guarantee that a program will execute *somehow*, - with a behavior as criticized in the section on *"Errors in dialog design"*.

The realistic answer is that *standardization of execution control of interactive programs has not even begun yet*. There is substantial experience with individual components of dialog control, but no unity about how to present these to the user.

4.3 The most frequent dialog control commands and their hardware realization

At the Yokosuka Electrical Communication Laboratory of NTT an experiment is underway for standardizing dialog control on a network of heterogeneous home computers. The idea is to freeze half a dozen of the most important dialog control commands, universal commands interpreted by the operating system, not by the application, into hardware. These commands include *motion and selection* (changing data and command environment), *viewing* (control over how much of an object is to be seen), and *operations on the history* (undo). The standard hardware is a *"mighty mouse"*, with five keys and a thumbwheel, tuned to the agility of a person's hand.

The motivation for developing this mouse needs explaining, since it points in the opposite direction of commercial "mouse architecture". The mouse popularized by the Xerox Alto computer has 3 keys, but its successor on the commercial Xerox Star has only 2, and the mouse of the Apple Lisa has a single key. This according to the "theory" that the user gets confused by many keys. However, the software developed for these computers tells a different story: often, different logical commands are multiplexed onto these one or two keys, and coded by such gimmicks as a "fast double click" (different from two slow clicks), or clicking while some key is depressed on the keyboard.

These examples lead to the conclusion that it is *not the number of keys that confuses the user, but the changing semantics assigned to the keys!* If a physical input device is assigned only one function, then we can reasonably use more than one such device. It's the idea of "modeless", not applied to the entire system, but restricted to a few universal dialog control commands.

Physically, the "mighty mouse" has five keys. Thumb and index finger are agile, so they control analog inputs. The other three fingers control 0-1 inputs. The thumb moves with greater freedom than any other finger, so its key can be switched into two different positions; in addition, the thumb glides sideways to turn a thumbwheel.

Logically, the following functions have been assigned to these keys. The thumb key realizes a linear motion command - gas pedal and gear shift in one. Varying pressure causes variable speed forward or backward, depending on the position of the key. A hard click means forward or backward by a unit.

The index finger controls a "show me"-key. Increasing pressure asks for increasing depth of detail, for example through zooming. A hard click selects this object, that is, turns it into the active data environment, the object of the following commands.

An object is identified by moving the cursor to it. For graphical object, built from intersecting lines, an oriented cursor is useful. It determines x, y, and an angle coordinate, which is set by turning the thumbwheel.

The remaining keys are conventional. The middle finger controls a pop-up menu, the ring finger an *"undo"*, the small finger an *"exit"* into the hierarchically next higher data or command environment.

A user must memorize these meanings once and for all - thereafter, he should be able to explore the entire system, even on unfamiliar hardware. In order to master an application program, he will undoubtedly need to learn commands specific to the application- but in order to browse in it and determine whether he really wants to use it, the universal commands of the "mighty mouse" might suffice.

5. Programming the man-machine interface

Programming methodology has emphasized different criteria for judging the quality of software during the three decades of its existence. In the early days of scarce computing resources the most important aspect of a program was its *functionality* - what it can do, and how efficiently it works. In the second phase of *structured programming* the realization that programs have a long life and need permanent adaptation to changing demands and environments led to the conclusion that functionality is not enough - a *program must be understandable*, so that its continued utility is not tied to its "inventor". Today, in the age of interactive computer usage and the growing number of casual users, we begin to realize that *functionality and good structure are not enough - good behavior towards the user is just as important.*

Thus the traditional programmer, trained to analyze systems of great logical complexity in detail, is confronted with an entirely new demand. He is called upon to design expressive pictures and lay them out on the screen, to formulate clear phrases and assemble them into an understandable presentation - skills that demand the creative-artistic flair of a graphic designer and author. It is satisfying to observe that many programmers have recently become concerned with the quality of the man-machine dialog that their interactive programs conduct, as judged from the user's point of view. Well they might, for the computer user population is changing rapidly. With the spread of low-cost single-user computers casual and occasional users abound, and for them the quality of the man-machine interface is crucial - an interactive system is only useful if the learning effort is commensurate with the brevity of the task they want to accomplish. It is for their benefit that computer professionals should start paying as much attention to their communicative skills as writers have always done.

References

[Be 82] G. Beretta, H. Burkhart, P. Fink, J. Nievergelt, J. Stelovsky, H. Sugaya, J. Weydert, A. Ventura, *XS-1: An integrated interactive system and its kernel,* 340-349, Proc. 6-th International Conference on Software Engineering, Tokyo, IEEE Computer Society Press, 1982.

[Ki 82] Y. Kitahara, **Information Network System · Telecommunications in the 21st century,** The Telecommunications Association, Tokyo, 1982.

[Mi 56] G. A. Miller, *The magical number seven, plus or minus two: Some limits on our capacity for processing information,* Psych. Review, Vol 63, No 2, 81-96, March 1956.

[NW 80] J. Nievergelt and J. Weydert,
Sites, Modes, and Trails: Telling the user of an interactive system where he is, what he can do, and how to get places, in **Methodology of Interaction**, R. A. Guedj (ed), 327-338, North Holland 1980.

[Ni 82] J. Nievergelt,
Errors in dialog design, and how to avoid them,
in **Document Preparation Systems · A Collection of Survey Articles,**
J. Nievergelt et al. (eds), North Holland Publ. Co., 1982.

[No 81] D. A. Norman, *The trouble with UNIX,* 139-150, Datamation, Nov 1981.

[St 84] J. Stelovsky,
XS-2: The user interface of an interactive system, ETH dissertation 7425, 1984.

CAVIAR
A Case Study in Specification

Bill Flinn
Standard Telecommunication Laboratories
Ib Holm Sørensen
Oxford University Computing Laboratory

Abstract

This paper describes the specification, written in the specification language known as Z, of a reasonably complex software system. Important features of the Z approach which are highlighted in this paper include the interleaving of mathematical text with informal prose, the creation of parametrised specifications, and use of the Z schema calculus to construct descriptions of large systems from simpler components.

0. Introduction

This paper presents a case study in system specification. The notation used to record the system's properties is known as Z [1, 2, 3]. Z is based on set theory, and its use as a specification language has been developed at the Programming Research Group at Oxford University. Some important aspects of the Z approach are illustrated in this paper.

As is well known, software development can be divided into several phases; requirements analysis, specification, design and implementation. Z can be applied in both the specification and design phases; however, in this paper we will address the specification phase only.

We view a specification as having a two-fold purpose: firstly, to give a formal (mathematical) system description which provides a basis from which to construct a design. Such a mathematical description is essential if we are to prove formally that a design meets its specification. Secondly, to give an informal statement of the system's properties, in order that the specification can be tested (validated) against the (usually informal) statement of requirements. Thus the Z approach is to construct a specification document which consists of a judicious mix of informal prose with precise mathematical statements. The two parts of the document are complementary in that the informal text can be viewed as commentary for the formal text. It can be consulted to find out what aspects of the real world are being described and how it relates to the informally stated requirements. The formal text on the other hand provides the precise definition of the system and hence can be used to resolve any ambiguities present in the informal text. A beneficial side effect for practitioners writing such documents is that their understanding of the system in question is greatly helped by the process of constructing both the formal and the informal descriptions.

It is often the case that the process of abstraction used to construct a specification results in structures which are more general than those actually required for the system being considered. It is part of the Z approach to identify and desribe such general structures. These descriptions can be placed in a specification library. Particular cases of these general components can then be used later, either as part of the current system or in subsequent projects.

This specification case study develops a number of general systems which are subsequently constrained and combined to form the complete system description.

1. The Case Study

This specification of a Computer Aided Visitor Information And Retrieval system resulted from the analysis of a manual system concerned with recording and retrieval of data about arrangements for visitors and meetings at a large industrial site. Standard Telecommunications Laboratories (U.K.) sponsored the study in order to investigate the feasibility of converting to a computer based solution. Of particular concern were the interrelation of the stored information, the quality of the user interface and the volume of data which was required to be processed. The customer provided as input to the study an informal requirements document. We attempt to provide in this paper an outline of the steps involved in development of the eventual formal specification. It is important to stress at the outset that we view the task of constructing such a specification to be an iterative process, involving several attempts at construction of a model for the system interspersed with frequent dialogues with the customer to clarify details which are ambiguous or undefined in the initial requirements document, and frequent redrafting to clarify the structure of the document.

At an early stage in the analysis it became clear that the CAVIAR system consisted of several highly independent subsystems. Each subsystem records important relationships within the complete system and these separate subsystems are themselves related according to some simple rules. Most of the operations to be provided in the user interface can be explained as functions which transform one particular subsystem only, leaving the others invariant. These observation led to the decision to first define the subsystems in isolation and then to describe the complete system by combining the definitions of the subsystems. Once this decision had been taken, it also became clear that each of the individual subsystems, when viewed at an appropriate level of abstraction, was a particular instance of a general structure. From this vantage point it was natural to specify each of the subsystems by "refining" a specification which describes the underlying general system.

The process of analysis as presented here begins with an identification of the sets which appear to be important from the customer's point of view. Next the relationships between these sets are investigated and a preliminary classification of the subsystems follows. The third phase consists of developing an appropriate general mathematical structure in which to place these subsystems. Various ways of *specialising* (restricting) the general structure are then investigated and particular subsystems are modelled by *instantiation*. Finally the subsystem models are combined.

2. Identification of the Basic Sets

We now present a brief account of the existing system, emphasizing the important concepts in boldface. **Visitors** come to the site to attend **meetings** and/or consult Company employees. A visitor may require a **hotel reservation** and/or **transport reservation**. Each meeting is also required to take place in a designated **conference room**, at a certain **time**. A meeting may require the use of a **dining room** for lunch, on a particular date. Booking a dining room requires **lunch information** including the number of places needed. Each conference room booking requires **session information** about resources required for use in the meeting, e.g., viewgraphs, projectors. The main operations required at the user interface can briefly be described as facilities for booking, changing and cancelling the use of resources. We list below the sets together with the names that we shall adopt for referring to them.

Set	Name
Meetings	M
Visitors	V
Conference Rooms	CR
Dining Rooms	DR
Lunch Information	LI
Session Information	SI
Hotel reservation	HR
Transport reservation	TR

The informal interpretation of these sets is straight forward and for the purpose of this specification no further detail is necessary. Note that the question of modelling time remains to be resolved; at this point we simply observe that hotel reservations are made for particular *dates*, transport reservations are made for certain *times* on particular dates, and conference room bookings are made for *sessions* on particular dates. We shall not specify the term *session* further apart from noting that a date is always associated with a session; it could, for example, denote complete mornings or afternoons, or hourly or half-hourly intervals, depending on the way conference rooms are allocated.

The notion of time and the relationship between the different units of time used within the system can be formalised by asserting the existence of three sets as follows:

```
Date
Session
Time
```

together with two total functions

```
date-of-session : Session → Date
date-of-time    : Time    → Date.
```

3. The Subsystems of CAVIAR

The first approach to a mathematical model stems from the realisation that several of the sets listed above can be viewed as *resources* and other sets viewed as *users* of those resources. We can identify the following subsystems of CAVIAR in this framework (i. e., Resource-User systems). Observe that in different subsystems the same set may appear in differing roles.

System	Resources	Users
CR-M	Conference rooms	Meetings
DR-M	Dining rooms	Meetings
M-V	Meetings	Visitors
HR-V	Hotel reservations	Visitors
TR-V	Transport reservations	Visitors

Once we have made this mathematical abstraction it seems worthwhile to develop a general theory of such resource-user systems for the following reasons:

1. A specification of such a general system would be more useful as part of a "specification library" than a specific instance of such a system. Re-usability is much more likely to be achieved by having *generic* specifications available which can be *instantiated* to provide particular systems.

2. Particular subsystems of the general system can be constructed as *special cases* of the general specification in various ways. This will amply repay care and time spent on the general case. Furthermore, such instantiation may well result in a more compact implementation.

4. A General Resource-User System

We consider a system parametrised over three sets;

$$[\ T, \ R, \ U \ ]$$

Informally, T is to be thought of as a set of *time slots*, R is a set of *resources* and U is a set of *users*. We describe a general resource-user system as a function from T to the set of relations between R and U. Thus we have a rather general framework: for each time slot $t \in T$, some users are occupying or using some resources. The set T will later be instantiated with different sets in the various applications. Notice that considering *relations* between R and U allows us the possibility of a user occupying several different resources simultaneously, as is shown informally in the following diagram:

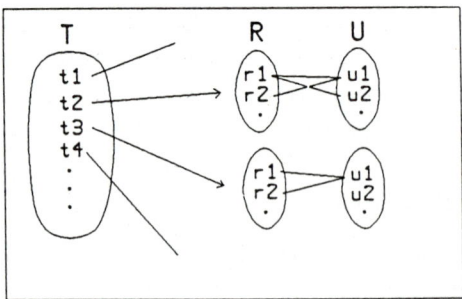

Formally, the structure we are describing is captured by a function of type

$$T \rightarrow (R \leftrightarrow U)$$

We shall now incorporate this into a *schema definition*. This schema is parametrised over the sets T, R and U, and contains some useful ancillary concepts in addition to the function ru above which will be useful in later analysis. In Z specifications it is common to introduce such derived components: as specifiers of software we are neither in the position of a pure mathematician looking for a particularly sparse set of concepts and axioms with which to define a mathematical structure, nor are we in the position of an implementor trying to minimise storage. The component in-use, which gives the set of resources in use at any point of time, will be useful in contexts where we are not concerned with the user component of the system state. The function users, which gives the users occupying resources at any point of time, will be used in situations where we do not require the information about resources. We also note that there may be occasions when we wish to consider the set of *inverse* relations generated by ru; we call this function ur.

$$
\begin{array}{|l}
\hline
\text{R-U} \\
\hline
\text{ru} \quad : T \rightarrow (R \leftrightarrow U) \\
\text{in-use} : T \rightarrow \mathbb{P}\, R \\
\text{users} \quad : T \rightarrow \mathbb{P}\, U \\
\text{ur} \quad : T \rightarrow (U \leftrightarrow R) \\
\hline
\forall t : T \bullet \\
\quad \text{in-use}(t) = \text{dom}(\text{ru}(t)) \land \\
\quad \text{users}(t) = \text{rng}(\text{ru}(t)) \land \\
\quad \text{ur}(t) = (\text{ru}(t))^{-1} \\
\hline
\end{array}
$$

The *initial state* of this system is defined by making ru(t) the empty relation for each t.

$$\text{Init-R-U} \;\hat{=}\; [\; \text{R-U} \mid \text{rng}(\text{ru}) = \{\, \{\}\, \} \;]$$

Our first theorem proves that such an inital state is reasonable and assures us of the consistency of the definition of R-U.

Theorem 1.
$$\vdash \; \exists\, \text{R-U} \bullet \text{Init-R-U}$$

In the interests of readability we have not given proofs of theorems stated in this paper.

We continue by defining the appropriate operations for this structure. The first step is to identify *commonalities*. For our purposes, the operations that we wish to consider on this structure are concerned with making a new booking, i.e., adding a new pair (r, u) to an existing relation at some time t, cancelling an existing booking, i.e., removing such an (r, u) pair, or modifying in some other way the relation that exists at some particular time. In fact we shall be a little more general and define a class of operations on R-U which allows the image of a *set* of time values to be altered. This is because we anticipate such operations as booking a conference room for a meeting which lasts for several time slots. Of course a booking which involves only a single time slot is a particular case.

Thus we may summarise the common part of all the operations as follows. Their description involves: a state before, R-U which introduces ru, in-use, users and ur; a state after, R-U′ which introduces ru′, in-use′, users and ur; a set of time values, t? which denotes an input. The operations always leave the function ru unchanged except for times in t?. Formally this is captured by

```
┌─ ΔR-U ─────────────────────────────┐
│  R-U                               │
│  R-U′                              │
│  t? : P T                          │
│ ───────────────────────────────    │
│  t? ◁ ru′  =  t? ◁ ru              │
└────────────────────────────────────┘
```

We now have a successful *booking* operation defined as follows

```
┌─ R-U-Book ─────────────────────────┐
│  ΔR-U                              │
│  r? : R                            │
│  u? : U                            │
│ ───────────────────────────────    │
│  ∀ t : t? •                        │
│      (r?, u?) ∉ ru(t) ∧            │
│      ru′(t) = ru(t) ∪ { (r?, u?) } │
└────────────────────────────────────┘
```

Thus R-U-Book inherits all the properties of ΔR-U. Furthermore, it takes two additional (input) parameters r?:R and u?:U, and is constrained by a predicate which imposes a requirement on the input parameters and also further relates the before and after states.

Notice that we are making the predicate

$$\forall\, t: t? \bullet (r?, u?) \notin ru(t)$$

a pre-condition for a successful booking. In fact, we can show that this condition is sufficient for performing a successful booking, i. e., if we are in a valid system state with the required input parameters of the correct type available and furthermore the above condition holds, then there exists a resulting valid system state which is related to the starting state according to the R-U-Book schema. Formally, this is the content of the following result:

Theorem 2
$$R\text{-}U \wedge [\,t?: \mathbf{P}\, T;\ r?: R;\ u?: U\ |\ \forall\, t:\, t? \bullet (r?, u?) \notin ru(t)\,]$$
$$\vdash$$
$$\exists\, R\text{-}U' \bullet R\text{-}U\text{-}Book$$

A successful *cancellation* operation may be defined via

```
┌─ R-U-Cancel ──────────────────────────────────┐
│  ΔR-U                                         │
│  r? : R                                       │
│  u? : U                                       │
├───────────────────────────────────────────────┤
│  ∀ t : t? •                                   │
│      (r?, u?) ∈ ru(t) ∧                       │
│      ru'(t) = ru(t) − { (r?, u?) }            │
└───────────────────────────────────────────────┘
```

The pre-condition for successful cancellation is that the pair $(r?, u?)$ is related by $ru(t)$ for all time values t in $t?$; i. e., the following theorem holds.

Theorem 3
$$R\text{-}U \wedge [\,t?: \mathbf{P}\, T;\ r?: R;\ u?: U\ |\ \forall\, t:\, t? \bullet (r?, u?) \in ru(t)\,]$$
$$\vdash$$
$$\exists\, R\text{-}U' \bullet R\text{-}U\text{-}Cancel$$

So far we have only specified successful operations; thus these descriptions are incomplete. We could at this stage define robust operations by introducing appropriate error recovery machinery. In the interests of simplicity we shall not give a general treatment of errors; however we shall indicate in a later section how the descriptions of the operations at the user interface can be completed.

We shall define two further operations on this structure. The first involves deleting a resource and all use of that resource. This is an operation to be treated with caution: see Theorem 7 below.

```
┌─ R-U-Del-Res ─────────────────────────┐
│ ΔR-U                                  │
│ r? : R                                │
├───────────────────────────────────────┤
│ ∀ t : t? •                            │
│     r? ∈ dom ru(t) ∧                  │
│     ru'(t) = { r? } ⩤ ru(t)           │
└───────────────────────────────────────┘
```

Informally, this operation may be described as follows. Consider each element t in t? and the corresponding relation ru(t) in turn. All elements (r?, u) are to be removed from ru(t).

Theorem 4
    R-U ∧ [t? : P T; r? : R | ∀ t: t? • r? ∈ dom ru(t)]
    ⊢
    ∃ R-U' • R-U-Del-Res

Corresponding to deleting a resource there is an operation which, given a user value u?, deletes all pairs (r, u?) from the relations associated with time values in t?. This is defined as follows:

```
┌─ R-U-Del-User ────────────────────────┐
│ ΔR-U                                  │
│ u? : U                                │
├───────────────────────────────────────┤
│ ∀ t : t? •                            │
│     u? ∈ rng ru(t) ∧                  │
│     ru'(t) = ru(t) ⩥ { u? }           │
└───────────────────────────────────────┘
```

Theorem 5
    R-U ∧ [t? : P T; u? : U | ∀ t : t? • u? ∈ rng ru(t)]
    ⊢
    ∃ R-U' • R-U-Del-User

So far we have listed theorems that a specifier is obliged to prove; viz the result that the initial state satisfies the required definition (and therefore that the specification is consistent) and the theorems that explicitly give the pre-conditions for each operation.

For the specifications that we shall develop from now on these theorems have been omitted in the interests of brevity.

In addition to these obligatory results, there are other "optional" theorems that are a consequence of the specification, and which often give insight into the structure being developed.

Two such results for our system are as follows:

Theorem 6
$$\text{R-U-Book} \; \text{\textbf{;}} \; \text{R-U-Cancel} \quad \vdash \quad ru' = ru.$$

Informally, this theorem states that if we make a booking and follow it immediately by a cancellation using the same input parameters, then the state of the system does not change.

Theorem 7
R-U-Del-Res
$$\vdash$$
$$\text{in-use}' = \text{in-use} \oplus (\lambda t:t? \bullet \text{in-use}(t) - \{ r? \}) \land$$
$$\text{users}' = \text{users} \oplus$$
$$(\lambda t:t? \bullet \text{users}(t) - \{ u : U \mid ur(t)(\!(\{u\})\!) = \{r?\} \})$$

This theorem makes precise the informal comment made earlier about the need for caution with the R-U-Del-Res operation. This theorem shows that resources are removed from the system structures, which we do expect, but furthermore the operation can also remove existing users.

There is a similar result concerning the R-U-Del-User operation.

5. Specialisation of the General R-U System

We shall now *specialise* the general R-U system into particular classes of the system. These specialisations are motivated by the observation that for some of the instances listed earlier, at any given time a resource may be related to only one user, or a user may occupy only one resource, or both.

5.1 An R-U system where resources cannot be shared

The first case we define is the class where each resource may be utilised by at most one user, but each user may occupy several resources. We denote this system by R⟩U (where "⟩" is just a character in the name) and define it formally by

$$R{\rangle}U \,\,\triangleq\,\, [\,\, R{-}U \,\,|\,\, rng(ru) \subseteq R \twoheadrightarrow U \,\,]$$

The initial state of this system is given by the same condition as for Init-R-U; thus we have

$$Init{-}R{\rangle}U \,\,\triangleq\,\, [\,\, R{\rangle}U \,\,|\,\, rng(ru) = \{\,\{\}\,\} \,\,]$$

All operations are described in terms of

$$\Delta R{\rangle}U \,\,\triangleq\,\, R{\rangle}U \wedge R{\rangle}U'$$

The operations on this system may be defined as special cases of the general operations for R-U. We first consider the booking operation.

$$R{\rangle}U{-}Book \,\,\triangleq\,\, \Delta R{\rangle}U \wedge [\,\, R{-}U{-}Book \,\,|\,\, \forall\, t: t? \bullet r? \notin dom\, ru(t) \,\,]$$

The qualifying predicate is included in indicate that there is a further pre-condition for booking a resource in a R⟩U system.

We now have two parts to the pre-condition for this operation; firstly this qualifying predicate, and secondly the pre-condition arising from R-U-Book. In fact the former implies the latter, as is easily checked.

The cancellation operation is defined as follows:

$$R{\rangle}U{-}Cancel \,\,\triangleq\,\, R{-}U{-}Cancel \wedge \Delta R{\rangle}U$$

On considering the two deletion operations defined for R-U, we observe that R-U-Del-Res is equivalent to a cancellation in our present context, because the resource is associated with only one user. We therefore need only the operation which deletes a user.

$$R{\rangle}U{-}Del{-}User \,\,\triangleq\,\, R{-}U{-}Del{-}User \wedge \Delta R{\rangle}U$$

5.2 An R-U system where each user may occupy at most one resource

The second case we define is the class where each user may occupy at most one resource but resources may be shared amongst users. We denote this system by R≲U and define it formally by

$$R{\leqslant}U \;\triangleq\; [\; R\text{-}U \;|\; rng(ur) \subseteq U \twoheadrightarrow R \;]$$

The initial state of this system is also given by the predicate for Init-R-U. We have

$$Init\text{-}R{\leqslant}U \;\triangleq\; [\; R{\leqslant}U \;|\; rng(ru) = \{\;\{\}\;\} \;]$$

The operations are described in terms of

$$\Delta R{\leqslant}U \;\triangleq\; R{\leqslant}U \wedge R{\leqslant}U$$

We now define the booking operation for the system.

$$R{\leqslant}U\text{-}Book \;\triangleq\; \Delta R{\leqslant}U \wedge [\; R\text{-}U\text{-}Book \;|\; \forall\, t : t? \bullet u? \notin rng\; ru(t) \;]$$

As before, a qualifying predicate is needed and again as before the constraint given here implies the earlier pre-condition for the general R-U-Book operation.

The cancellation operation is defined as follows:

$$R{\leqslant}U\text{-}Cancel \;\triangleq\; R\text{-}U\text{-}Cancel \wedge \Delta R{\leqslant}U$$

On considering the two deletion operations defined for R-U, we observe that this time R-U-Del-User is equivalent to a cancellation in our present context, because a user may be associated with only one resource. We therefore need only the operation which deletes a resource.

$$R{\leqslant}U\text{-}Del\text{-}Res \;\triangleq\; R\text{-}U\text{-}Del\text{-}Res \wedge \Delta R{\leqslant}U$$

5.3 An R-U system where a user occupies at most one nonsharable resource

The third and last specialisation we define shares all the properties of the systems defined in the preceding two sections. It is therefore defined as the *conjunction* of the two schemas above. In this system each user may occupy at most one resource and each resource may be occupied by at most one user. Formally we have

$$R\overline{\subset}U \triangleq R\overline{>}U \wedge R\overline{<}U$$

The initial state of this system is clearly defined by

$$\text{Init-}R\overline{\subset}U \triangleq [\ R\overline{\subset}U\ |\ rng(ru) = \{\ \{\}\ \}\ ]$$

The operations on this system are given by the conjunction of the operations defined for each of the two earlier systems. For this system we require only the booking and cancellation operations. Thus we have

$$R\overline{\subset}U\text{-Book} \triangleq R\overline{>}U\text{-Book} \wedge R\overline{<}U\text{-Book}$$

$$R\overline{\subset}U\text{-Cancel} \triangleq R\overline{>}U\text{-Cancel} \wedge R\overline{<}U\text{-Cancel}$$

5.4 The specification library

We have now constructed four specifications which might be considered to form the nucleus of a specification library for resource-user systems. We may summarise the relationships between the four classes of system schematically as follows:

```
              R-U ───── Most general
             ↗ ↑ ↖
           ╱   │   ╲
        R>U   ·   R<U
           ╲   │   ╱
            ↖ │ ↗
              R⊂U ───── Most constrained
```

6. Classification and Instantiation

6.1 Some laws for CAVIAR

In this section, in order to illustrate the clarification process which took place during requirements analysis, we list some observations about the CAVIAR system which emerged during dialogue with the customer. We formalise the important constraints as *laws* which need to be taken account into account in the development which follows.

1. At any time a conference room is associated with only one meeting.

2. At any time a meeting may be associated with *more than one* conference room.

Law 1 is reasonably obvious: it would be difficult to hold more than one meeting in a given room. Law 2 is not obvious: it was unclear from the informal description whether or not a meeting could occupy more than one room. In fact the customer believed initially that a meeting could only take up one room, but a counter-example was found amongst the supporting documentation.

3. At any time a meeting is associated with only one dining room.

4. At any time participants from several meetings can occupy the same dining room.

These laws followed from the informal information provided that all visitors in a particular meeting would go to lunch in the same dining room. It was further established that all seats in a dining room were treated as indistinguishable, so further meetings could be accommodated if enough seats were available. Further clarification was necessary regarding lunch times: it transpired that there were "early" and "late" lunches; however this was handled by "doubling up" each dining room. For example, a booking would be made for "DR 1, early" and this was a different dining room from "DR 1, late."

5. At any time a visitor is associated with only one meeting.

6. At any time a meeting may involve several visitors.

Law 5 had to be checked out with the customer.

7. At any time a hotel room is associated with only one visitor and vice versa.

8. At any time a transport reservation is associated with only one visitor and vice versa.

Law 7 was natural, but law 8 was less so. It was established that even if the transport department decided to use a minibus, a separate transport reservation would be issued to each visitor.

6.2 Matching system with models

In this section we first consider each CAVIAR subsystem in turn and match it to the appropriate model. In fact we have enough structure available to define two subsystems directly and we do this in the remainder of this section.

(1) We first consider the conference room - meeting system CR-M.

From laws 1 and 2 we see that CR-M is an instance of the R$>$U subsystem.

(2) The dining room - meeting subsystem DR-M.

Applying laws 3 and 4 we find that DR-M is an instance of R$<$U.

However this system does not contain any information about numbers of seats or the lunch details, so we will need to extend this system later.

(3) The meeting - visito subsystem M-V.

From laws 5 and 6 M-V is an instance of R$<$U.

However we have not documented the fact that meetings have to be created before visitors can be attached to them; this will also be done later.

(4) The hotel reservation - visitor subsystem HR-V, and the transport reservation - visitor subsystem TR-V, both have the property that each resource is occupied by only one user and vice versa. Therefore both these systems are instances of R$=$U.

In fact this model is sufficient to define HR-V and TR-V completely, by *instantiation*, as we now show.

6.3 The hotel reservation subsystem - HR-V

We define HR-V as follows:

$$\text{HR-V} \triangleq \text{R}=\text{U}_{\text{HR-V}}[\text{Date, HR, V}]$$

This object is a *decorated* instance of the R$=$U schema, with its parameter sets instantiated by the sets Date, HR and V introduced in section 2. To be more explicit, the definition above is shorthand for the following:

┌─ HR-V ───┐
│ ru_{HR-V} : Date → (HR ↔ V)   │
│ $in\text{-}use_{HR-V}$: Date → $\mathbb{P}$ HR   │
│ $users_{HR-V}$: Date → $\mathbb{P}$ V   │
│ ur_{HR-V} : Date → (V ↔ HR)   │
│   │
│ $rng(ru_{HR-V}) \subseteq HR \twoheadrightarrow V$ ∧ │
│ $rng(ur_{HR-V}) \subseteq V \twoheadrightarrow HR$ ∧ │
│ (∀t: Date; r: HR •   │
│   r ∈ $in\text{-}use_{HR-V}(t)$ ⇔ r ∈ $dom(ru_{HR-V}(t))$) ∧ │
│ (∀t: Date; u: V •   │
│   u ∈ $users_{HR-V}(t)$ ⇔ u ∈ $ran(ru_{HR-V}(t))$) ∧ │
│ (∀t: Date • $ur_{HR-V}(t) = (ru_{HR-V}(t))^{-1}$) │
└──┘

Thus each component of the schema is given the decoration in the definition, and each occurrence of the parametrised sets is instantiated as shown above. From now on we shall use such decoration without further comment.

The initial state of HR-V is given by

$$\text{Init-HR-V} \triangleq \text{Init-R}\Xi U_{HR-V}[\text{Date, HR, V}]$$

and the operations are given by

$$\text{Book-Hotel-Room}_0 \triangleq \text{R}\Xi\text{U-Book}_{HR-V}[\text{Date, HR, V}]$$

and

$$\text{Cancel-Hotel-Room}_0 \triangleq \text{R}\Xi\text{U-Cancel}_{HR-V}[\text{Date, HR, V}]$$

6.4 The transport reservation subsystem - TR-V

This subsystem is essentially the same as the HR-V subsystem except for the parametrisation. The instances of the parameters are denoted respectively Time, TR and V, where once again the sets TR and V are as in section 2. We shall not specify the set Time further, except to repeat that it contains a Date component (see section 2). Thus we have

$$\text{TR-V} \triangleq \text{R}\Xi U_{TR-V}[\text{Time, TR, V}]$$

with initial state given by

$$\text{Init-TR-V} \;\triangleq\; \text{Init-R}\Xi\text{U}_{TR-V}[\text{Time, TR, V}]$$

and operations given by

$$\text{Book-Transport}_O \;\triangleq\; \text{R}\Xi\text{U-Book}_{TR-V}[\text{Time, TR, V}]$$

and

$$\text{Cancel-Transport}_O \;\triangleq\; \text{R}\Xi\text{U-Cancel}_{TR-V}[\text{Time, TR, V}]$$

7. The Meeting Attendance Subsystem

We now turn our attention to what is necessary in order to complete a model for M-V. Booking and cancelling operations have been defined already but so far we have not taken account of the fact that before bookings can be made the system has to "create" meetings. The question of exactly which objects are "currently defined" at any particular time is important because in several cases only those objects known to the system (i. e., those objects that have been created but not yet destroyed) can book resources, etc.

7.1 A pool system

We can model this situation with a simple structure which we term a Pool. This schema is parametrised over the set T and an arbitrary set X. There are only two operations to be defined; namely those that add an object to, and delete an object from, the pool, over a specified time period.

Formally we have

$$[\,T,\,X\,]$$
$$\begin{array}{|l}\hline \text{Pool} \\ \hline \text{exists} : T \rightarrow \mathbb{P}\,X \\ \hline \end{array}$$

with initial state given by

$$\text{Init-Pool} \;\triangleq\; [\;\text{Pool} \;|\; \text{rng}(\text{exists}) = \{\,\{\}\,\}\;]$$

For later use we define

$$\equiv\text{Pool} \;\triangleq\; [\;\Delta\text{Pool} \;|\; \text{Pool}' = \text{Pool}\;]$$

Given

$$\Delta\text{Pool} \triangleq \text{Pool} \wedge \text{Pool}'$$

The operations are given by

```
┌─ Create ─────────────────────────────────────────────────┐
│ ΔPool                                                    │
│ t? : P T                                                 │
│ x? : X                                                   │
├──────────────────────────────────────────────────────────┤
│ exists' = exists ⊕ (λ t : t? • exists(t) ∪ { x? } )      │
└──────────────────────────────────────────────────────────┘
```

and

```
┌─ Destroy ────────────────────────────────────────────────┐
│ ΔPool                                                    │
│ t? : P T                                                 │
│ x? : X                                                   │
├──────────────────────────────────────────────────────────┤
│ exists' = exists ⊕ (λ t : t? • exists(t) − { x? } )      │
└──────────────────────────────────────────────────────────┘
```

We could have included in the Create operation the pre-condition that the object x? not already exist for any of the times in t?. However we make a deliberate decision here to omit this - having in mind the situation where an object may already exist for some of the times in t? and its existence needs to be extended to all of t?. A similar remark applies to the Destroy operation.

7.2 The meeting - visitor subsystem

To construct the model for the M-V system we combine the Pool and R≼U structures.

```
┌─ M-V ────────────────────────────────────────────────────┐
│ R≼U_{M-V}[Session, M, V]                                 │
│ Pool_M[Session, M]                                       │
├──────────────────────────────────────────────────────────┤
│ ∀t : T •                                                 │
│    in-use_{M-V}(t) ⊆ exists_M(t)                         │
└──────────────────────────────────────────────────────────┘
```

Thus we have combined an M-V instance of an R≼U system and a meeting instantiation of a Pool system (with the parameter sets as shown). The predicate assures that visitors can only attend existing meetings.

The initial state is given by

$$\text{Init-M-V} \triangleq \text{Init-R}{\leqslant}\text{U}_{M-V}[\text{Session}, M, V] \wedge \text{Init-Pool}_M[\text{Session}, M]$$

We now define the operations on M-V in terms of

$$\Delta\text{M-V} \triangleq \text{M-V} \wedge \text{M-V}'$$

The first operation is concerned with adding a visitor to a meeting.

$$\text{Add-Visitor-to-Meeting}_0 \triangleq$$
$$\Delta\text{M-V} \wedge \equiv\text{Pool}_M[\text{Session}, M] \wedge \text{R}{\leqslant}\text{U-Book}_{M-V}[\text{Session}, M, V]$$

When an operation is "promoted" in this way, its new pre-condition is determined as follows: the "old" pre-condition (i. e., that arising from its definition) must be conjoined with a further predicate which arises from the new invariant of the larger state. Here, for example, the pre-condition for the earlier booking operation is given in section 5.2: namely

$$\forall\ t\ :\ t?_{M-V} \bullet u?_{M-V} \notin rng(ru_{M-V}(t))$$

and this must be conjoined with

$$\forall\ t\ :\ t?_{M-V} \bullet r?_{M-V} \in exists_M(t).$$

This second predicate is a consequence of the M-V invariant.

Thus the complete pre-condition for the Add-Visitor-to-Meeting operation is given by

$$\forall\ t\ :\ t?_{M-V} \bullet u?_{M-V} \notin rng(ru_{M-V}(t)) \wedge r?_{M-V} \in exists_M(t)$$

which states that the visitor ($u?_{M-V}$) is not already attending a meeting at that time and that the meeting he is going to attend actually exists.

The second operation removes a visitor from a meeting.

$$\text{Remove-Visitor-from-Meeting}_0 \triangleq$$
$$\Delta\text{M-V} \wedge \equiv\text{Pool}_M[\text{Session}, M] \wedge \text{R}{\leqslant}\text{U-Cancel}_{M-V}[\text{Session}, M, V]$$

It is easy to check that the pre-condition for the Remove-Visitor-from-Meeting operation is simply the predicate which is inherited from the initial R-U-Cancel operation; namely

$$\forall\ t\ :\ t?_{M-V}\ \bullet\ (r?_{M-V}, u?_{M-V})\ \in\ ru_{M-V}(t)$$

We now define the operations which create and cancel meetings as follows:

$\text{Create-Meeting}_0 \triangleq$
$\quad \Delta M\text{-}V\ \land\ \equiv R\leqslant U_{M-V}[\text{Session}, M, V]\ \land\ \text{Create}_M[\text{Session}, M]$

For the creation there is no pre-condition.

```
┌─ Cancel-Meeting₀ ─────────────────────────────────┐
│ ΔM-V                                              │
│ R≼U-Del-Res_{M-V}[Session, M, V]                  │
│ Destroy_M[Session, M]                             │
├───────────────────────────────────────────────────┤
│ t?_M = t?_{M-V} ∧                                 │
│ x?_M = r?_{M-V}                                   │
└───────────────────────────────────────────────────┘
```

The pre-conditions for cancelling a meeting arise from the original R-U-Del-Res operation, i. e., that

$$\forall\ t\ :\ t?_{M-V}\ \bullet\ r?_{M-V}\ \in\ \text{dom}(ru_{M-V}(t))$$

and secondly from the identifications required for the input parameters.

8. The Meeting Resource Subsystems

We are left with the systems CR-M and DR-M to define. We observe that both of these have further information associated with the resource-user relationship, so in order to capture this facet in our model we introduce the concept of a *diary* system.

8.1 A diary system

The diary is to record information about some elements of a set. We denote the set in question by X and the associated information by I_X. For each t, the set of elements of X for which we have information is defined as $\text{recorded}(t)$. Once again this system is dependent on time, T.

$$[\ T,\ X,\ I_X\ ]$$

```
┌─ Diary ─────────────────────────────
│  info     : T → (X ↦ I_X)
│  recorded : T → P X
│─────────────────────────────────────
│  ∀ t : T • recorded(t) = dom(info(t))
└─────────────────────────────────────
```

with initial state given by

$$\text{Init-Diary} \;\triangleq\; [\ \text{Diary} \mid \text{rng}(\text{info}) = \{\ \{\}\ \}\ ]$$

The two operations to be defined both involve a change over a particular time period. Note that we are motivated to make this definition in order to maintain compatibility with existing systems. Formally we define

$$\Delta\text{Diary} \;\triangleq\; \text{Diary} \wedge \text{Diary}' \wedge [\ t?\ :\ P\ T\ ]$$

```
┌─ Add ───────────────────────────────────────────────────
│  ΔDiary
│  x? : X
│  i? : I_X
│─────────────────────────────────────────────────────────
│  (∀ t : t? • x? ∉ recorded(t)) ∧
│  info' = info ⊕ (λ t : t? • info(t) ⊕ { x? ↦ i? })
└─────────────────────────────────────────────────────────
```

The complementary erasure operation would remove one element (and the information associated with it) from info(t). However we note that this is a special case of the following more powerful operation.

```
┌─ Erase ─────────────────────────────────────────────────
│  ΔDiary
│  x? : T ↦ P X
│─────────────────────────────────────────────────────────
│  dom(x?) = t? ∧
│  (∀ t : t? • x?(t) ⊆ recorded(t)) ∧
│  info' = info ⊕ (λ t : t? • x?(t) ⩤ info(t))
└─────────────────────────────────────────────────────────
```

8.2 The conference room booking subsystem

We are now in a position to fully specify the subsystem CR-M, by instantiation as follows:

```
┌─ CR-M ──────────────────────────────────┐
│ R⟩U_{CR-M}[Session, CR, M]              │
│ Diary_{CR}[Session, CR, SI]             │
│─────────────────────────────────────────│
│ in-use_{CR-M} = recorded_{CR}           │
└─────────────────────────────────────────┘
```

with initial state given by

$$\text{Init-CR-M} \triangleq \text{Init-R}\rangle\text{U}_{CR-M}[\text{Session, CR, M}]$$
$$\wedge\ \text{Init-Diary}_{CR}[\text{Session, CR, SI}]$$

It would be more correct to regard the session information SI as being related to a meeting rather than a conference room. The reason for associating SI with conference rooms is that it contains information which is issued to the department supplying equipment for meetings, and they are concerned with the *venue* rather than what is to take place there.

The operations that we require for CR-M are given below. Information is recorded about each resource when it is booked, and must be erased when a cancellation takes place. The definitions use

$$\Delta\text{CR-M} \triangleq \text{CR-M} \wedge \text{CR-M}'$$

```
┌─ Book-Conf-Room_0 ──────────────────────┐
│ ΔCR-M                                   │
│ R⟩U-Book_{CR-M}[Session, CR, M]         │
│ Add_{CR}[Session, CR, SI]               │
│─────────────────────────────────────────│
│ t?_{CR-M} = t?_{CR} ∧                   │
│ r?_{CR-M} = x?_{CR}                     │
└─────────────────────────────────────────┘
```

```
┌─ Cancel-Conf-Rooms_0 ───────────────────┐
│ ΔCR-M                                   │
│ R⟩U-Del-User_{CR-M}[Session, CR, M]     │
│ Erase_{CR}[Session, CR, SI]             │
│─────────────────────────────────────────│
│ t?_{CR-M} = t?_{CR} ∧                   │
│ (∀t: t?_{CR-M} • x?_{CR}(t) = ur_{CR-M}(t)⦇{u?_{CR-M}}⦈) │
└─────────────────────────────────────────┘
```

The cancellation operation here deletes all conference rooms associated with a particular meeting over the specified time period. This is the operation which is most compatible with the Cancel-Meeting operation defined for M-V. However, if required, we could also define the operation that cancels just one conference room - meeting pairing.

8.3 The dining room booking subsystem

The final subsystem that we need to consider is DR-M.

The analysis so far does not take account of the fact that dining rooms have a finite capacity, so we need to extend out model. We suppose that we have been given a function

$$\text{max-no} : DR \rightarrow N$$

which records this capacity and we record the number of seats in each dining room which have been reserved already.

The DR-M system is defined formally as follows:

```
┌─ DR-M ─────────────────────────────────────
│ R≼U_{DR-M}[Date, DR, M]
│ Diary_{DR}[Date, M, LI]
│ rsvd : T → (DR ↛ N)
│───────────────────────────────────────────
│ users_{DR-M} = recorded_{DR} ∧
│ (∀t:Date • dom(rsvd(t)) = in-use_{DR-M}(t) ∧
│    (∀r: in-use_{DR-M}(t) • rsvd(t)(r) ≤ max-no(r))
│ )
└───────────────────────────────────────────
```

Observe that in this case information is associated with each *user*, and therefore the diary system takes M as its main parameter. Dining rooms that are in use have a number of seats reserved, and this number has to be within the dining room's capacity.

The initial state of DR-M is given by

$$\text{Init-DR-M} \triangleq \text{Init-R≼U}_{DR-M}[\text{Date, DR, M}] \land \text{Init-Diary}_{DR}[\text{Date, M, LI}]$$

The two operations that we require for this structure are *booking* a (number of seats in a) dining room and *cancelling* a lunch booking for a particular meeting. In normal circumstances, a resource (dining room) will not be subject to being taken out of service (although this occurrence is clearly easy to model if required).

Both these operations leave rsvd unchanged for time values outside the period in question; we make this part of the operation invariant.

─── ΔDR-M ──────────────────────────────────
$\Delta R \!\succeq\! U_{DR-M}[\text{Date, DR, M}]$
$\Delta Diary_{DR}[\text{Date, M, LI}]$
amount? : $T \twoheadrightarrow \mathbb{N}$
─────────────────────────────────
$t?_{DR-M} = t?_{DR}$ $\wedge$
$dom(amount?) = t?_{DR-M}$ $\wedge$
$t?_{DR-M} \triangleleft rsvd' = t?_{DR-M} \triangleleft rsvd$
──────────────────────────────────

─── Book-Dining-Room$_0$ ───────────────────────────
ΔDR-M
$R \!\succeq\! U\text{-Book}_{DR-M}[\text{Date, DR, M}]$
$Add_{DR}[\text{Date, M, LI}]$
─────────────────────────────────
$x?_{DR} = u?_{DR-M}$ $\wedge$
$(\forall t : t?_{DR-M} \bullet$
  $rsvd(t)(r?_{DR-M}) + amount?(t) \leq max\text{-}no(r?_{DR-M})$ $\wedge$
  $rsvd'(t) = rsvd(t)$
     $\oplus \{ r?_{DR-M} \mapsto rsvd(t)(r?_{DR-M}) + amount?(t) \}$
)
──────────────────────────────────

─── Cancel-Dining-Room$_0$ ─────────────────────────
ΔDR-M
$R \!\succeq\! U\text{-Cancel}_{DR-M}[\text{Date, DR, M}]$
$Erase_{DR}[\text{Date, M, LI}]$
─────────────────────────────────
$(\forall t : t?_{DR-M} \bullet$
  $x?_{DR}(t) = \{ u?_{DR-M} \}$ $\wedge$
  $rsvd'(t) = rsvd(t)$
     $\oplus \{ r?_{DR-M} \mapsto rsvd(t)(r?_{DR-M}) - amount?(t) \}$
)
──────────────────────────────────

8.4 The visitor pool - V-P

From the informal requirements we find that visitors must be "legitimate" before they are allowed to attend meetings or have resources booked on their behalf. This requirement is easily met by introducing a visitor Pool structure, with actual parameters Date and V. Thus we define V-P as

$$\text{V-P} \triangleq \text{Pool}_V[\text{Date}, V]$$

with initial state given by

$$\text{Init-V-P} \triangleq \text{Init-Pool}_V[\text{Date}, V]$$

The operations that we require on this structure are simply those of creation and destruction of visitors. Formally we have

$$\text{Create-Visitor}_0 \triangleq \text{Create}_V[\text{Date}, V]$$

and

$$\text{Destroy-Visitor}_0 \triangleq \text{Destroy}_V[\text{Date}, V]$$

8.5 The construction process

In this section we summarise the constructions we have used to build the individual CAVIAR components.

In sections 7 and 8 we added *pool* and *diary* components to our basic library in section 5.4. We now have a library which consists of the 6 components R-U, R$\geqslant$U, R$\leqslant$U, R$\sqsupseteq$U, Pool and Diary. We indicate in the following diagram how each subsystem has been constructed using components from the library.

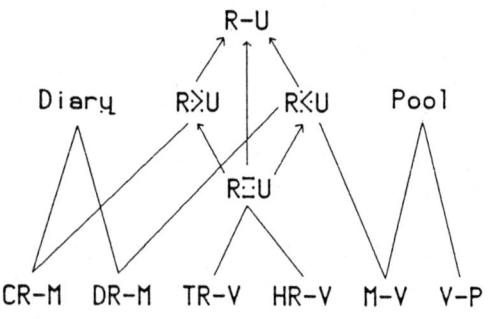

9. The Complete CAVIAR System

We have now achieved our first goal of specifying all constituent subsystems of CAVIAR. We have yet to combine the subsystems into a coherent whole. This is now a comparatively easy task, once we have observed a few extra constraints.

9.1 Combining subsystems to form the system state

We define the visitor part of the system as follows:

```
┌─ V-SYS ──────────────────────────────────────────────────────┐
│  V-P                                                          │
│  HR-V                                                         │
│  TR-V                                                         │
│  ──────────────────────────────────────────                   │
│  (∀d : Date • users_HR-V(d) ⊆ exists_V(d) ) ∧                 │
│  (∀t : Time • users_TR-V(t) ⊆ exists_V(date-of-time(t)) )     │
└───────────────────────────────────────────────────────────────┘
```

The invariant states that visitors that have hotel or transport reservations must be known.

The meeting part of the system is defined by

```
┌─ M-SYS ──────────────────────────────────────────────────────┐
│  M-V                                                          │
│  CR-M                                                         │
│  DR-M                                                         │
│  ──────────────────────────────────────────                   │
│  (∀s:Session • users_CR-M(s) ⊆ exists_M(s) ) ∧                │
│  (∀d:Date •                                                   │
│      users_DR-M(d) ⊆                                          │
│      ∪ { s:Session | date-of-session(s) = d • exists_M(s) }   │
│  )                                                            │
└───────────────────────────────────────────────────────────────┘
```

The invariant states that meetings which are occupying conference rooms or dining rooms must be known to the system at that time.

These two subsystems are now combined to form the CAVIAR system.

```
┌─ CAVIAR ─────────────────────────────────────────────────────┐
│  V-SYS                                                       │
│  M-SYS                                                       │
│  ─────────────                                               │
│  ∀s : Session • users_{M-V}(s) ⊆ exists_V(date-of-session(s))│
└──────────────────────────────────────────────────────────────┘
```

Informally, the invariant states that all visitors who are attending meetings must be known to the system.

The initial state of the system is given by the conjunction of all the initialisations. It is easy to verify that this conjunction satisfies the invariant.

$$\text{Init-CAVIAR} \;\hat{=}\; \text{Init-HR-V} \wedge \text{Init-TR-V} \wedge \text{Init-M-V} \wedge$$
$$\text{Init-CR-M} \wedge \text{Init-DR-M} \wedge \text{Init-V-P}$$

9.2 Operations on CAVIAR

The operations on CAVIAR may be divided naturally into three groups.

9.2.1 Operations which involve meetings only

These operations are concerned with M-SYS only and leave V-SYS unchanged. We denote this by

$$\text{M-OP} \;\hat{=}\; \Delta\text{CAVIAR} \wedge \equiv\text{V-SYS}$$

where

$$\Delta\text{CAVIAR} \;\hat{=}\; \text{CAVIAR} \wedge \text{CAVIAR}'$$

and

$$\equiv\text{V-SYS} \;\hat{=}\; [\, \text{V-SYS} \wedge \text{V-SYS}' \mid \text{V-SYS} = \text{V-SYS}' \,]$$

(Note: in the following similar definitions of ≡CR-M, ≡DR-M, etc. are omitted.)

The first operation is to construct a meeting

$$\text{Create-Meeting} \;\hat{=}\; \text{M-OP} \wedge \text{Create-Meeting}_0 \wedge \equiv\text{CR-M} \wedge \equiv\text{DR-M}$$

This operation has no pre-condition (there is no pre-condition for Create-Meeting$_0$), so it is total. The next operation is to cancel a meeting.

$$\text{Cancel-Meeting}_1 \triangleq \text{M-OP} \wedge \text{Cancel-Meeting}_0 \wedge \equiv\text{CR-M} \wedge \equiv\text{DR-M}$$

We can determine the pre-condition for this operation as follows: first we establish the constraint arising from the system invariant. The operation removes an element from exists$_M$ so this element cannot be a user in CR-M or DR-M during the period $t?_M$. Formally, we require that

$$\forall t : t?_M \bullet r?_{M-V} \notin \text{users}_{CR-M}(t) \cup \text{users}_{DR-M}(\text{date-of-session}(t))$$

The second part of the pre-condition arises from the earlier pre-condition for Cancel-Meeting$_0$. This is precisely

$$t?_M = t?_{M-V} \wedge x?_M = r?_{M-V} \wedge (\forall t : t?_{M-V} \bullet r?_{M-V} \in \text{dom}(ru_{M-V}(t))).$$

We shall at this point fulfil the promise made in section 4.1: indicating how to define the corresponding total operation. This is formed by the *disjunct* of the successful operation with the schema which takes as its qualifying predicate the *negation* of the pre-condition established above.

```
┌─ Cancel-Meeting-Fail ────────────────────────────────────────────┐
│ ≡CAVIAR                                                          │
│ t?_{M-V} : P Session                                             │
│ t?_M     : P Session                                             │
│ x?_M     : M                                                     │
│ r?_{M-V} : M                                                     │
├──────────────────────────────────────────────────────────────────┤
│   (∃ t : t?_M •                                                  │
│         r?_{M-V} ∈ users_{CR-M}(t) ∪ users_{DR-M}(date-of-session(t)) )
│ ∨ t?_M ≠ t?_{M-V}                                                │
│ ∨ x?_M ≠ r?_{M-V}                                                │
│ ∨ (∃t : t?_{M-V} • r?_{M-V} ∉ dom(ru_{M-V}(t)) )                 │
└──────────────────────────────────────────────────────────────────┘
```

$$\text{Cancel-Meeting} \triangleq \text{Cancel-Meeting}_1 \vee \text{Cancel-Meeting-Fail}$$

Informally, if the required pre-condition for the meeting cancellation is not satisfied, the system is unchanged. In practice we would require an appropriate error message to be output.

For the sake of brevity, we shall present the remainder of the operations without going through this process.

The next two operations add visitors to, and delete visitors from, a meeting.

$\quad$ Add-Visitor-to-Meeting $\triangleq$
$\qquad$ M-OP $\wedge$ Add-Visitor-to-Meeting$_0$ $\wedge$ $\equiv$CR-M $\wedge$ $\equiv$DR-M

$\quad$ Remove-Visitor-from-Meeting $\triangleq$
$\qquad$ M-OP $\wedge$ Remove-Visitor-from-Meeting$_0$ $\wedge$ $\equiv$CR-M $\wedge$ $\equiv$DR-M

The pre-conditions for these operations are straightforward to determine in the usual way and we shall omit them and also those for the remaining operations.

The next two operations deal with conference rooms.

$\quad$ Book-Conf-Room $\quad\triangleq\quad$ M-OP $\wedge$ $\equiv$M-V $\wedge$ Book-Conf-Room$_0$ $\quad\wedge$ $\equiv$DR-M

$\quad$ Cancel-Conf-Room $\triangleq$ M-OP $\wedge$ $\equiv$M-V $\wedge$ Cancel-Conf-Room$_0$ $\wedge$ $\equiv$DR-M

We now have the two operations concerning dining rooms.

$\quad$ Book-Dining-Room $\quad\triangleq$ M-OP $\wedge$ $\equiv$M-V $\wedge$ $\equiv$CR-M $\wedge$ Book-Dining-Room$_0$

$\quad$ Cancel-Dining-Room $\triangleq$ M-op $\wedge$ $\equiv$M-V $\wedge$ $\equiv$CR-M $\wedge$ Cancel-Dining-Room$_0$

There is one final operation to be defined in this section: namely the cancellation of both dining room and conference room(s) associated with a particular meeting. This is **not** the conjunct of the two cancellation operations already given because each of these leaves the components it is not acting on **fixed**. Hence we need a different operation defined by

$\quad$ Cancel-Meeting-Arrangements $\triangleq$
$\qquad$ M-OP $\wedge$ $\equiv$M-V $\wedge$ Cancel-Conf-Room$_0$ $\wedge$ Cancel-Dining-Room$_0$

9.2.2 Operations which involve visitors only

This section contains operations which involve V-SYS only and leave M-SYS unchanged. We denote this group by

$$\text{V-OP} \triangleq \Delta\text{CAVIAR} \land \equiv\text{M-SYS}$$

The first pair of operations introduce visitors to and remove visitors from the visitor system.

$$\text{Create-Visitor} \triangleq \text{V-OP} \land \text{Create-Visitor}_0 \land \equiv\text{HR-V} \land \equiv\text{TR-V}$$

$$\text{Destroy-Visitor} \triangleq \text{V-OP} \land \text{Destroy-Visitor}_0 \land \equiv\text{HR-V} \land \equiv\text{TR-V}$$

The Caviar invariant induces the following pre-condition for the Destroy operation.

$$\forall t : t?_V \bullet x?_V \notin \text{users}_{HR-V}(t)$$
$$\cup \cup \{ t : \text{date-of-time}^{-1}(\!(t?_V)\!) \bullet \text{users}_{TR-V}(t) \}$$
$$\cup \cup \{ s : \text{date-of-session}^{-1}(\!(t?_V)\!) \bullet \text{users}_{M-V}(s) \}$$

The two operations concerned with hotel rooms are as follows:

$$\text{Book-Hotel-Room} \triangleq \text{V-OP} \land \equiv\text{V-P} \land \text{Book-Hotel-Room}_0 \land \equiv\text{TR-V}$$

$$\text{Cancel-Hotel-Room} \triangleq \text{V-OP} \land \equiv\text{V-P} \land \text{Cancel-Hotel-Room}_0 \land \equiv\text{TR-V}$$

The two operations concerned with transport reservations are

$$\text{Book-Transport} \triangleq \text{V-OP} \land \equiv\text{V-P} \land \equiv\text{HR-V} \land \text{Book-Transport}_0$$

$$\text{Cancel-Transport} \triangleq \text{V-OP} \land \equiv\text{V-P} \land \equiv\text{HR-V} \land \text{Cancel-Transport}_0$$

9.2.3 A general visitor removal operation

Finally we define an operation which removes a visitor entirely from the system for a particular set of dates.

$$
\begin{array}{|l}
\underline{\text{Delete-Visitor}} \\
\Delta\text{CAVIAR} \\
\equiv\text{CR-M} \\
\equiv\text{DR-M} \\
\text{Cancel-Hotel-Room}_O \\
\text{Cancel-Transport}_O \\
\text{Remove-Visitor-from-Meeting}_O \\
\text{Destroy-Visitor}_O \\
\hline
x?_V = u?_{HR-V} = u?_{TR-V} = u?_{M-V} \land \\
t?_V = t?_{HR-V} \land \\
t?_{TR-V} = \{\, d : t?_V;\ t : \text{Time} \mid \text{date-of-time}(t) = d \\
\qquad\qquad\qquad\qquad \land\ u?_{TR-V} \in \text{users}_{TR-V}(t) \bullet t\,\} \land \\
t?_{M-V} = \{\, d : t?_V;\ s : \text{Session} \mid \text{date-of-session}(s) = d \\
\qquad\qquad\qquad\qquad \land\ u?_{M-V} \in \text{users}_{M-V}(s) \bullet s\,\}
\end{array}
$$

10. Conclusion

This specification has created a conceptual model for the CAVIAR system which provides a precise description of the system state and its external interface, together with an exact functional specification of every operation. The subtle inter-relationships between constituent subsystems are described in the predicates which constrain the combination of these subsystems, and these have been taken into account in the specification of the operations. The system designer can now concentrate on the important parts of the design task: namely selecting appropriate data structures and algorithms, without having to be simultaneously concerned with the complexity of subsystem interactions. This reflects the classical principle of *separation of concerns*.

It may be argued that a specification such as we have given above is a long way from an actual software product. Experience shows however that minimal effort is required to develop software once such a specification has been constructed. For example, in the case of CAVIAR, a Pascal implementation was constructed directly and quickly from the specification.

11. Acknowledgements

A formal specification of CAVIAR was given in 1981 by J.-R. Abrial. This work was carried out at the Programming Research Group at Oxford University in collaboration with B. Sufrin, T. Clement and one of the co-authors. T. Clement implemented a prototype version of the specification on a ITT-2020 computer in UCSD Pascal. J.-R. Abrial's original specification document listed most of the properties of the system that appear in this document, though the style of the presentation, the notation, and the conventions used in this paper have since been developed by members of the Programming Research Group.

We would like to thank J.-R. Abrial for his original contribution, I. Hayes for editing this paper and all those involved in helping with the project, particularly the personnel in the Visitor Services Department of STL, who willingly provided the team with information about the current manual system in operation at that time.

We would also like to thank Bernie Cohen, Tim Denver and Tom Cox for their initial effort in setting up this collaborative effort between STL and the Programming Research Group and their continuing interest.

12. References and Related Work

1. Abrial, J.-R. The specification language Z: Basic library. *Oxford University Programming Research Group internal report*, (April 1980).

2. Morgan, C. C. Schemas in Z: A preliminary reference manual. *Oxford University Programming Research Group Distributed Computing Project report*, (March 1984).

3. Sufrin, B. A., Sørensen, I. H., Morgan, C. C., and Hayes, I. J. Notes for a Z Handbook. *Oxford University Programming Research Group internal report*, (July 1985).

4. Morgan, C. C., and Sufrin, B. A. Specification of the UNIX file system. *IEEE Transactions on Software Engineering, Vol. 10, No. 2*, (March 1984), pp. 128-142.

5. Hayes, I. J. Specification Case Studies. *Oxford University Programming Research Group Monograph, PRG-46*, (July 1985).

Z Reference Card
Mathematical Notation
Version 2.2

Programming Research Group
Oxford University

1. Definitions and declarations.
Let x, x_k be identifiers and T, T_k sets.

LHS $\cong$ RHS	Definition of LHS as syntactically equivalent to RHS.
$x: T$	Declaration of x as type T.
$x_1: T_1;\ x_2: T_2;\ \ldots;\ x_n: T_n$	List of declarations.
$x_1, x_2, \ldots, x_n : T$	$\cong x_1:T;\ x_2:T;\ \ldots;\ x_n:T.$
$[A, B]$	Introduction of generic sets.

2. Logic.
Let P, Q be predicates and D declarations.

true, false	Logical constants.
$\neg P$	Negation: "not P".
$P \wedge Q$	Conjunction: "P and Q".
$P \vee Q$	Disjunction: "P or Q".
$P \Rightarrow Q$	Implication: "P implies Q" or "if P then Q".
$P \Leftrightarrow Q$	Equivalence: "P is logically equivalent to Q".
$\forall x : T \bullet P$	Universal quantification: "for all x of type T, P holds".
$\exists x : T \bullet P$	Existential quantification: "there exists an x of type T such that P".
$\exists! x : T \bullet P_x$	Unique existence: "there exists a unique x of type T such that P". $\cong (\exists x : T \bullet P_x \wedge \neg(\exists y:T \mid y \neq x \bullet P_y))$
$\forall x_1:T_1;\ x_2:T_2;\ \ldots;\ x_n:T_n \bullet P$	"For all x_1 of type T_1, x_2 of type $T_2, \ldots$, and x_n of type T_n, P holds.
$\exists x_1:T_1;\ x_2:T_2;\ \ldots;\ x_n:T_n \bullet P$	Similar to $\forall$.
$\exists! x_1:T_1;\ x_2:T_2;\ \ldots;\ x_n:T_n \bullet P$	Similar to $\forall$.
$\forall D \mid P \bullet Q$	$\cong (\forall D \bullet P \Rightarrow Q)$.
$\exists D \mid P \bullet Q$	$\cong (\exists D \bullet P \wedge Q)$.
$t_1 = t_2$	Equality between terms.
$t_1 \neq t_2$	$\cong \neg(t_1 = t_2)$.

3. Sets.
Let S, T and X be sets; t, t_k terms; P a predicate and D declarations.

$t \in S$	Set membership: "t is an element of S".
$t \notin S$	$\cong \neg(t \in S)$.
$S \subseteq T$	Set inclusion: $\cong (\forall x : S \bullet x \in T)$.
$S \subset T$	Strict set inclusion: $\cong S \subseteq T \wedge S \neq T$.
$\{\}$	The empty set.
$\{t_1, t_2, \ldots, t_n\}$	The set containing $t_1, t_2, \ldots$ and t_n.
$\{x : T \mid P\}$	The set containing exactly those x of type T for which P holds.
$(t_1, t_2, \ldots, t_n)$	Ordered n-tuple of $t_1, t_2, \ldots$ and t_n.
$T_1 \times T_2 \times \ldots \times T_n$	Cartesian product: the set of all n-tuples such that the kth component is of type T_k.
$\{x_1:T_1;\ x_2:T_2;\ \ldots;\ x_n:T_n \mid P\}$	The set of n-tuples $(x_1, x_2, \ldots, x_n)$ with each x_k of type T_k such that P holds.
$\{D \mid P \bullet t\}$	The set of t's such that given the declarations D, P holds.
$\{D \bullet t\}$	$\cong \{D \mid \text{true} \bullet t\}$.
$\mathbb{P}\ S$	Powerset: the set of all subsets of S.
$\mathbb{F}\ S$	Set of finite subsets of S: $\cong \{T: \mathbb{P}\ S \mid T \text{ is finite}\}$.
$S \cap T$	Set intersection: given $S, T: \mathbb{P}\ X$, $\cong \{x:X \mid x \in S \wedge x \in T\}$.

$S \cup T$	Set union: given $S, T: \mathbf{P}\ X$, $\triangleq \{\ x:X\ \|\ x \in S \vee x \in T\ \}$.	$x \mapsto y$	$\triangleq (x, y)$
$S - T$	Set difference: given $S, T: \mathbf{P}\ X$, $\triangleq \{\ x:X\ \|\ x \in S \wedge x \notin T\ \}$.	$\{\ x_1 \mapsto y_1,\ x_2 \mapsto y_2,\ \ldots,\ x_n \mapsto y_n\ \}$	The relation $\{\ (x_1, y_1),\ \ldots,\ (x_n, y_n)\ \}$ relating x_1 to $y_1, \ldots,$ and x_n to y_n.

$\cap SS$ Distributed set intersection: given $SS: \mathbf{P}\ (\mathbf{P}\ X)$, $\triangleq \{x:X\ \|\ (\forall S:SS \bullet x \in S)\}$.

$\cup SS$ Distributed set union: given $SS: \mathbf{P}\ (\mathbf{P}\ X)$, $\triangleq \{x:X\ \|\ (\exists S:SS \bullet x \in S)\}$.

$|S|$ Size (number of distinct elements) of a finite set.

$\#S$ $\triangleq |S|$.

4. Numbers.

$\mathbf{N}$ The set of natural numbers (non-negative integers).

$\mathbf{N}^+$ The set of strictly positive natural numbers: $\triangleq \mathbf{N} - \{\ 0\ \}$.

$\mathbf{Z}$ The set of integers (positive, zero and negative).

$m..n$ The set of integers between m and n inclusive: $\triangleq \{\ k:\mathbf{Z}\ \|\ m \leq k \wedge k \leq n\ \}$.

$\min S$ Minimum of a set, $S : \mathbf{F}\ \mathbf{N}$. $\min S \in S \wedge (\forall x : S \bullet x \geq \min S)$.

$\max S$ Maximum of a set, $S : \mathbf{F}\ \mathbf{N}$. $\max S \in S \wedge (\forall x : S \bullet x \leq \max S)$.

5. Relations.

A relation is modelled by a set of ordered pairs hence operators defined for sets can be used on relations.

Let X, Y, and Z be sets; $x : X$; $y : Y$; and $R : X \leftrightarrow Y$.

$X \leftrightarrow Y$ The set of relations from X to Y: $\triangleq \mathbf{P}\ (X \times Y)$.

$x\ R\ y$ x is related by R to y: $\triangleq (x, y) \in R$.

$\mathrm{dom}\ R$ The domain of a relation: $\triangleq \{x:X\ \|\ (\exists y:Y \bullet x\ R\ y)\}$.

$\mathrm{rng}\ R$ The range of a relation: $\triangleq \{y:Y\ \|\ (\exists x:X \bullet x\ R\ y)\}$.

$R_1\ \mathbf{;}\ R_2$ Forward relational composition: given $R_1: X \leftrightarrow Y;\ R_2: Y \leftrightarrow Z$, $\triangleq \{\ x:X;\ z:Z\ \|\ (\exists y:Y \bullet x\ R_1\ y \wedge y\ R_2\ z\ )\}$.

$R_1 \circ R_2$ Relational composition: $\triangleq R_2\ \mathbf{;}\ R_1$.

R^{-1} Inverse of relation R: $\triangleq \{\ y:Y;\ x:X\ \|\ x\ R\ y\ \}$.

$\mathrm{id}\ X$ Identity function on the set X: $\triangleq \{\ x : X \bullet x \mapsto x\ \}$.

R^k The relation R composed with itself k times: given $R : X \leftrightarrow X$, $R^0 \triangleq \mathrm{id}\ X,\ R^{k+1} \triangleq R^k \circ R$.

R^* Reflexive transitive closure: $\triangleq \cup\ \{\ n: \mathbf{N} \bullet R^n\ \}$.

R^+ Non-reflexive transitive closure: $\triangleq \cup\ \{\ n: \mathbf{N}^+ \bullet R^n\ \}$.

$R(\!|S|\!)$ Image: given $S : \mathbf{P}\ X$, $\triangleq \{y:Y\ \|\ (\exists x :S \bullet x\ R\ y)\}$.

$S \triangleleft R$ Domain restriction to S: given $S: \mathbf{P}\ X$, $\triangleq \{x:X; y:Y\ \|\ x \in S \wedge x\ R\ y\}$.

$S \triangleleft\!\!\!- R$ Domain subtraction: given $S: \mathbf{P}\ X$, $\triangleq (X - S) \triangleleft R$.

$R \triangleright T$ Range restriction to T: given $T: \mathbf{P}\ Y$, $\triangleq \{x:X; y:Y\ \|\ x\ R\ y \wedge y \in T\}$.

$R \triangleright\!\!\!- T$ Range subtraction of T: given $T: \mathbf{P}\ Y$, $\triangleq R \triangleright (Y - T)$.

$R_1 \oplus R_2$ Overriding: given $R_1, R_2 : X \leftrightarrow Y$, $\triangleq (\mathrm{dom}\ R_2 \triangleleft\!\!\!- R_1) \cup R_2$.

6. Functions.

A function is a relation with the property that for each element in its domain there is a unique element in its range related to it. As functions are relations all the operators defined above for relations also apply to functions.

$X \nrightarrow Y$ The set of partial functions from X to Y:
$\triangleq \{ f: X \leftrightarrow Y \mid$
     $(\forall x: \text{dom } f \bullet$
         $(\exists ! y: Y \bullet x \, f \, y)) \}$.

$X \rightarrow Y$ The set of total functions from X to Y:
$\triangleq \{ f: X \nrightarrow Y \mid \text{dom } f = X \}$.

$X \rightarrowtail\mkern-18mu\nrightarrow Y$ The set of one-to-one partial functions from X to Y:
$\triangleq \{ f: X \nrightarrow Y \mid$
     $(\forall y: \text{rng } f \bullet$
         $(\exists ! x: X \bullet x \, f \, y)) \}$.

$X \rightarrowtail Y$ The set of one-to-one total functions from X to Y:
$\triangleq \{ f: X \rightarrowtail\mkern-18mu\nrightarrow Y \mid \text{dom } f = X \}$.

$f \, t$ The function f applied to t.

$(\lambda \, x : X \mid P \bullet t)$
   Lambda-abstraction:
the function that given an argument x of type X such that P holds the result is t.
$\triangleq \{ x: X \mid P \bullet x \mapsto t \}$.

$(\lambda \, x_1: T_1; \ldots ; x_n: T_n \mid P \bullet t)$
$\triangleq \{ x_1:T_1; \ldots ; x_n:T_n \mid P \bullet$
    $(x_1, \ldots, x_n) \mapsto t \}$.

7. Orders.

partial_order X
   The set of partial orders on X.
$\triangleq \{ R: X \leftrightarrow X \mid \forall x, y, z: X \bullet$
   $x \, R \, x \, \wedge$
   $x \, R \, y \wedge y \, R \, x \Rightarrow x = y \, \wedge$
   $x \, R \, y \wedge y \, R \, z \Rightarrow x \, R \, z$
$\}$.

total_order X
   The set of total orders on X.
$\triangleq \{ R: \text{partial\_order } X \mid$
   $\forall x, y: X \bullet$
     $x \, R \, y \vee y \, R \, x$
$\}$.

monotonic $X <_X$
   The set of functions from X to X that are monotonic with respect to the order $<_X$ on X.
$\triangleq \{ f : X \rightarrow X \mid$
   $x <_X y \Rightarrow f(x) <_X f(y)$
$\}$.

8. Sequences.

seq X   The set of sequences whose elements are drawn from X:
$\triangleq \{ A: \mathbb{N}^+ \nrightarrow X \mid$
   $\text{dom } A = 1..|A| \}$.

$|A|$ The length of sequence A.
$[\,]$ The empty sequence $\{\}$.
$[a_1, \ldots, a_n]$
   $\triangleq \{ 1 \mapsto a_1, \ldots, n \mapsto a_n \}$.
$[a_1, \ldots, a_n] \frown [b_1, \ldots, b_m]$
   Concatenation:
   $\triangleq [a_1, \ldots, a_n, b_1, \ldots, b_m]$,
   $[\,] \frown A = A \frown [\,] = A$.

head A $\triangleq A(1)$.
last A $\triangleq A(|A|)$.
tail $[x] \frown A \triangleq A$.
front $A \frown [x] \triangleq A$.
rev $[a_1, a_2, \ldots, a_n]$
   Reverse:
   $\triangleq [a_n, \ldots, a_2, a_1]$,
   rev $[\,] = [\,]$.

$\frown$/AA   Distributed concatenation:
given AA : seq(seq(X)),
$\triangleq AA(1) \frown \ldots \frown AA(|AA|)$,
$\frown/[\,] = [\,]$.

$;$/AR   Distributed relational composition:
given AR : seq $(X \leftrightarrow X)$,
$\triangleq AR(1) \, ; \, \ldots \, ; \, AR(|AR|)$,
$;/[\,] = \text{id } X$.

disjoint AS   Pairwise disjoint:
$$\text{given AS: seq } (\mathbb{P}\ X),$$
$$\hat{=}\ (\forall\ i, j : \text{dom AS} \bullet i \ne j$$
$$\Rightarrow AS(i) \cap AS(j) = \{\}).$$
AS <u>partitions</u> S
$$\hat{=}\ \text{disjoint AS}$$
$$\wedge\ \cup\ \text{ran AS} = S.$$
A <u>in</u> B   Contiguous subsequence:
$$\hat{=}\ (\exists C, D: \text{seq } X \bullet$$
$$C \frown A \frown D = B).$$
squash f   Convert a function, $f: \mathbb{N} \twoheadrightarrow X$, into a sequence by squashing its domain.
$$\text{squash } \{\} = [\ ],$$
and if $f \ne \{\}$ then
$$\text{squash } f =$$
$$[f(i)] \frown \text{squash}(\{i\} \triangleleft f)$$
where $i = \min(\text{dom } f)$ e.g.
$$\text{squash } \{2 \mapsto A,\ 27 \mapsto C,\ 4 \mapsto B\}$$
$$= [A, B, C]$$
S ↾ A   Restrict the sequence A to those items whose index is in the set S:
$$\hat{=}\ \text{squash}(S \triangleleft A)$$
A ↾ T   Restrict the range of the sequence A to the set T:
$$\hat{=}\ \text{squash}(A \triangleright T).$$

9. Bags.

bag X   The set of bags whose elements are drawn from X:
$$\hat{=}\ X \twoheadrightarrow \mathbb{N}^+$$
A bag is represented by a function that maps each element in the bag onto its frequency of occurrence in the bag.

[]   The empty bag $\{\}$.

[x_1, x_2, ..., x_n]   The bag containing x_1, x_2, ... and x_n with the frequency they occur in the list.

items s   The bag of items contained in the sequence s:
$$\hat{=}\ \{\ x: \text{rng } s \bullet$$
$$x \mapsto |\{i: \text{dom } s\ |\ s(i) = x\}|$$
$$\}$$

Z Reference Card
Schema Notation
[For details see "Schemas in Z"]

Programming Research Group
Oxford University

Schema definition: a schema groups together some declarations of variables and a predicate relating these variables. There are two ways of writing schemas: vertically, for example

$$\begin{array}{|l}\hline S \\ \hline x : \mathbb{N} \\ y : \text{seq } \mathbb{N} \\ \hline x \le |y| \\ \hline \end{array}$$

or horizontally, for the same example
$$S \hat{=} [\ x: \mathbb{N};\ y: \text{seq } \mathbb{N}\ |\ x \le |y|\ ].$$
Use in signatures after $\forall, \lambda, \{\ldots\}$, etc.:
$$(\forall S \bullet y \ne [\ ]) \hat{=} (\forall x: \mathbb{N};\ y: \text{seq } \mathbb{N}\ |$$
$$x \le |y| \bullet y \ne [\ ]).$$

tuple S   The tuple formed of a schema's variables.

pred S   The predicate part of a schema: e.g. pred S is $x \le |y|$.

Inclusion   A schema S may be included within the declarations of a schema T, in which case the declarations of S are merged with the other declarations of T (variables declared in both S and T must be the same type) and the predicates of S and T are conjoined. e.g.

$$\begin{array}{|l}\hline T \\ \hline S \\ z : \mathbb{N} \\ \hline z < x \\ \hline \end{array}$$

is

$$[\,x, z : N;\ y : \text{seq } N \mid x \leq |y| \wedge z < x\,]$$

S | P The schema S with P conjoined to its predicate part. e.g. (S | x>0) is
[x:N; y:seq N | x≤|y| ∧ x>0].

S ; D The schema S with the declarations D merged with the declarations of S. e.g. (S ; z : N) is
[x, z:N; y:seq N | x≤|y|]

S[new/old] Renaming of components: the schema S with the component old renamed to new in its declaration and every free use of that old within the predicate. e.g. S[z/x] is
[z:N; y:seq N | z ≤ |y|]
and S[y/x, x/y] is
[y:N; x:seq N | y ≤ |x|]

Decoration Decoration with subscript, superscript, prime, etc.; systematic renaming of the variables declared in the schema. e.g. S′ is
[x′:N; y′:seq N | x′≤|y′|]

¬S The schema S with its predicate part negated. e.g. ¬S is
[x:N; y:seq N | ¬(x≤|y|)]

S ∧ T The schema formed from schemas S and T by merging their declarations (see inclusion above) and and'ing their predicates. Given
T ≙ [x: N; z: P N | x∈z],
S ∧ T is

$$[\,x : N;\ y : \text{seq } N;\ z : P\,N \mid x \leq |y| \wedge x \in z\,]$$

S ∨ T The schema formed from schemas S and T by merging their declarations and or'ing their predicates. e.g. S ∨ T is

$$[\,x : N;\ y : \text{seq } N;\ z : P\,N \mid x \leq |y| \vee x \in z\,]$$

S ⇒ T The schema formed from schemas S and T by merging their declarations and taking pred S ⇒ pred T as the predicate. e.g. S ⇒ T is similar to S ∧ T and S ∨ T except the predicate contains an "⇒" rather than an "∧" or an "∨".

S ⇔ T The schema formed from schemas S and T by merging their declarations and taking pred S ⇔ pred T as the predicate. e.g. S ⇔ T the same as S ∧ T with "⇔" in place of the "∧".

S \ (v_1, v_2, ... , v_n)
Hiding: the schema S with the variables v_1, v_2, ..., and v_n hidden: the variables listed are removed from the declarations and are existentially quantified in the predicate. e.g. S \ x is
[y:seq N | (∃x:N • x≤|y|)]

A schema may be specified instead of a list of variables; in this case the variables declared in that schema are hidden. e.g. (S ∧ T) \ S is

$$[\,z : P\,N \mid (\exists\, x : N;\ y : \text{seq } N \bullet x \leq |y| \wedge x \in z)\,]$$

S ↾ (v$_1$, v$_2$, ... , v$_n$)
    Projection: The schema S with any variables that do not occur in the list v$_1$, v$_2$, ..., v$_n$ hidden: the variables removed from the declarations are existentially quantified in the predicate.
    e.g. (S ∧ T) ↾ (x, y) is

    ┌─────────────────────┐
    │ x : N   │
    │ y : seq N   │
    ├─────────────────────┤
    │ (∃ z : P N •   │
    │   x ≤ |y| ∧ x ∈ z) │
    └─────────────────────┘

    The list of variables may be replaced by a schema as for hiding; the variables declared in the schema are used for the projection.

The following conventions are used for variable names in those schemas which represent operations:
undashed   state before the operation,
dashed   state after the operation,
ending in "?"   inputs to the operation, and
ending in "!"   outputs from the operation.

The following schema operations only apply to schemas following the above conventions.

pre S   Precondition: all the state after components (dashed) and the outputs (ending in "!") are hidden. e.g. given
    S ┌─────────────────────┐
      │ x?, s, s', y! : N   │
      ├─────────────────────┤
      │ s' = s - x? ∧ y! = s│
      └─────────────────────┘

    pre S is

    ┌─────────────────────┐
    │ x?, s : N   │
    ├─────────────────────┤
    │ (∃ s', y! : N •   │
    │   s' = s-x? ∧ y! = s)│
    └─────────────────────┘

post S   Postcondition: this is similar to precondition except all the state before components (undashed) and inputs (ending in "?") are hidden.

S ⊕ T   Overriding:
         ≙ (S ∧ ¬pre T) ∨ T.
         e.g. given S above and
    T ┌─────────────────────┐
      │ x?, s, s' : N   │
      ├─────────────────────┤
      │ s < x? ∧ s' = s   │
      └─────────────────────┘

    S ⊕ T is

    ┌──────────────────────────┐
    │ x?, s, s', y! : N   │
    ├──────────────────────────┤
    │ (s' = s-x? ∧ y! = s ∧   │
    │ ¬(∃ s' : N •   │
    │   s < x? ∧ s' = s))   │
    │ ∨ (s < x? ∧ s' = s)   │
    └──────────────────────────┘

    The predicate can be simplified:

    ┌──────────────────────────┐
    │ x?, s, s', y! : N   │
    ├──────────────────────────┤
    │ (s' = s-x? ∧ y! = s   │
    │   ∧ s ≥ x?)   │
    │ ∨   │
    │ (s < x? ∧ s' = s)   │
    └──────────────────────────┘

S ⨟ T   Schema composition: if we consider an intermediate state that is both the final state of the operation S and the initial state of the operation T then the composition of S and T is the operation which relates the initial state of S to the final state of T through the intermediate state. To form the composition of S and T we take the state after components of S and the state before components of T that have a basename* in common, rename both to new variables, take the schema "and" (∧) of the resulting schemas, and hide the new variables.

e.g. S ⨾ T is

---
x?, s, s', y! : N
(∃ s_0 : N . $s_0 = s-x? \wedge y! = s \wedge$ $s_0 < x? \wedge s' = s_0$)
---

* basename is the name with any decoration ("'", "!", "?", etc.) removed.

S >> T   Piping: this schema operation is similar to schema composition; the difference is that rather than identifying the state after components of S with the state before components of T, the output components of S (ending in "!") are identified with the input components of T (ending in "?") that have the same basename.

Measurement of SQL
Problems and Progress

Phyllis Reisner
IBM Almaden Research Center

ABSTRACT: Since development of the database query language, SQL, began, there have been behavioral experiments to measure and improve its ease of use. These experiments are important in their own right. In addition, they are examples of how methodology in the new field of "software human factors" is developing. The present paper, written for the 17th annual IBM Computer Science Symposium at Nemu-no-sato, Japan, uses the experiments with SQL as a springboard to illustrate both the progress that has been made in this new field and the problems that still remain.

INTRODUCTION

It is almost eight years since the first study measuring the usability of SQL appeared in the literature. It is almost 10 years since that first study began. On the occasion of the 17th Annual Computer Science Symposium at Nemu-no-sato, Japan, it is appropriate to review both the progress that has been made and the problems that still remain in this area.

These experiments are important in two ways. First, of course, they illustrate various aspects of SQL's ease-of-use. In addition, they illustrate the progress that has been made in the development of a very new part of computer science. This is the field that is generally becoming known as "software human factors".

Software human factors is a "baby" field that is just beginning to take shape. The first professional meeting in this field took place as recently as 1981. This was the meeting of Human Factors and Computer Science in Gaithersburg, Maryland (5). Also, the very first issue of Computing Surveys devoted entirely to this field appeared as recently as 1981 (6). In addition, in June of this year (1983) the prestigious National Academy of Sciences convened a workshop on Software Human Factors (14). This workshop started to synthesize some of the knowledge about methodology that has been developing in this field. It is clear, from the dates above, that software human factors is a very new, emerging field.

There are two themes that I want to illustrate in this talk. First of all, when the SQL work began, many people were saying that software ease of use could not be measured. They felt that the entire area was just too nebulous. The experiments described in this talk show this is not true. It is indeed possible to be precise and to measure software ease of use.

The second theme is that, yes it is possible to measure. But it is also very hard to do so. It is extremely hard to do so well. While it is possible to measure, and to get numbers as the result of that measurement, the numbers must be interpreted with a great deal of caution. Very serious errors can be made if the numbers are used blindly, without understanding.

The talk is organized as follows. First, I will give some background: a very brief introduction to SQL, to the state of the art when this work began, and to the goals of the work. Then will come discussion of the early SQL experiments. Following that I will discuss experiments that compare two or more query languages for ease of use. These experiments compare SQL with another query language, QBE. I will discuss the work of Thomas and Gould (1975), of Greenblatt and Waxman (1978), and of Boyle, Bury and Evey (1983). Following that will come experiments to improve SQL ease-of-use. Here I will discuss some other aspects of my own early work, and the more recent work of Welty (1981, 1983). Notice again the dates of these experiments.

BACKGROUND

SQL

SQL was a database query language, originally called "SEQUEL". It was developed at the IBM Research laboratory in San Jose, California. SQL was based on Ted Codd's relational model (3). It was not intended originally to be used by programmers. Rather, it was intended for professionals who were not programmers, such as doctors, lawyers, accountants, teachers, etc. The main idea was the following. Stored in the computer was a data base. A data base is, conceptually, a collection of tables with information in them, such as the names of people working for a company, the names of the departments each one works in, their salaries, etc. A very simplified example of a database is shown in Figure 1.

EMP	NAME	DEPTNO	SAL

Figure 1. A sample data base with one table.

Users of this data base would want to find information in these tables. They would have questions such as the following:

1. Find the names of employees in department 50.
2. Find the names of employees who work in Stockton.
3. Find the departments with average salary over $15,000.

To find the answers to such questions, the user would type his question into the computer in a particular standardized way. A particular question asked in this standardized way is a SQL query. The collection of such standardized ways of asking questions of a database is the database query language.

Some sample SQL queries are shown in Table I.

TABLE I
Queries in SQL and in QBE

Query Language	Example query for the question " Find the names of employees in department 50".
SQL	SELECT NAME FROM EMP WHERE DEPTNO=50
QBE	<table><tr><td>EMP</td><td>NAME</td><td>DEPTNO</td><td>SAL</td></tr><tr><td></td><td>p. Brown</td><td>50</td><td></td></tr></table>

The words SELECT, FROM, and WHERE, are keywords. For example, to find the names of employees in department 50, the user would type SELECT, then the name of the column in the table that has the information he wants. In this case, the column is NAME. Then the user would type FROM, and the name of the table that had the information (EMP). Then comes the WHERE clause, the condition (that the department number is 50). More complicated queries are described in (2).

STATE OF THE ART IN 1975

The developers of SQL, in 1975, wanted to find out whether SQL was indeed as easy to use as they (just as developers everywhere) felt it was. However, the state of the software human factors art, in 1975, was very primitive indeed. Hardware human factors was relatively well-developed. We knew, in general, how to test hardware (such as keyboards). Not only did we know, but this knowledge was even used when time and resources were available. However, software was another story. There had only been a few scattered attempts at testing software, and many people felt the problems were too fuzzy to be amenable to experimentation.

GOALS

A major goal of the experimental work with SQL, therefore, was simply to learn how to think about the problems of testing software. Primarily, the problem was to learn what questions made sense to ask: what did "ease-of-use" mean? Secondly, it was important to learn how to actually do the experiments, and third, to actually do them. The goal of this early work, therefore, was twofold: (1) to learn, by doing it, how to do this kind of experimentation, and (2) to actually use the methodology that was developed to measure the ease-of-use of SQL.

EARLY EXPERIMENTS WITH SQL

There were many questions to be asked about SQL in these early experiments: how easy was it to use, was it easier than another language, and how could it be made easier. In this section will will talk about the first of these. Analyses of the data to see how the language could be improved will be discussed in a later section. In the early SQL experiments, there was another language taught and tested at the same time, called SQUARE, which was used for comparison. However, I will discuss comparative experiments in a later section of this talk, using experiments comparing SQL with QBE.

APPROACH

The main approach to SQL testing was to teach surrogate users the language and then give them various kinds of tests to see how well they could use it (7,8). The tests were graded, just as ordinary classroom-type tests. However, the purpose and focus of the testing was different. In a classroom situation, the focus is on the students. The language material is considered "constant" and is used to measure the students' performance. In testing SQL, however, just the reverse was true. The focus was on the students, not the language. In a sense, the students were considered "constant", and their performance was used to measure the language.

SUBJECTS

In testing SQL, the people taught and tested were similar in background to the kinds of people who would be using the language in real life. Both programmers and non-programmers were taught. It may seem strange that programmers were taught, when the language was intended to be used by non-programmers. There was, however, a sensible reason for this. We were not sure at that time whether even programmers could use the language. If programmers could not do so, then almost certainly non-programmers could not. In addition, teaching both non-programmers and programmers was a check on the methodology. Programmers would be expected to perform better than non-programmers - at least in very early stages of the teaching. If this did not occur, one would have to start wondering why not.

TESTS

In order to decide what tests make sense, it is first necessary to decide what is meant by the ease-of-use of a language. There were two "dimensions" of ease-of-use I chose to consider: the task and the stage of use.

The most obvious task is writing. The user should be able to write queries in the language. Another important task is reading. The user should be able to read back what he or someone else has written (to modify or debug a query, for example). The user should also understand what he has written, not just write queries mechanically, by rote.

Each of these tasks should be performed at different stages of using a language: when the person is learning the language, when he has actually learned it and is using it do to work (productive stage), when he has not used the language for a while and may have forgotten some of it (memory stage), and when he must relearn what he has forgotten. I will talk only about the learning tests. Some tests of memory were done. They follow a similar pattern.

Writing Tests

For the writing tests, people were given English sentences. They were then expected to write the corresponding SQL query.

For example,

    Given: Find the names of employees in department 50.

    Write: SELECT NAME
           FROM EMP
           WHERE DEPT = 50

The students were then graded on whether or not they wrote the query correctly.

Reading Tests

In the reading tests, the task was reversed. The students were given a query in SQL and asked to write, in English, what it meant. For example, they were given:

    Given: SELECT NAME
                FROM EMP
              WHERE DEPTNO = 50

    Write: Find the names of employees in department 50.

Deciding whether or not students could read the queries correctly from these statements is not completely straightforward. Students sometimes wrote very strange English sentences. From these strange sentences, it was hard to tell whether they really did not know how to read the query or whether their English was careless. The reading tests thus gave some data, but it was not as clean as the data from the writing tests.

However, by putting the results of the reading tests together with the results of the "understanding" tests which I will discuss next, it was possible to make a reasonable judgement about whether they could read the SQL queries.

Understanding Tests

To find out whether students understood what they were doing or were simply following learned patterns, they were given tests of understanding. For the understanding tests, they were given a SQL query, plus a data base (tables with data) and they were asked to "play computer". That is, they were asked to make believe they were computer and actually find the data that the computer would find as an answer to the query.

However, there were also problems in interpreting the test results with these tests of understanding. When a student made a mistake, it was sometimes not clear whether he really did not understand the query or whether he was simply making a clerical error.

To mitigate these problems of interpretation, I combined the reading and the understanding tests. I gave students a SQL query and asked them to do two things for that same query - write an English sentence and "play computer". From the results of these combined tests, it was possible to make a better judgement about their understanding.

Subjective Opinions

We are dealing, in these tests, with people, and we want to find out what they think. There was therefore another kind of test that was very important: subjective opinion questionnaires. These questionnaires asked, for example, whether the language was hard or easy, what was hard about it, why it was hard, which functions were hardest, etc. These tests are an important supplement to the more objective tests described above. However, they are not a substitute for the more objective tests. Some of the students, for example, would tell me that something was very easy. However, when I looked at the corresponding objective tests, it was clear that they understood very little indeed.

Some Details

Some details about the testing are the following.

I obtained students from a local university, San Jose State University. I taught them for two hours a day over a two-week span. This made a total of 12-14 hours of classroom time. There were 64 students. They had a wide variety of backgrounds. The teaching was an ordinary classroom situation (teacher and students working at a blackboard). There was one teacher (myself). Testing was done with paper and pencil, not with an implemented computer system.

It will be clear from later portions of this talk, and from Mr. Lambeck's talk (see elsewhere in these Proceedings), that on-line versions of languages can be tested, and that much of the testing can be automated. However, it is important to understand that on-line testing is not necessary for learning whether or not subjects understand the <u>concepts</u> of a query language. Testing an implemented version is important to understand both the benefits and problems of the man-machine interaction. However, for testing basic understanding, pencil and paper can be very efficient.

RESULTS

Writing Tests

Results of the writing test can be see in Table II. They show that people could actually write queries correctly. While these results are important for what they tell us about SQL, they are more important because they show that we can, rigorously, measure a query language. However, such measurement is not straightforward. Decisions had to be made about what was, and was not a correct query. For example, is a query with a small mistake in spelling correct or not? To settle such questions a classification system had to be developed and applied.

Table II. Mean percentage of correct answers on writing exam

	SQL	(SQUARE)
Non-programmers	65.0	(54.7)
Programmers	77.5	(77.7)

For those interested in statistics, the results were as follows. The difference between programmers and non-programmers was significant. This difference is not surprising. However, it is a way to check the methodology. If results that one normally expects to find do not occur, this suggests there may be some flaw in the methodology. Differences between SQL and SQUARE were significant, but only for non-programmers.

Reading and Comprehension Tests

Results of the reading test and of the comprehension test can be found in (8).

Distribution of Scores

Not only is it possible to get average values for the group, but it is also possible, of course, to get distributions of values. Figure 2 is a cumulative frequency distribution, showing percentage of correct answers on the writing exam. The importance of this curve is not so much the

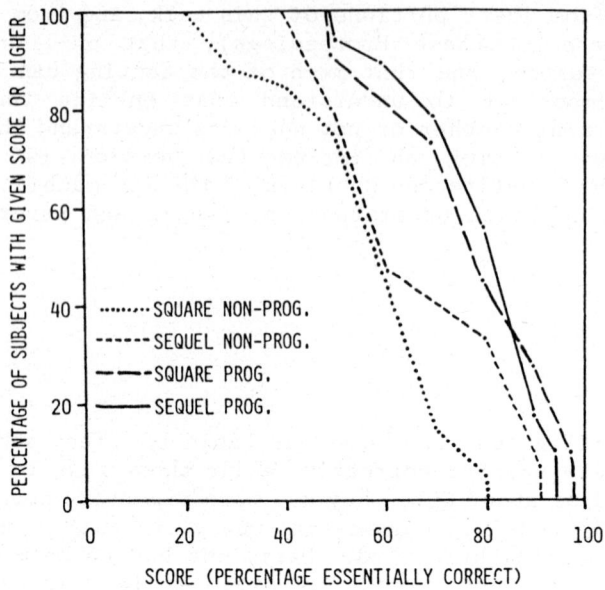

Figure 2. Cumulative frequency distribution of subjects on final exam

particular form of the curve, as the demonstration that one can get data in this field.

Learning Curves

Figure 3 shows another kind of information, a learning curve. This curve shows hours of instruction vs. performance for a "simple mapping", one of the simplest kinds of SQL query. The dip you can observe is the weekend.

DISCUSSION

It was clear that measurement of a query language is possible. Furthermore, we now had some experience with such measurement. We knew that we could do this very complicated kind of testing in a reasonable amount of time, and still get numbers that could be trusted. We knew how to think about the problems, what the problems were. For example, one problem that arose in planning the experiment was how to think about the teaching. How well people could learn the language depended very much on the adequacy of teaching. So what were we testing, after all, the teaching or the language? Furthermore, we also wanted to find out where people had problems with the language so that it could be improved. But if we developed a very fine teaching course, one that circumvented the language problems that we thought were there, then we would not be able to confirm the existence of these problems!

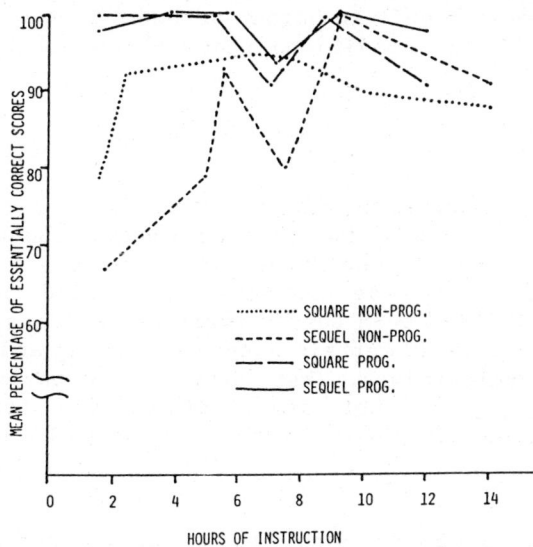

Figure 3. Learning curve for simple mapping

In addition to learning how to test languages, we also had some clear results about SQL. We knew that it could be learned by both programmers and non-programmers. We claimed that it could be used by the kinds of people it was intended for after about 12-14 hours of instruction.

Cautions

However, claims such as the above can be misleading. It is necessary to look very carefully at the methodology used for testing. This is true both for the SQL tests and for other tests of this kind.

First, only some aspects of ease of use were measured. Learning was tested, and memory as well. But productivity after a period of time was not measured, nor was relearning. Furthermore, these were paper and pencil test. Paper and pencil tests are appropriate and can be very efficient for finding out whether people understand the concepts of a language. However, this clearly does not say anything about the advantages or disadvantages of on-line use. In addition, the test was not standardized. The test itself had not been tested. Nor was the teaching completely reproducible. A detailed curriculum had been developed and was used for teaching. However, classroom teaching adapts to minute to minute classroom needs and will differ from class to class.

One particularly important point is the following. The claim that the language could be learned only holds for the given conditions. One of those conditions was that a Ph.D researcher was the teacher. It is not clear whether this would result in higher or lower tests scores, but it is not representative of the conditions under which SQL could reasonably expect to be taught.

Progress and Problems

It was clear that SQL measurement was possible, and we had experience with one test to support this claim. We had some results about SQL itself. We knew that it could be learned under the given conditions. But there were many problems remaining. (These are sometimes called "opportunities"). Only some aspects of SQL had been tested (writing, reading, comprehension) and these were tested only at certain stages (learning and memory, but not productive usage or relearning). There had been no on-line testing, and the tests themselves were not standardized. The teaching was not completely reproducible. Therefore, much remained to be done.

EARLY EXPERIMENTS WITH QBE (Thomas and Gould)

At approximately the same time that SQL was being developed at the IBM Research Laboratory in San Jose, California, another query language was being developed on the opposite coast of the United States, at the IBM Research Center in Yorktown Heights, New York. This language, QBE (Query-by-Example) was also based on Codd's relational model. QBE was the brainchild of Moshe Zloof (5). Like SQL, it was also intended for non-programmer professionals.

An obvious question was: which is easier to use, SQL or QBE.

QBE

An example of a QBE query is shown in Table I. In QBE, templates (table outlines with column names) were displayed on the computer console. The user had to enter data in these columns by typing at the computer keyboard. In this example, the user entered "50", the condition that he knew, in the column DEPTNO. Then, to get the information he wanted, he would enter his request in the NAME column, which had the answers. "Brown" is an "example element". Any example could have been used. The "p." stands for "print".

QBE TESTING

A full discussion of QBE testing can be found in the paper by Thomas and Gould (12). I will concentrate only on the particular points I wish to make.

The basic approach to testing QBE was the same as that of SQL, teach the language, then give people tests in the language. However, there were differences in details. The SQL tests included tests of writing and understanding, which the QBE tests did not. On the other hand, the QBE tests included "ill-formulated questions", which the SQL tests did not. Ill-formulated questions were not precise questions such "Find the names.....". Rather, they described a situation, perhaps in three or four sentences, and the user would have to formulate the question from this longer description.

Details

Some details of the testing are the following.

Schedule: Students were taught for one hour and 45 minutes, then tested on the material that had just been covered. Then they were taught for 70 minutes more, and tested on the new material.

Subjects: Fifteen of the subjects were college students or recent college graduates. Twenty-four were high school graduates. They were taught in four groups, ranging in size from 4 to 12.

Teaching: Both Thomas and Gould were teachers. They used an overhead projector in a classroom-type situation.

Testing: Testing was with pencil and paper.

RESULTS

Results of the tests, as given by Thomas and Gould, were:

   o Percentage of correct queries: 67%.

   o Mean training time: 2-3 hours.

WHICH IS EASIER: SQL or QBE

Table III summarizes data for SQL and QBE. The reader is invited to look at this table, and decide whether SQL or QBE is the easier query language.

Table III: WHICH IS EASIER?

	SQL EXPERIMENT	QBE EXPERIMENT
Training Time	12-14 hrs	2-3 hrs
Correctness	65%	67%

The answer to the question is that, based on these two experiments, WE DON'T KNOW.

First, we do not know whether subjects could perform the same tasks. We have information about reading and understanding in SQL, but not in QBE. And we have information about ill-formulated questions in QBE, but not in SQL.

Were the subjects equivalent? We don't really know. The SQL subjects were all college students, without programmer training. Some of the QBE subjects were only high school students, but some did have a little programming experience. Which subject group was more likely to succeed, if either? The college students might possibly do better than the high school ones, but the students with some programming experience might do better than those without.

Was the teaching method the same, or not? One important facet of teaching method is the ratio of students to teacher: the fewer students per teacher, the better. This ratio was 15:1 in the SQL tests, but 5:1 for QBE.

When was the test given:? In SQL testing, it was given after all functions had been taught and well practiced. In QBE testing, there were two separate tests - each one immediately after some functions had been taught.

Were the test questions equally difficult? This has not been assessed.

There is another crucial point which is frequently misunderstood. Teaching time in these tests was not an experimental result, it was part of the experimental procedure. The experimenters decided approximately how to teach, and from that came the teaching time. For example, I decided to teach to the slower members of the class, and not move on until they understood. I also decided to use blackboard work, and to make sure students knew what they were doing, not write queries by rote. All this affects the teaching time. But this is a decision about procedure. It is not a result of the experiment. Suppose, for example, I had decided to teach whatever I could in three hours? Would the percentage of correct queries have been substantially different than they were? Suppose 12-14 hours had been spent on QBE testing. Would the results have differed markedly? Without doing experiments which control teaching time, we do not know.

It is clear that some aspects of the testing might seem to favor QBE, others, SQL. The result is that from these two separate tests, we simply can not tell whether QBE or SQL is easier to use. The results shown in Table III were deliberately misleading. Such comparisons, of tests made under different conditions, do not tell us which language is easier to use.

DISCUSSION

Cautions

TO COMPARE LANGUAGES, it is necessary to have COMPARABLE EXPERIMENTS. The following apects of the tests must be equivalent:

- o kinds of tasks (writing, reading...)
- o kinds of test questions (well-formulated, ill-formulated...)
- o test question difficulty
- o subjects
- o teaching methods
- o testing "situation" (e.g. instructions to subjects)
- o experimenter attitude

This list is far from complete.

Experimenter attitude, in particular, is very important. If the experimenter wants the results to show that a language is easy to use, this can influence the data he/she gets. This is true even though the experimenter is attempting to run a careful, unbiased experiment. There are subtle clues one can sometimes use to judge an experimenter's attitude. If the report of the experiment says "we want to validate (i.e. prove) the ease of use of the language", this is very different from saying "we want to find out whether the language is easy to use". It turns out that many psychological experiments are sensitive to experimenter attitude. Even in experiments with rats, the experimenter's attitude can unconsciously influence the way he/she handles the animal, and thus affect the results. Reference (9) discusses comparative experiments in more detail.

Progress and Problems

What then was learned from the QBE experiment, and what remains to be done? Basically, the results are similar to those for SQL. It was clear that such experimentation was possible, and there were results of the one test to prove it. It was also clear that QBE could be learned in "reasonable" time under the given conditions.

But the same kinds of limitations that held for the SQL tests also hold for QBE. Only certain tests were done, but not others (e.g. no tests of productivity, reading, understanding). There was no on-line testing. The test itself is not standardized. The "given conditions" included two Ph.D researchers as teachers, hardly a typical situation.

Thus, while much was learned, much remained to be done.

COMPARISON OF QBE and SQL (Greenblatt and Waxman)

In order to compare languages, it is necessary to run a <u>controlled</u> experiment. This is what was done by two professors at Queens College, New York (4). In a controlled experiment, one holds constant those factors that might make a difference in the experimental results. One can also examine other factors that might make a difference in the results, but have not been controlled, to see whether experimental results could have come from these "uncontrolled" factors.

Greenblatt and Waxman ran a controlled experiment comparing QBE and SQL. They held constant a number of factors (e.g. the teaching method, the exam questions, the exam procedure, the scoring technique, etc.) They also gathered data on the students' background: age, sex, number of computer courses, high school grade average, etc. This data is shown in Table IV. There are some differences in the number of subjects in each class, and some differences in the amount of computer training they had. There are also differences in the percentage of male and female students in each class. (I personally do not think the different proportion of males and females makes a difference, but that is a personal bias. Such questions can be tested experimentally).

Results of Greenblatt and Waxman's test can be seen in Table V. Notice that only the last two questions were statistically significant. Very loosely, for those unfamiliar with statistics, significance means that these are the only results you can trust.

On the basis of this data, Greenblatt and Waxman claimed that QBE was "superior to SQL in learning and application ease". Let us see whether this claim is indeed supported by this experiment.

DISCUSSION

Cautions

What have we learned from this experiment? The only things that we know to be true are that people wrote queries faster in QBE than in SQL and that they were more confident that their answers were correct. It is not surprising that they wrote faster. They had less to write. Confidence is interesting, but is much less basic than correctness, for example.

Furthermore, even though this was a controlled experiment, some of the uncontrolled factors might have favored one language or the other. The number of subjects per class might have favored QBE. The source of questions (they were largely Thomas and Gould's questions) might have favored QBE. The subjects' computer background might have favored SQL. Thus, the end result is still somewhat unclear.

TABLE IV
Subject background, adapted from Greenblatt and Waxmann

	QBE	SQL
No. of Subjects	8	17
Mean Age	19.3	24.8
Mean High School Average	90.8	82.8
Mean College GPA	3.3	3.4
Mean No. Computer courses	1.0	1.9
Percent Male	71.4	52.9

TABLE V
Overall Results, adapted from Greenblatt and Waxman

	QBE	SQL
Training Time (hrs:min)	1:35	1:40
Mean Total Exam Time (min)	23.3	53.9
Mean % Correct Queries	75.2	72.8
Mean Time/Query (min)	.9	2.5 ***
Mean Confidence/Query (1 = very sure correct, 5 = very sure incorrect)	1.6	1.9 ***

*** Statistically significant ($p < .001$).

Progress and Problems

The progress that has been made is that a controlled experiment has been run. This gives some experience in the running of such experiments. And there is some data on the SQL-QBE comparison.

However, there is still much to learn. The important questions (e.g. percentage of correct queries) did not yield statistically significant results. And even more experimental control would be useful.

QMF (Boyle, Bury and Evey)

This year (1983) my colleagues, Boyle, Bury and Evey, of the IBM Human Factors Center in San Jose, California, ran a controlled experiment comparing QBE and SQL, using implemented versions of the languages. A recently announced IBM product, QMF (Query Management Facility) has options for use of both QBE and SQL. During the development of QMF, the question of which language was easier to use arose. Boyle, Bury and Evey decided to find out (1).

SUBJECTS

As subjects, Boyle, Bury and Evey used twelve business students and eight secretaries. Each subject learned both languages. When one person learns both languages, there clearly is less variability in the results because of differences between people.

TRAINING METHOD

Training method for this experiment was no longer a human teacher, as in the previous experiments. Instead, students learned from self-study manuals. This is clearly a more reproducible form of teaching than learning from a human tutor. With this method, it is also possible to measure the amount of learning time.

TEST METHOD

Instead of paper-and-pencil tests, Boyle, Bury and Evey used a special facility they had for on-line testing. The subject took the test at a computer terminal. The computer directed the presentation of the experimental tasks and collected data on time and errors.

Language coverage: This experiment was less compehensive in its coverage of the various functions of the languages than some of the previous experiments. Only 20% of the functions were tested. However, this 20% is supposed to account for 80% of usage.

RESULTS

Data was obtained on training time, on writing queries, and on subjective opinions. Results showed that neither language had a clear advantage over the other: SQL was better in some ways, QBE in others. The training time was faster for some QBE functions (e.g. joins); but it was faster for some SQL functions (e.g. arithmetic). For query writing, there were few strong differences. Again, QBE was faster for some functions (e.g. multiple tables) and SQL for others (e.g. arithmetic). Subjective opinion, too, were divided: some of the subjects preferred QBE, others SQL.

As a consequence of these results, the experimenters' recommendation was: offer both languagues in QMF. Let the user choose. This, in fact, is what was done.

DISCUSSION

What progress had been made, and what remains to be done? More experimental control has been introduced. There is some automation of the testing. This is both good and bad (fewer students can be tested at a time unless you have many computer terminals). The training method is reproducible. But only 20% of the language has been tested.

Notice that the recommendation (to use both QBE and SQL) has <u>itself</u> not been tested. Such a test might involve comparing QBE with QMF, and also comparing SQL with QMF. The major issue here is whether offering subjects the option of using either of the two languages will confuse them more than the benefit they would get from having the choice.

FINDING USER PROBLEMS: EARLY SQL EXPERIMENTS

We turn now to another topic. In addition to measuring ease of use and to comparing two languages for it, we also want to help in the development of languages. We want to provide feedback about the problems people are having with using a language so that the problems can be fixed. It is also possible to use measurement of ease of use to help with such feedback.

IDENTIFYING DIFFICULT FUNCTIONS

In order to improve the ease of use of a language, one thing to do is try to determine which functions of the language are particularly difficult. To do this, one can measure the ease of use of the individual functions, using the same kinds of techniques that are used for the overall language evaluations. Table VI shows the results of such measurement from the early SQL experiments (8). This table shows the different functions, such as the simple mapping illustrated in Table I, as well as more complex functions.

It is clear that there is a range of difficulty of the functions. This is hardly surprising. In any language, it is highly likely that some functions will be harder than others. What is important is that the measurement identifies the ones which are difficult. It is then possible to look at them, try to determine why they are difficult, and try to fix them.

IDENTIFYING COMMON USER ERRORS

There is another kind of feedback one can supply for improving the ease of use of a language. We would like to know what specific kinds of errors people make. The point is the following. If one person makes a silly

TABLE VI.
Mean percentage essentially correct scores on SQL final exam

	Non-programmers	Programmers
MAPPING	91	98
SELECTION	87	89
PROJECTION	73	100
ASSIGNMENT	87	94
BUILT-IN FUNCTIONS	88	89
AND/OR	77	82
SET OPERATIONS	70	88
COMPOSITION	53	74
GROUP BY	46	61
CORRELATION VAR.	12	33
COMPUTED VAR.	7	44

mistake, that is just a silly mistake. If two people make the same silly mistake, that is just a coincidence. But when you find twenty people making the same silly mistake, it is highly likely that a problem in the language is causing it. Table VII shows some of the more frequent kinds of errors that subjects made in using SQL. They made some of the same kinds of mistakes you would expect them to make in using a natural language query system. As is clear from the table, they could not spell. They could not spell even when the words, with the correct spelling, were directly in front of them. They would see the word "personnel", for example, spelled correctly in the data base table, yet spell it with one "n" and two "ll"s. They also made errors with endings. For example, If the question were: find the names etc. (with an "s" ending) and the correct SQL query was SELECT NAME (without the "s"), they would write SELECT NAMES.

TABLE VII
Percentage of subjects making at least one error on final exam

	Non-programmers	Programmers
ENDING	60	39
SPELLING	53	61
SYNONYM	33	50
QUOTATION MARK	47	50
OTHER PUNCTUATION	20	22

As a result of these and other kinds of errors, a number of recommendations for improving SQL were made. For example, it was recommended that the SQL processor automatically correct some unambiguous syntactic errors. It should add parentheses where needed, correct minor errors in spelling, change some keywords, such as the word WHERE to HAVING, when this was required and unambiguous.

DISCUSSION

There are a number of cautions to be aware of in understanding this work. Experiments like these can _find_ problems. However, they can not _explain_ them. They don't say why people are making these mistakes. Sometimes the reasons are obvious, but often they are not, and more work is needed before solutions can be suggested. Furthermore, these kinds of experiments do not suggest _solutions_ to the problems found. Again, sometimes the solutions are obvious and simple. However, sometimes they are far from obvious, and perhaps not even possible without total redesign of the language.

Testing, therefore, is not a bandaid that one can simply take and put at the end of the cycle of developing a query language and say "we will do a little testing and patch things up". Testing to find errors is just a start. There can be a great deal of work beyond that if we truly want to develop languages that are easy to use. Furthermore, even when there are suggested redesigns, the redesigns themselves should be tested. Suggesting improvements to a language is not enough. It is then necessary to find out whether the suggested improvements are correct.

Progress and Problems

Some progress had clearly been made. We knew how to go about trying to identify user problems in a controlled way. Various kinds of problems with SQL had been found. In fact, some corrections had been suggested. However, the suggested corrections themselves had not been tested.

TESTING "USER-FRIENDLY SQL (Welty)

The next step, clearly, was to develop an improved SQL, and then to test it to see whether there really was an improvement. This is what Welty did. Welty was a long distance student of mine at the University of Massachusetts at Amherst. His Ph D. dissertation was a comparison of SQL with another query language, TABLET. In his tests, he, too, found numerous problems with SQL that he felt needed to be corrected.

Welty is now a professor at the University of Southern Maine. At Maine, he developed a version of SQL, User-Friendly SQL, which corrected some of the errors that people had made. For example, User-Friendly SQL corrected mistakes that people made in spelling, corrected some errors in punctuation, and corrected some keyword errors. (User-Friendly SQL is not yet implemented).

Welty then ran a comparative test to see whether the improved SQL really resulted in better human performance than the original version (13). As subjects, he used college students. His tests were paper-and-pencil type tests. He tested query writing with a final exam and memory (short term retention) tests.

His results did show improvement with User-Friendly SQL. There was an 8.2% improvement in the percentage of errors made on the final exam, and a 26% improvement on the retention exam.

DISCUSSION

What progress has been made and what remains to be done? This is the first test we know of which does before-and-after testing. It takes some suggested improvements to a language and tries to determine whether there really is improvement. However, there was no long-term testing, and no on-line testing.

SUMMARY AND GENERAL DISCUSSION

What then, have we learned about query language testing, in general, and about SQL ease of use, in particular?

Clear progress has been made in measurement of ease of use. We know that it is possible to do this kind of testing. Ease of use is not just a fuzzy notion. It can be defined, and we can be somewhat precise in our measurements. The experiments themselves are becoming more controlled, and there is some tendency towards using the computer for testing when it is appropriate. I want to be clear that it is _not_ always appropriate to do so. We have some data about the ease of use of SQL, about its ease of use relative to that of other query languages, and about some of the problems that people can encounter in using it.

Much, however, remains to be done. To the best of my knowledge, there has been no testing of SQL over a long stretch of time. Furthermore, the tests themselves have not been standardized. In educational testing, for example, there are groups that spend a lot of time and effort preparing standardized tests. We know, then, which are the hard questions on these tests, and which are the easy ones; which questions can be answered by most people and which by only a few. Furthermore, we have ironed out problems in the tests themselves that might cause erroneous results. To truly compare query languages, we would need such standardized tests. It may not be feasible to develop them, however, for quite some time.

We need more comprehensive testing, and we need to be able to do the testing faster than we can today. This is not a criticism of the people doing the tests today. Experimenters are usually very well aware that their tests are not as comprehensive as they would like them to be, that they have had to make choices about what to test and what to eliminate. These experiments are so time consuming that it is frequently not possible to do everything you would like to do. Designing these experiments is analogous to a problem in system design. You have a limited amount of resources, and you must decide what is most important, given those resources.

Another problem is the lack of trained people in this field to do the experimenting. A degree, even a Ph.D in psychology is not enough. People in this field have basically trained themselves, and then the experimenters are training others. Some schools are now starting to train people, too. But there is not yet a large enough pool of people that know how to do this work and know how to do it well. This is one of the crucial needs in this very new, "baby" field.

A GLIMPSE OF THE FUTURE

What then, is needed for the future? So far we have been talking about behavioral tests. Behavioral tests require the administrative task of obtaining subjects, the task of preparing test materials, possibly the task of preparing teaching materials, giving tests, taking measurements, interpreting the results. Such testing is time consuming and possibly costly. Frequently, it is done late in the development cycle, too late to be truly useful. Testing to find usability problems should be done very early in the development cycle, while there is still time to rethink the issues and fix the problems. Prototyping tools can be used to test earlier in the development cycle, and they should be used. However, they still require the overhead of behavioral tests: bringing in subjects, preparing and evaluating tests, etc. Paper and pencil tests, too, can sometimes be run early in the development cycle. In fact, most of the tests discussed in this talk are paper and pencil tests. They, too, however, are behavioral tests and require the same overhead. We need <u>analytic tools</u> to use in addition to the behavioral ones we have.

Let me give you an analogy. Suppose I wanted to build a bridge and I wanted to find out whether that bridge would support the weight of people who were to cross it. Or suppose I had two different bridge designs, and I wanted to find out which one would support more weight. What would I do if I were an experimental psychologist ? I would run behavioral experiments. I would send people across the bridge, wait till it fell down, then redesign the bridge, and try again. Eventually, I would get a design that worked. This is a very costly and time consuming procedure. It would be better if I could describe the design of a bridge on paper, abstractly, and then analyze that design on paper. I could then predict, from the analysis, that this bridge would not fall down or that this bridge design would support more weight than that one. We need that same kind of analytic tool for software human factors. This is the kind of work I am currently doing (10, 11).

A FINAL WORD

There is one final word which applies to behavioral testing and to analytic tools and to everything else we do in this field: MEASUREMENT IS NOT THE END, IT IS THE BEGINNING.

REFERENCES

1. Boyle, J. M., Bury, K. F., and Evey, R. J., Two studies evaluating learning and use of QBE and SQL, Proceedings of the Human Factors Society, 27th Annual Meeting, 1983, pp. 663-667.

2. Chamberlin, D. D., Astrahan, M. M., Eswaran, K. P., Griffiths, P. P. Lorie, R. A. Mehl, J. W., Reisner, P., and Wade, B. W., Sequel 2: a unified approach to data definition, manipulation and control, IBM Journal of Research and Development, 20, Nov., 1976, pp. 560-575.

3. Codd, E. F. A relational model of data for large shared data bases, Communications of the ACM, 13, 2, June 1970, pp. 377-397.

4. Greenblatt, D. and Waxman, J. A study of three database query languages, in Databases: Improving usability and responsiveness (B. Shneiderman, ed.), Academic Press, New York, 1978.

5. Human Factors in Computer Systems, Proceedings, March 15-17, Gaithersburg, Md.

6. Moran, T. P. (ed.). The psychology of human-computer interaction, special issue of ACM Computing Surveys, 13, 1981.

7. Reisner, P., Boyce, R. F., and Chamberlin, D. D., Human factors evaluation of two data base query languages - SQUARE and SQL, in Proceedings of the National Computer Conference, AFIPS Press, Arlington, Va., 1975, pp. 447-452.

8. Reisner, P. Use of psychological experimentation as an aid to development of a query language, IEEE Transactions on Software Engineering, SE-3, May 1977, pp. 218-229.

9. Reisner, P., Human factors studies of database query languages: a survey and assessment, Computing Surveys, 13 (1), March 1981, pp.13-31.

10. Reisner, P., Formal grammar and human factors design of an interactive graphics system, <u>IEEE Transactions on Software Engineering,</u> 1981, SE-7, pp. 229-240.

11. Reisner, P. Formal grammar as a tool for analyzing ease-of-use: some fundamental concepts, <u>Human Factors in Computer Systems,</u> (Thomas, J. and Shneider, M. eds), Ablex, 1984.

12. Thomas, J. C., and Gould, J. D., A psychological study of query by example, <u>Proceedings of the National Computer Conference,</u> AFIPS Press, Arlington, Va., 1975, pp. 439-445.

13. Welty, C., Correcting user errors in SQL, <u>International Journal of Man-Machine Studies</u> (in press).

14. Workshop on Software Human Factors, report (to be published)

15. Zloof, M. M., Query by example, <u>Proceedings National Computer Conference,</u> AFIPS Press, Arlington, Va, May, 1975, pp. 431-437.

Human Factors in Electronic Musical Instruments

Yasunori Mochida
Nippon Gakki Co., Ltd.

1.   Introduction

This paper is presented to participate in a symposium under the theme of "Human Factors in Design". The title "Human Factors in Design" refers to the research on machine/device designs and studies based on considerations given to the people who use these machines/devices.

Past studies on this subject have been concentrated in the fields of aviation, motor vehicles, and trains, and various papers discussing the adaptability between people and machines have been published. However, it is felt that other topics of different kinds of relationships between people and machines, such as harmonization, duty sharing, and the master-servant relationship, should also be discussed in addition to adaptability.

I have dedicated my business career to the development of musical instruments, particularly electronic musical instruments. And in dosing so, I have experienced various problems in the relationships between people and the instruments in terms of both software and hardware. In this discussion, I would like to present these experiences for reference.

2.   Harmonization

When considering the subject of people and machines, it is necessary to understand that emphasis is placed on harmonization between the two.

When considering the phenomenon of performance ("playing music"), the instrument represents the hardware and the musician the software. The hardware and software are combined to produce output in the form of music. Consequently, to produce better music, both the hardware and software must be improved. Hardware improvements involve improved functions, while software improvements involve improved techniques.

3.   Hardware Aspects of Electronic Musical Instruments

3.1 Progress

During the development of electronic musical instruments, the technological advances made in hardware have been remarkable. A general assumption might be that this progress concerns the development of fully automatic performance functions or improved instrument operability. Although these are, in fact, important factors of progress, the "tone making" factor is of greater importance. The "tone making" factor, the ability of synthesized electric signal waves to be heard as good tones, is the most highly desired factor.

Because music is constituted by a timewise alignment of tones, the tones themselves must be of good quality. Thirty years ago, during the early stages of electronic musical instruments development, one may recall the all too "electrical" quality of tones produced by electronic musical instruments. Compared to these, the tones produced by recently developed electronic musical instruments have a wider variety and a much more natural quality.

3.2 Tone making

Tones are considered to generally consist of three elements: pitch (tone height), timbre (tone color) and loudness (tone intensity). These terms refer to the frequency, harmonic structure, and amplitude, respectively. However, good tones for music cannot be made simply by applying these three elements. It is necessary to add the dimension of timewise variation. For example, if the frequency is alternately fluctuated up and down, and the so-called vibrato effect is imparted, the resulting tones sound more natural than those at constant and continuous frequencies. If tones are suddenly generated with a certain amplitude and are sustained for a certain period, and are then suddenly discontinued, the resulting tones take on the same abrupt effect as Morse code signals or the sound of a cheap car horn. However, if an envelope on the tones is imparted to give a slight roundness in their build-up portion (attack), and is then gradually reduced (release), the resulting tones are of a much better quality.

In addition, when natural tones were studied in more detail, it was noted that they were irregular or slightly unstable in their build-up portion. As a result, these elements were also incorporated in the tone making. The music synthesizer rendered fluctuation in the build-up portion of the pitch or timbre by timewisely varying the frequency of the signal or the filter characteristics of the electronic circuit.

Non-vibrato

Vibrato

Attack and
Sustain Imparted

Attack Portion   Release Portion (Sustain Effect)

(Although other portions were treated here, the build-up portion was focused on). It was determined that, although the variation occurs in a split second, human ears are sensitive enough to detect such variation.

Such findings were subsequently incorporated in the development of electronic musical instruments. However, the developmental progress in the quality of timbre lagged behind. This was not only because no practical circuits were available at a feasible price, but because the previous development of tone making had only focused on the amounts of the respective harmonic components.

Simple electric signal waves generated by an ordinary electronic circuit contain harmonic components at pure integer ratios. However, it was determined that such tones, which included components at pure integer ratios, are not good quality sounds.

Similar to the fact that delicious water is not distilled water, but is spring water that contains suitable amounts of calcium and carbonate, the tones of a musical instrument sound more natural and pleasant when they are not only familiar, but when they contain their appropriate share of impurities.

The impurities mentioned here refer to the harmonic components at ratios deviating from the integer relations. The tones with harmonic structures slightly deviating from the integer relations were determined to be those that sound the most beautiful. This is called the beauty of inharmonicity. On the other hand, if the deviation is too great, problems may arise. If it is asked what determines this beauty of inharmonicity, the only possible reply is that the refined auditory sense of a person must be the judge.

In the past, electrical signal waves with inharmonically related harmonic structures could not be easily generated by conventional circuits. This task has been made possible in recent years by using a method introduced by Doctor Chowning of Stanford University. This method is based on an FM (frequency modulation) method that incorporates the use of LSI.

In the conventional FM method used in broadcasting, the carrier signal of a radio frequency is modulated by a modulating signal of an audio frequency. Then, based on this carrier frequency, a plurality of frequency components are sequentially distributed above and below the carrier frequency at intervals of the modulating frequency.

The magnitude of each components in this case is represented by a value of the Bessel function of each order. However, to simplify the explanation of inharmonicity, only the frequencies of each component will be noted here.

For example, if a carrier wave of 500 Hz is frequency-modulated by a wave of 100 Hz, then components of 400 Hz, 300 Hz, 200 Hz, 100 Hz and so on are produced below the carrier, and components of 600 Hz, 700 Hz, 800 Hz, 900 Hz and so on are produced above.

Since 100 Hz is the lowest frequency, it can be regarded as the fundamental component wave, while 200 Hz, 300 Hz, etc., correspond to the harmonic component of two times, three times, etc. Here, each harmonic component is precise in the integer relation.

However, if a carrier wave of 508 Hz is frequency-modulated by a wave of 102 Hz, then components of 406 Hz, 304 Hz, 202 Hz, 100 Hz and so on

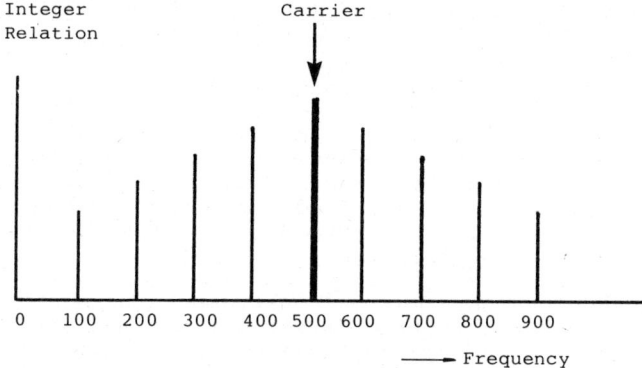

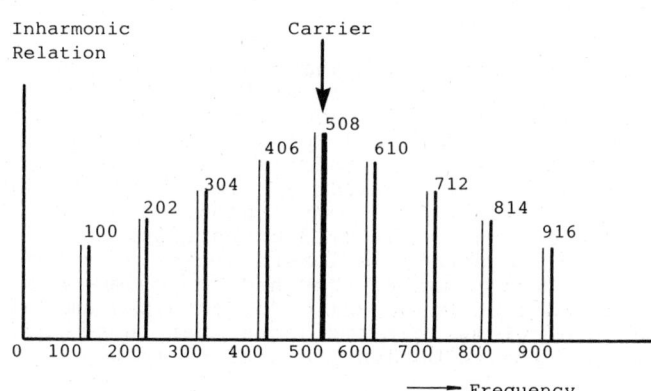

are produced below the carrier, and components of 610 Hz, 712 Hz, 814 Hz, 916 Hz and so on are produced above the carrier. Because 100 Hz is the lowest frequency, it can also be regarded as the fundamental component wave, while 202 Hz, 304 Hz, etc., correspond to the harmonic components of 2.02 times, 3.04 times, etc.

In this case, each harmonic component is not precise in the integer relation. According to this principle, the harmonic components exhibiting an inharmonically related harmonic structure are produced.

3.3 Response

Just like a musician who expects certain tone responses from his musical instrument, a computer operator must also obtain the responses he expects from the computer. If the response is not what was expected, the operator will not be able to perform his job. In the computer, a certain time interval (lag) is required from input to output. However, an interval that is too long or inconsistent will cause problems for the operator. Because a computer is used for a variety of purposes, multiple jobs are controlled according to priority, and, within one job, a certain response is maintained according to the arrangement of such resources as disks, circuits, and memory.

The importance of a suitable interval can be explained by the following analogy. If a person using an automatic vending machine receives a carton of milk or pack of cigarettes in the dispenser immediately after inserting a coin, this person may momentarily think that the item is what was left behind by a previous user. In the same way that a suitable interval is set for vending machines, a processing interval is set between computer input and output.

In the case of musical instruments, however, there is a slight difference. Because sounds are instantaneously aligned in succession to constitute tones during musical instrument performance, exact real-time processing is required. The expected performance cannot be obtained unless the hardware can freely respond to the fast and slow movements required for the performance.

In the days when electronic musical instruments were not yet digitalized (1970s), when each instrument circuit was individually connected by electrical wire on one-to-one basis with its respective circuit, the instrument's performance, in effect the operation, responded in real time.

Contrary to the assumption that a sharper response could be more easily obtained once these instruments were digitalized, it became more difficult for engineers to improve response because of the time sharing process. In other words, with the introduction of digitalization and the greater degree of freedom in information processing, the engineers had set new goals that were previously unimaginable, such as the improved ability to conduct more complicated changes during short intervals of the aforementioned build-up portion, and adding individual control to harmonic components in inharmonicity. In order for the engineers to attain their goals, they had to achieve more detailed processing in a limited time frame using an increased amount of information. It was under these circumstances that these engineers continued their efforts to obtain the desired response speeds.

3.4 Automatic performance functions

Automatic performance functions are added to the electronic musical instrument to simplify operation. These functions expand the system to improve the effects of the performance. If a player desires to give a complicated performance in which he is not able to perform all the necessary movements, automatic performance functions can be used to assist in the performance of necessary operations in place of a human player.

Instruments may be equipped with any combination of such functions. These include auto-rhythm performance with percussion sounds and the automatic addition of arpeggio performance, duet tones, and counter melody lines.

4. Human Engineering in Electronic Instrument Construction

Electronic musical instruments that include manual keyboards and pedal keyboards have a variety of operation levers and buttons arranged for easy operation.

The manual keyboard, which is the most frequently operated mechanism, is used to select the pitch of tones. The keys are arranged in a single row of several octaves, each octave consisting of twelve tones comprising seven white keys and five black keys. This pitch-wise arrangement provides a great advantage and allows the player to select notes intuitively. Because the entire arrangement is such that the

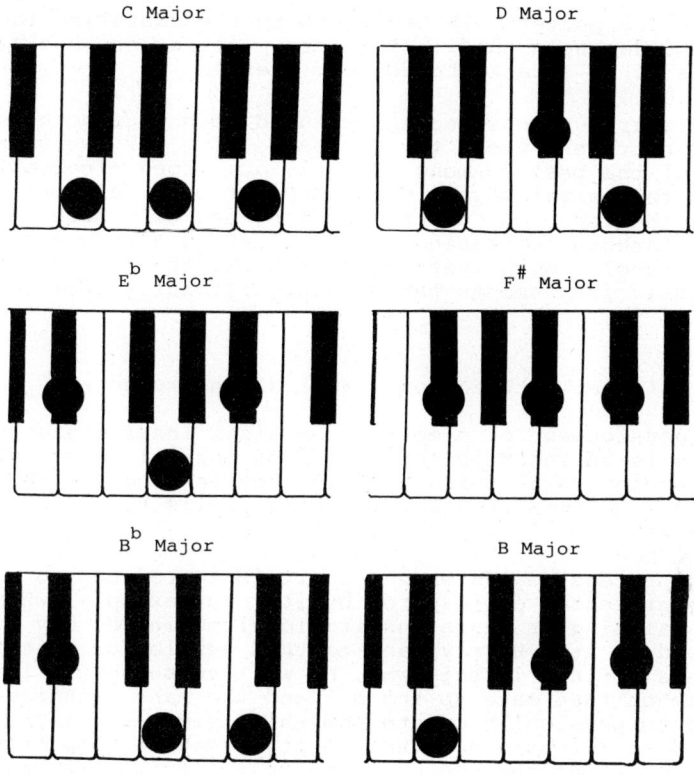

Fingering Patterns for Chords

black keys are arranged alternately in groups of twos and threes, the fingering pattern varies depending on the chords. Even for the major cords, the fingering pattern differs from that for C major, D major, Eb major, F# major, Bb major and B major.

A variety of proposals had been made in the past to change the existing key arrangement, such as to arrange the black keys and white keys alternately throughout the entire keyboard, or to arrange buttons in several rows as in a typewriter. However, none of these proposals ever replaced the current arrangement. The reason for this was not due to the conservative attitudes of players, but because of the ideal nature of the existing arrangement that provides easy operation during instrument performance. Because the black keys are placed in alternating groups of twos and threes, the locations of the desired keys can be found in a single glance or detected by the feel of the fingers without even looking at the keyboard. If all the keys had been arranged uniformly, a need would have arisen to provide some indication, either by a mark or a raised surface, of a set reference point such as the C key. By using the existing arrangement, the desired chords can also be easily pressed using various fingering patterns, and the chord being played can be easily identified by the player. Although some time is required for a player to master these fingering patterns, this existing arrangement was found to be the most convenient. Mastering these various fingering patterns is analogous to the mastering the reading and writing of Japanese Kanji. The more patterns one knows, the easier and faster is is to select the correct ones.

Although there are various mechanisms used to select notes depending on the kind of instrument used, the keyboard, with its twelve-note octaves, is one of the best. Among recently developed electronic musical instruments, the manual keyboard is used not only for note selection but also for the tone control of intensity and/or timbre. This is accomplished through depressing force, which is also called the touch-responsive control. With these developments, the expression ability of electronic music instruments has been significantly improved.

5.   Software Aspect of Electronic Musical Instruments

Because the development of electronic musical instruments is relatively recent, there is an insufficient number of musical scores available for these instruments. With this insufficiency in mind, we should examine the educational methods used for teaching pupils how to play the musical instruments.

The learning process begins when a pupil is introduced to a certain art and becomes interested or inspired by it. For example, when a pupil sees a fine painting or hears inspirational music, he may be emotionally moved by it. He may want to then develop or imitate works of others in this field. In this way, he will subsequently learn the various elements that make up the art and the many techniques and methods used to physically create something similar. As he works through repeated imitations, a unique style may develop that deviates from simply imitating someone else, which relies on previous learning, to create unique expressions. Although such unique works may not be masterpieces, he will eventually improve, and will benefit from self-criticism or constructive criticism from others. In this way, the pupil refines his craft until he masters it.

If this learning method is applied to music, the pupil must proceed through these steps in three areas: melody, rhythm and harmony. The sequence of learning the sensations of melody and rhythm followed by the sensation of harmony is a normal human leaning experience. Although the conventional method used in Japan involves the faithful reproduction of the etudes of Beyer or Czerny, and is an effective method for learning melody, it is not always effective in learning harmony.

In addition, this conventional method teaches theories first. For example, it stresses that such melody notes as fa and la should be accompanied by a chord of the fourth degree. Also, if the preceding chord is a "Doppledominante", the succeeding chord should be a "Dominante". The method also teaches that two notes which are separated by an interval of the perfect fifth cannot move in parallel. This method equates learning music with hard work instead of something pleasureable. And, above all, it does not encourage creativity because is places too many restrictions on the pupil.

In teaching, creativity should be encouraged and restrictions should be minimized. It is important to encourage children to play in ways in which they feel are comfortable, so that they will learn music with their senses. In this way, they will be able to appreciate a melody as being interesting, or feel that a certain chord accompaniment is better as they discover and learn the rules along the way.

According to this basic concept, we at Nippon Gakki Co., Ltd., established Yamaha Music Foundation as a separate non-profit organization to nurture musicians who are able to understand and create music, as well as to develop newer educational methods and software.

To correct the common misconception that music is difficult to learn, we stand committed to teaching children the general generic basics, such as melody, rhythm, harmony and absolute hearing, by using educational methods that incorporate natural and compatible procedures.

At present, there are 9500 Yamaha Music Schools in Japan, as well as established schools in 250 cities worldwide. The current student enrollment is 650,000, and 3 million students have graduated from our schools.

6.   Conclusion

It is rather difficult to fully utilize the capacity of an electronic musical instrument, despite the various functions or utilities. To alleviate this problem, more advanced software and players with better technical skills are required.

These types of instruments are equipped with complex, high-speed LSI components that can process extremely large amounts of information in a very short time. These instruments are, in a broad sense, a type of special-purpose computer.

It can be said that, in the effort to improve the relationship between people and electronic musical instruments in terms of both hardware and software, our trials and tribulations have been quite similar to those of people engaged in the development of the computer.

To promote a better relationship between people and musical instruments, it is necessary to create new and better musical scores for these newly developed instruments and, on the human engineering level, to develop wider and deeper understanding of the instruments.

As explained earlier, there are various approaches to discussing the relationships between people and machines. And although this particular discussion focused on the subjects of tone making and educational methods, we hope active discussion will be approached from a wider range of perspectives.

Incorporating Human Engineering in Motor Vehicle Design

Shunji Tsuchiya
Toyota Motor Corporation

1. INTRODUCTION

Motor vehicles are a primary source of transportation for mankind. Consequently, "Human Factors" must be incorporated into motor vehicle design to ensure safety as well as comfort. The interaction between man and machine is diagrammed in the "Man-Vehicle System" proposed by Professor Hirao of Tokyo University (Fig. 1).

This system implies that in order to drive a vehicle, we must consider such environmental factors as road conditions and the actual course. Operational judgements are made according to feedback signals received through our recognition of the environmental factors. In this way, we steer our vehicle along a winding road without driving off the road. The driver must rely on personal knowledge, emotion, and will to ensure safe driving operation. Thus, the difference of driving pattern is depend upon the difference of control degree between each driver.

Therefore, when examining the operation of a motor vehicle, the correlation between these three driver factors with the vehicle and environment must be considered. In other words, a superior motor vehicle design must include these "Human Factors".

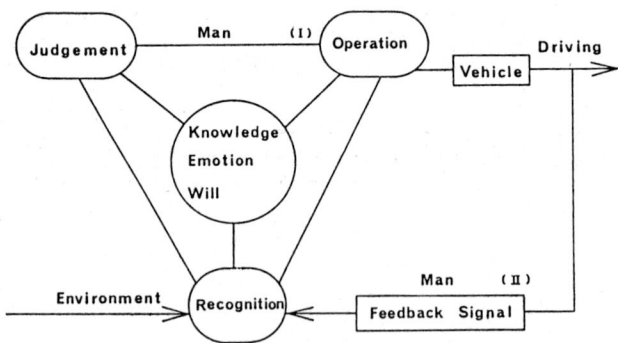

Fig. 1. Man-Vehicle System

2. HUMAN ENGINEERING APPROACH TO MOTOR VEHICLE DESIGN

The human engineering factors in motor vehicle design can be roughly divided into comfort and safety as shown in Fig. 2, although there are many ways to categorize these factors. Comfort includes habitability, operability, riding comfort, low noise, and ease of entry/exit, etc. Although riding comfort and low noise have no direct impact on driving, they are important factors in reducing driver fatigue and discomfort.

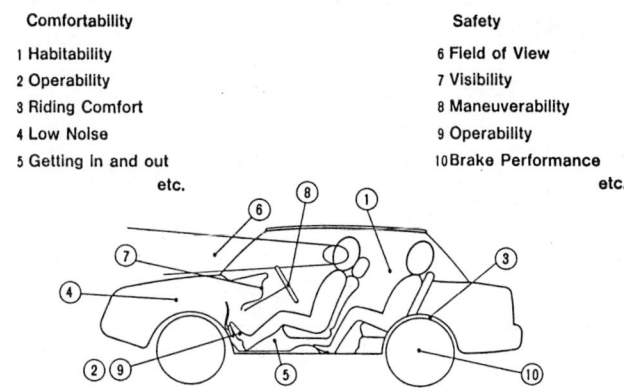

Fig. 2. Human factors on vehicles

Safety includes the field of view, instrument panel access, driving operability, and brake performance, etc. With regard to these factors, the major human engineering considerations, examples of typical values, and actual applications in motor vehicle design are explained as follows.

2.1 Habitability

Habitability includes the follows:
i)   Interior space
ii)   Arrangement of controls
iii) Seating comfort
iv)   Air conditioning

Only interior space, arrangement of controls, and seating comfort will be discussed.

Such factors as legroom, knee clearance, head clearance, and lateral space (e.g., shoulder room) must be considered when designing the habitable space of a vehicle.

Although the driver's physical size is a major consideration, the driving and seating postures are also important. It should be noted that driving and seating postures are often different even among persons having the same physical size. This variation can be easily understood by examining the many different angles of joints in the human body (Fig. 3).

The information in Fig. 4 is provided by Renault, and indicates the comfortable angles of driving posture. Although the range of angles is wide, the correlation is important. The driving or seating posture varies depending on physical size and sex. Driving postures are primarily affected by the characteristic of height. Figure 5 shows the driving posture of a rather short female. Figure 6 shows the driving posture of a tall male.

Figure 7 shows these two driving postures overlapped. By comparing the different angles of the joints (shoulders, elbows, knees, etc.), it is easy to understand why the postures differ.

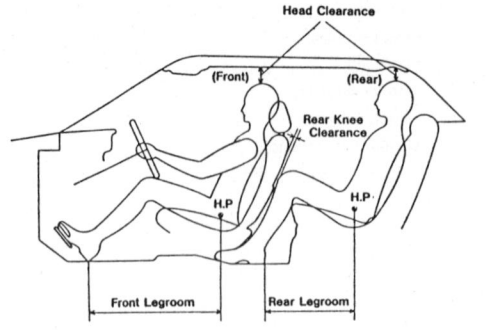

Fig. 3. Interior space

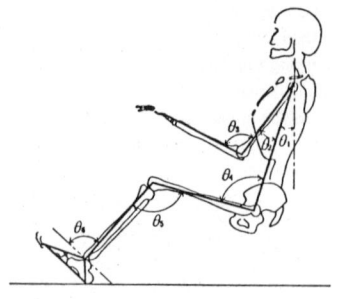

Fig. 4. Comfort angles

Fig. 5

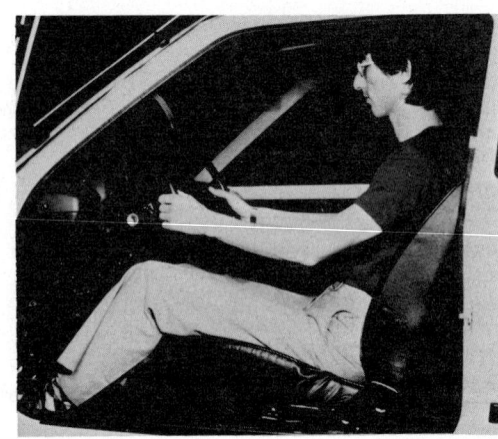

Fig. 6

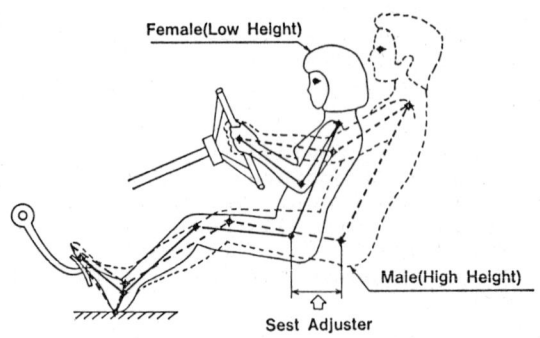

Fig. 7. Comparison of driving positions

Consequently, the interior of a vehicle must be designed to accommodate different physical sizes and postures. For this interior design, a prototype (three-dimensional model) is used to evaluate such items as the layout of the instrument panel, steering wheel position, foot pedals, gear shift, and seating.

Figure 8 shows a three-dimensional model being used to confirm that the interior space and arrangements meet the planned specifications. The primary characteristics that affect seating comfort are the final posture and distribution of body pressure. The final posture is the posture assumed after the driver is properly seated. For example, the stoop posture indicated in Fig. 9 (by the dashed-dotted line) is considered bad posture. On a well-designed seat, the center of the body pressure distribution should be at the lumber support near the bottom of the seat back, and at the seat cushion.

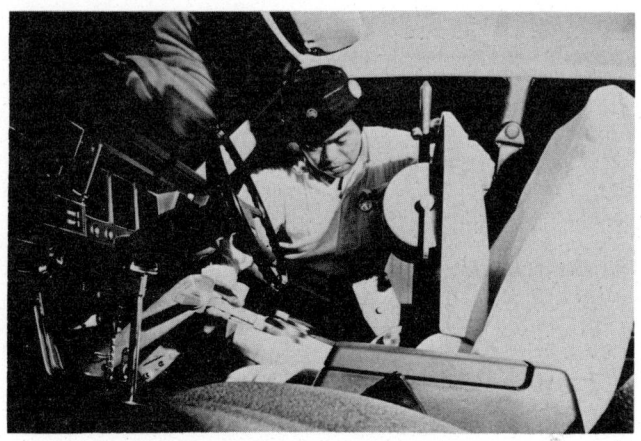

Fig. 8

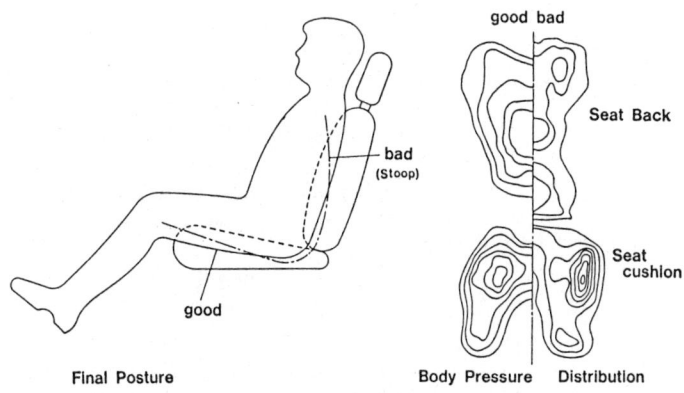

Fig. 9. Seat comfort

In the past several years, motor vehicles have been equipped with a variety of items designed to improve habitable space, ease of operating controls, and seating adjustment. Figure 10 shows some of these items.

The adjustable mechanisms shown above are provided to simplify the operation of various control devices and seating adjustment after obtaining the optimal field of view. Note that the position of the driver's eyes is essential in acquiring maximum driving visibility, which will be described later in detail.

As the number of required adjustments is increased, the level of complexity is also increased. To help alleviate this increasing complexity, a system has been developed in which desired adjustments are all stored in a computer. In this way, all adjustments can be automatically set by simply pushing a button.

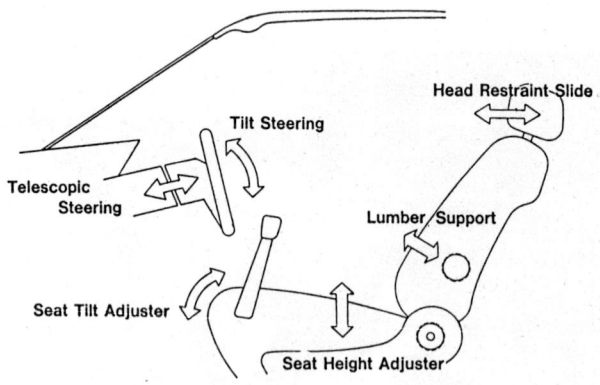

Fig. 10. Equipment to improve habitability

2.2 Field of View

The field of view is divided into the direct field of view and the indirect field of view. The components of these two fields are listed as follows:

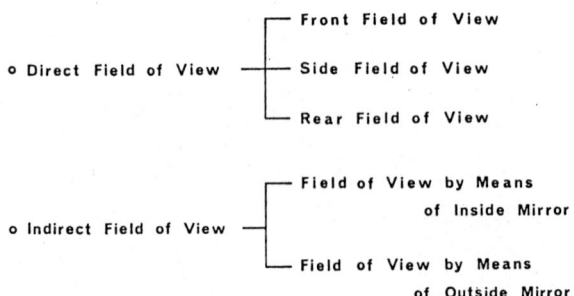

A wide-angle camera lens was set at the position of the driver's eyes to take photographs of the fields of view. In this way, all required field of view ranges could be evaluated. Figure 11 shows the front field of view at 180°. Figure 12 shows the field of view to the rear of the driver's position.

Fig. 11

Fig. 12

The projection chart of Fig. 13 was used to quantitatively represent the viewing positions, support frames, instrument panel, and ceiling of the vehicle. In this way, the front and rear fields of view could be studied by using the photographs taken.

To study the field of view, the position of the driver's eyes was used as the base. The ellipse shown in Fig. 14 indicates the position of the driver's eyes. This is referred to as the "eyellips". The "99 percentile" indicates that this applies to 99% of all people. Other important considerations include the minimum elevation to the ceiling (θ_1) in regard to the eyellips, and the minimum dip (θ_2) used to evaluate the front field of view.

As indicated previously, the indirect field of view is facilitated by the inside mirror and outer mirror. Note how the field of view differs when only the outer mirror is used.

Figure 15-a shows how distance is judged between vehicles, and how this judgement can be affected by changing the curvature of the outer mirror (by using concave mirrors).

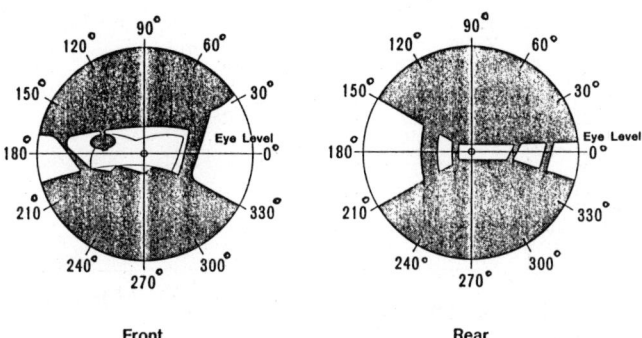

Fig. 13. Projection chart of front and rear field of view

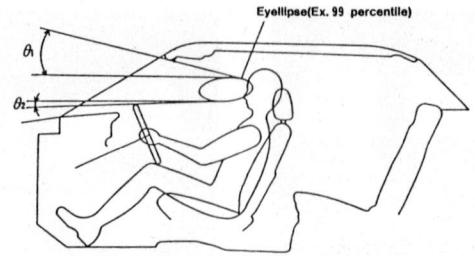

Fig. 14. Front field of view

Figure 15-b indicates how distance judgement can be impaired for given driving conditions using different mirrors. When the curvature of the mirror is more pronounced, or when the mirror is more rounded, a vehicle behind you appears to be farther away than it actually is. In other words, a more pronounced curvature reflects a wider range, but distorts images by making them appear smaller than they actually are. This significantly impairs distance judgement and can be very dangerous when driving.

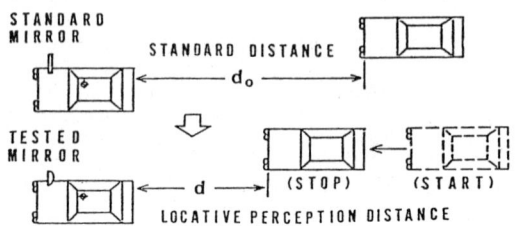

a

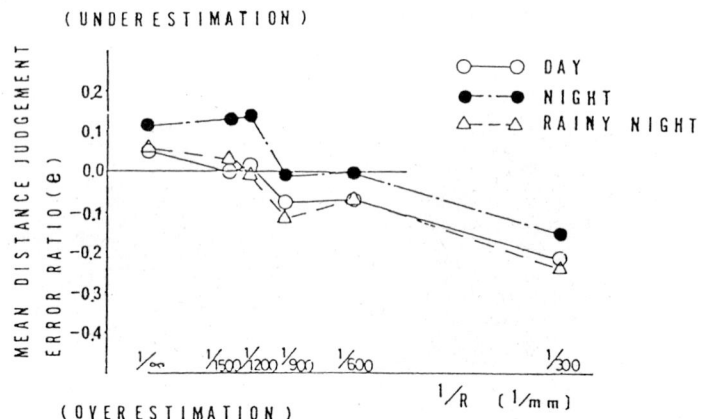

b

Fig. 15-a, b. Distance judgement

2.3 Visibility

The items to be studied when evaluating visibility are as follows:

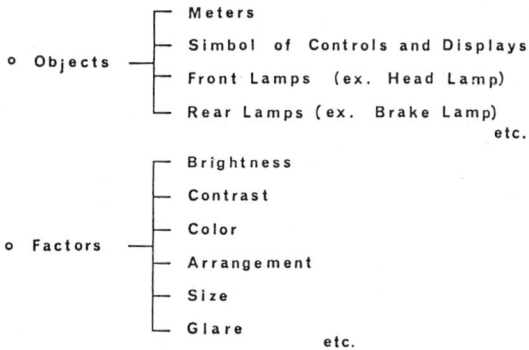

- Objects
  - Meters
  - Simbol of Controls and Displays
  - Front Lamps (ex. Head Lamp)
  - Rear Lamps (ex. Brake Lamp)
  - etc.

- Factors
  - Brightness
  - Contrast
  - Color
  - Arrangement
  - Size
  - Glare
  - etc.

Figure 16 shows an EOG (electrooculogram) measurement being conducted while the subject is observing a digital meter. The time required for visibility is obtained from the waveform being evaluated. In other words, such items as the size of digital characters being displayed, shape and location of the meter, time required to display characters, and the blinking of displayed characters affect this measurement.

Figure 17 shows measurement data on visual recognition times (provided by another company). This data is used for comparing the visibility of digital and analog meters. The axis of abscissas indicates the time required to correctly recognize speed, and the axis of ordinates indicates the frequency response in a percentage. In this example, the digital meter reflected a visual recognition time reduced by an average of 0.1 second.

Figure 18 shows the digital speedometer of a TOYOTA SOARER, which was designed based on the results of human engineering studies conducted using EOG and related measurements. Due to its high degree of visibility, this type of digital speedometer is becoming very popular in the automobile industry.

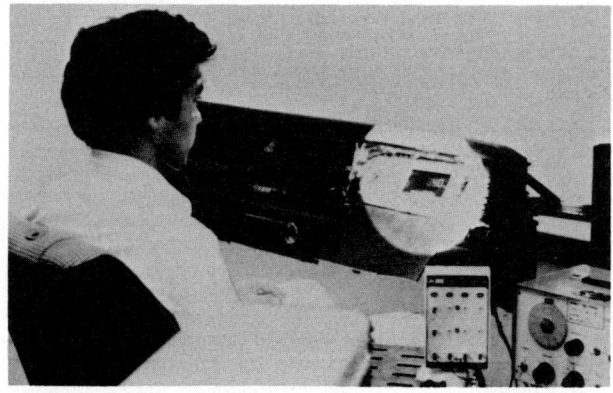

Fig. 16

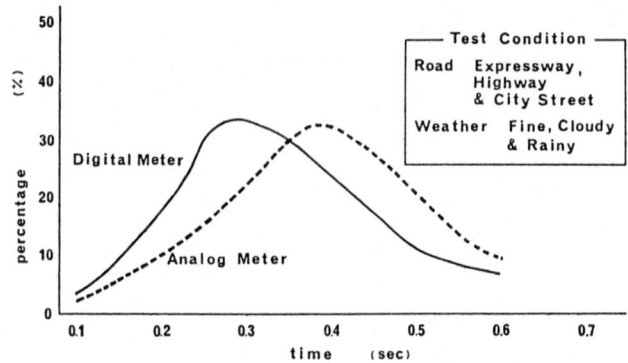

Fig. 17. Visual recognition time

Fig. 18

Another important factor that affects visibility is the lighting provided by headlamps. Figure 19 shows the visual area enabled by the passing headlamp beam for the vehicle on the right when the vehicle in the opposite lane is 100 m ahead. The curving line in the middle of the figure indicates the limit of the effective visual area.

The dots in Fig. 19 represent the following:
- visual area enabled by headlamps of subject vehicle
- area that cannot be seen
- visual area enabled by both headlamps of subject vehicle and vehicle in opposite lane

The desired light distribution can be determined by measuring the visual area through simulation, and by measuring the response of visual perception when the oncoming headlamps are initially seen.

Figure 20 shows an example of the required light distribution characteristics of four-beam headlamps. Note that these special headlamps are specifically designed to meet the demands of increased highway driving. These headlamps are similar to the three-beam headlamps discussed next, although four-beam headlamps have not been introduced commercially due to regulations governing headlamp specifications.

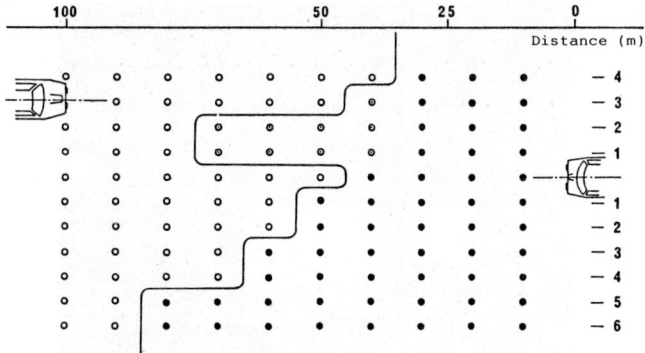

Fig. 19. Measurement of visual area by headlamps

Beam	Light Distribution
Highway High Beam (H·Hi)	
Highway Low Beam (H·Lo)	
Town High Beam (L·Hi)	
Town Low Beam (L·Lo)	

Fig. 20. 4-beam headlamps

Another reason that Toyota uses the three-beam headlamp system is that it is more practical than the four-beam headlamp system. The three-beam headlamp system is equipped on Toyota ESV-III models and was used to evaluate the required light distribution characteristics. Figure 21 shows the lighting conditions of the three-beam headlamp system.

In this headlamp system, the beams are automatically changed according to signals issued from the beam or illumination sensor to the CPU. The three-beam headlamp system represents a major step forward in the development of vehicle lighting systems of the future.

Taillights are another factor that significantly affect distance judgement between vehicles at night. Figure 22 shows the results of experiments conducted to determine the visibility at different distances. It shows the results of experiments conducted in a darkroom using a model lamp to test the effects of color, brightness, and size on the apparent distance of colored light.

A tendency was observed that when a greater or brighter light was used, the distance appeared to be closer than it actually was.

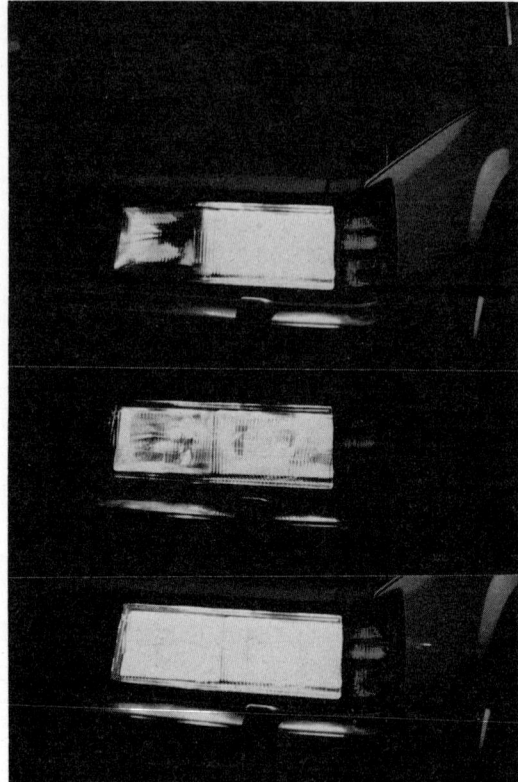

Fig. 21

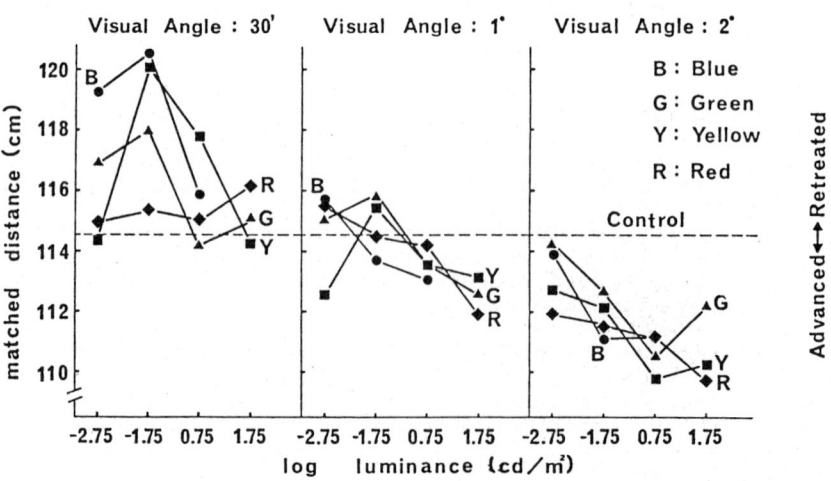

Fig. 22. Matched distance as function of log luminance

2.4 Operability

The components of operability and human engineering factors are listed as follows:

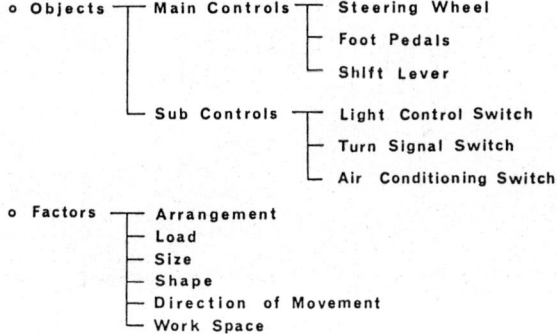

The most important consideration of operability is where the operating equipment (steering wheel, control knobs, gear shift, etc.) is located in relation to how far the driver must reach to gain access to this equipment. Figure 23 shows an experiment that measures the maximum reach by using a number of protruding bars. Of course, the maximum reach of individual drivers depends on such factors as physical height, sex, and arm length.

Operability is largely affected by the size, operational direction, and force required to operate the controls. Figure 24 shows a situation in which operability is evaluated for a relatively short woman in an actual vehicle.

The results of such studies have indicated that frequently used controls and controls required for safe operation (windshield wipers, etc.) should be located near the steering wheel to maximize operability. Studies are also being conducted to determine whether operability can be improved by mounting other switches on the steering wheel column.

Fig. 23

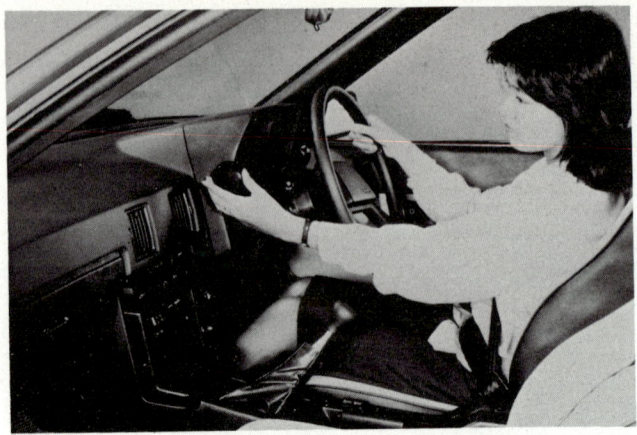

Fig. 24

2.5 Impaired Driver's Response

In previous discussions, studies were conducted based on the assumption that the driver was in a fully conscious state of mind. However, consideration must be given to the fact that driver response is seriously impaired by the following:
- Fatigue
- Drugs
- Lack of sleep
- Sickness
- Drinking

We are currently evaluating ways to detect such driver conditions in order to prevent tradgic accidents from occurring. A basic test to detect intoxicated drivers is explained as follows.

Figure 25 shows the relationship between the Blood-Alcohol Level (BAL) and corresponding values of the force required to push a floor-mounted pedal. Low scores indicate a normal response. Conversely, as scores increase, a significant variation is noted. This indicates an intoxicated condition. Note that some levels of intoxication indicate that this task adjustment is virtually impossible (scores of 1000 or more).

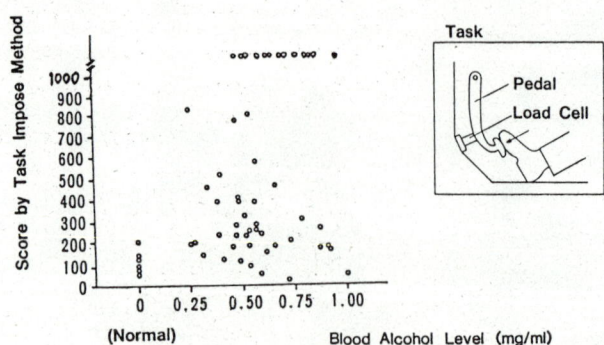

Fig. 25. Detection of intoxicated drivers

It should be noted that this method is not foolproof; some subjects being tested may be able to perform normally even if intoxicated. However, other factors such as the ability to distinguish and responsiveness are also being studied.

3. MANEUVERABILITY

Consideration must be given to motor vehicle dynamics by the human engineering approach to vehicle design. Two of the more important factors to be considered are stability and controllability. Figure 26 is a block diagram showing the various driving conditions and factors that must be considered when driving a motor vehicle.

o Controllability
>   Yaw Velocity Gain to Steering Wheel Torque
>   Yaw Velocity Response to Steering Wheel Angle

o Stability
>   US-OS Characteristics
>   Directional Stability

o Others
>   Steering Wheel Effort
>   Steering Wheel Returnability
>   Various Task Performances

The major items that must be evaluated regarding stability and controllability are listed above. In addition to the physically dynamic responsiveness and vehicle stability, a keen perception on the part of the driver is necessary for this evaluation. For purposes of brevity, only the US-OS characteristics will be briefly discussed because they are mentioned in subsequent discussion. The stability and control of a motor vehicle are largely dependent on the driver's skill and perception. Consequently, this evaluation is theoretically based not only on the physically dynamic stability of the vehicle, but also includes interaction with the driver under controlled conditions.

The US-OS characteristics are basic elements that indicate the stability and controllability of a motor vehicle. As shown in Fig. 27, the US (understeer) characteristics refer to when speed is gradually increased while turning at a constant steering angle. In this case,

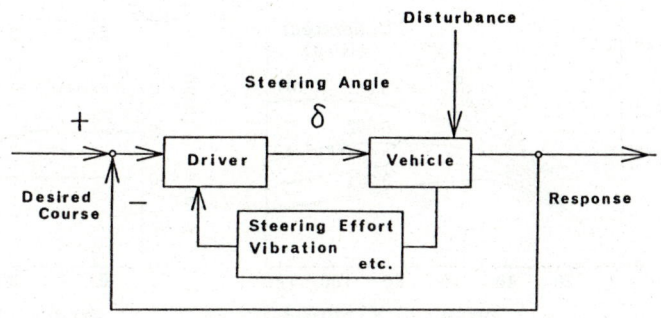

Fig. 26. Driver / vehicle model

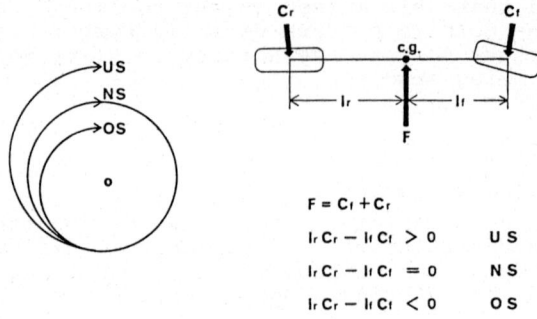

Fig. 27. US-OS characteristics

the turning radius will gradually increase. Consequently, the driver feels impelled to slightly increase the steering angle when rounding a curve. The OS (oversteer) characteristics are exactly the opposite. The driver feels compelled to decrease the steering angle.

These steering tendencies are indicated by the above simple expressions. In other words, for the US characteristics, the moment coefficient opposite to the turning direction remains constant at the moment of force. Conversely, for the OS characteristics, the moment coefficient in the turning direction remains constant. (Of course, the moment of force and moment coefficient are balanced by various dynamic force).

The degree of US-OS characteristics are indicated in units of K (stability factor). Figure 28 to the left shows a measurement example of US-OS characteristics. This measurement was conducted at slow driving speeds to enable lateral acceleration of up to 0.2 - 0.3 g. A motor vehicle is generally considered easy-to-steer when it has low US characteristics, or an approximate K value of 0.002.

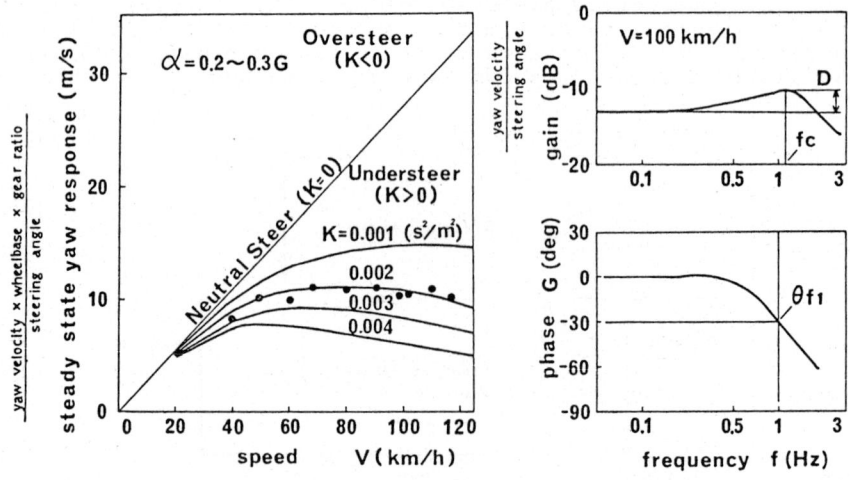

Fig. 28. Example of steering response

Figure 28 to the right shows the frequency characteristics of vehicle response to input signals. Values f_c, D, and 0_{f1} are closely related to precise response or convergence. Such physical quantities that correspond to steering tendencies are called the alternative characteristics.

Figure 29 shows the results of our studies. Note that the axis of ordinates represents the stability factor, and the axis of abscissas represents the yawing natural frequency. The plotted points shown in the diagram indicate the values of different vehicles.

Although there is no single desired value, an f_c value that is too low is generally considered unsatisfactory, while a K value of approximately 0.002 is considered satisfactory. It should be noted that most skilled drivers prefer a low K value.

Figure 30 shows another evaluation example of transient vehicle response characteristics to steering-step input. This figure also

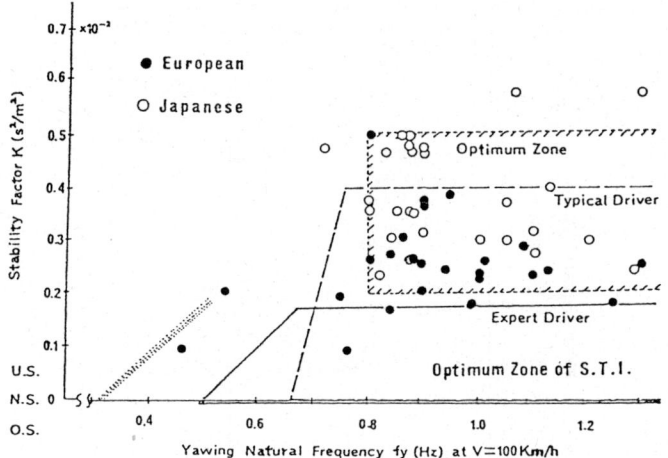

Fig. 29. Optimum zone of directional control characteristics

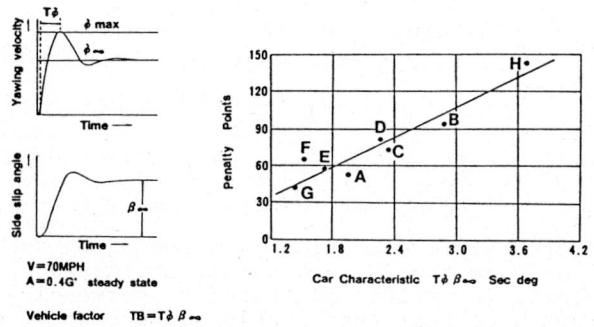

Fig. 30. Evaluation b task performance

shows the relationship between TB (product of time interval T∅ when the maximum yaw velocity and convergence value of side slip angle B are reached) and the driving performance. Note that a vehicle having a lower TB value has fewer penalty points, which indicates better driving performance.

Vehicle response was previously discussed where constant input was provided by the driver without any corrective steering action. Figure 31 shows the results of a study in which a vehicle was driven onto a circular course from a straight one. The purpose of this study was to indicate the relationship between the above mentioned US-OS characteristics and a normalized mean square value of course deviation. As noted in Figure 31, a lower stability factor is desired at lower speeds. Conversely, note that the vehicle cannot be driven at high speeds unless a specific high K value is obtained.

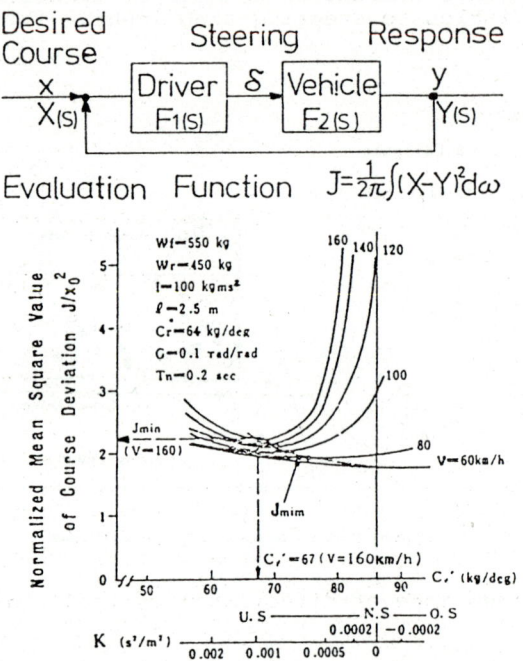

Fig. 31. Calculation of optimum control by a simple vehicle/driver model

4. VIBRATION

The degree of motor vehicle vibration is affected by such factors as road conditions, driving speeds, and number of passengers. Ideal vibration characteristics differ according to personal preference. Consequently, shock absorbers that enable adjustment of damping characteristics and suspension springs that enable adjustment of spring coefficient should be used so that the desired suspension can be set.

Figure 32 shows the variable adjustable shock absorber used in the Toyota Soarer. In addition to the switch that enables mode selection (Normal, Sport, Auto) according to personal preference and road conditions, the damping force is automatically increased to minimize vertical vehicle movement when cornering, braking, or accelerating. The damping force is automatically decreased to compensate for other conditions to ensure a smooth ride.

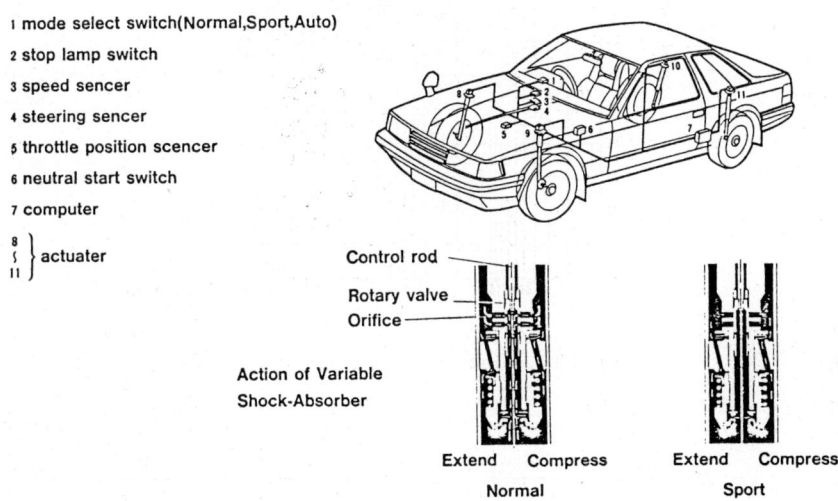

Fig. 32. TOYOTA electronic modulated suspension

5. NOISE

A variety of noise levels encountered in everyday life are listed in Fig. 33. The curves in the figure indicate weighing curves. In other words, the same noise level is heard at a sound pressure of 1,000 Hz/60 dB as 50 hz/90 dB.

The maximum noise level for a motor vehicle is during full acceleration at 7.5 m. Under normal driving conditions, the noise level is significantly lower. The noise levels of different motor vehicles running at full acceleration are summarized in Fig. 34.

Passenger cars usually fall within the lower range, while heavy-duty trucks fall within the upper range. Careful consideration must be given to internal vehicle noise. This is because even if the internal noise is minimized, it still bothers the driver.

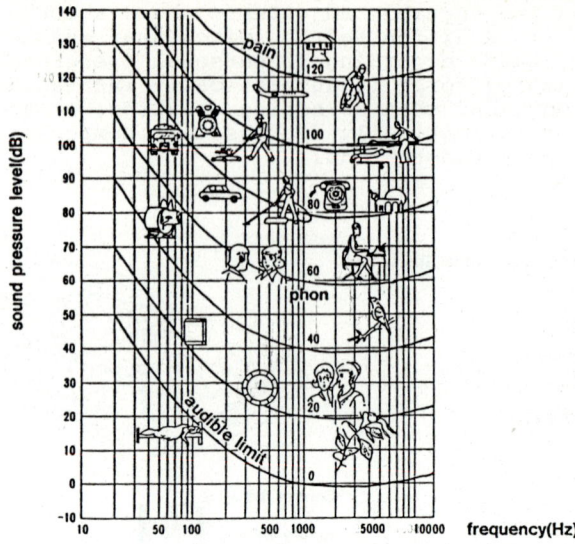

Fig. 33. Noises in everyday life

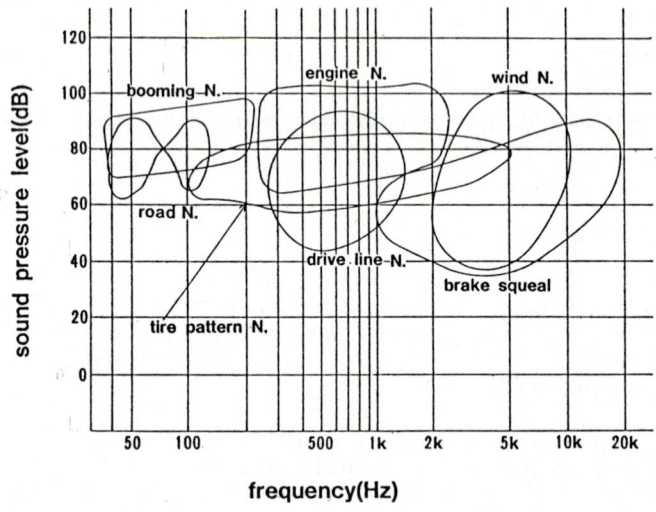

Fig. 34. Vehicle noises

6. MISCELLANEOUS ITEMS

In addition to those topics that have been previously discussed, there are other items in motor vehicle design that merit discussion. Among the more important are:

○ Acceleration	Response, Stretch
○ Drivability	Stumble, Surge
	Acceleration & Deceleration Shock
○ Brake	Pedal Effort, Pedal Stroke
	Effectiveness
○ Speed Change	Shift Shock (Automatic)
	Effort, Stroke (Manual)

To achieve high-quality vehicle design and to satisfy driver requirements, these items all need to be further improved. In pursuit of these objectives, we will continue our research efforts in basic human engineering, as well as conduct psychological surveys to determine such factors as personal preferences in motor vehicle design.

A special technique is employed that includes coordinating equations for such unrelated items and conditions. Our motor vehicle design process also relies heavily on computer-aided design (CAD) to provide the best possible mechanisms to satisfy the most-demanding road and temperature conditions.

7. APPLICATIONS

This report has outlined the incorporation of human engineering in motor vehicle design. Subsequent studies will be undertaken to further improve the quality of human life through this design process.

Figure 35-a shows the Toyota ESV-I model that was developed based on a variety of studies conducted to primarily improve safety. This safety improvement effort was undertaken to meet the standards set forth by the ESV (Experimental Safety Vehicle) Project, which was proposed by the U.S. Department of Transportation in 1973. The Toyota ESV-I embodies the results of lengthy research and development efforts represented by the components shown in Fig. 35-b.

For purposes of brevity, these components will not be individually discussed. Suffice it to say that they each resulted from extensive research and development activities. The Radar Sensor System, equipped on the Toyota ESV-I, is one of the many components resulting from our many development projects (Fig. 35-c). This system enables the detection of forward objects (other vehicles, etc.).

The Radar Sensor design incorporates a receiving antenna that receives reflected 10 GHz radar beams. These beams are analyzed by the computer to determine the distance to the vehicle ahead and relative speed from the Doppler shift. Warning alarms are then generated to facilitate automatic operation while maintaining a fixed distance from the vehicle ahead by controlling the accelerator and brakes via instructions issued by the computer. This detection capability is further enhanced by using the intensity of the reflected beams.

Other technologically advanced features that were introduced with the Toyota ESV-III model in November, 1982 are shown in Fig. 36. The Toyota ESV-III was primarily designed to reduce driver operational responsibilities, and to improve safety. This model includes such

a

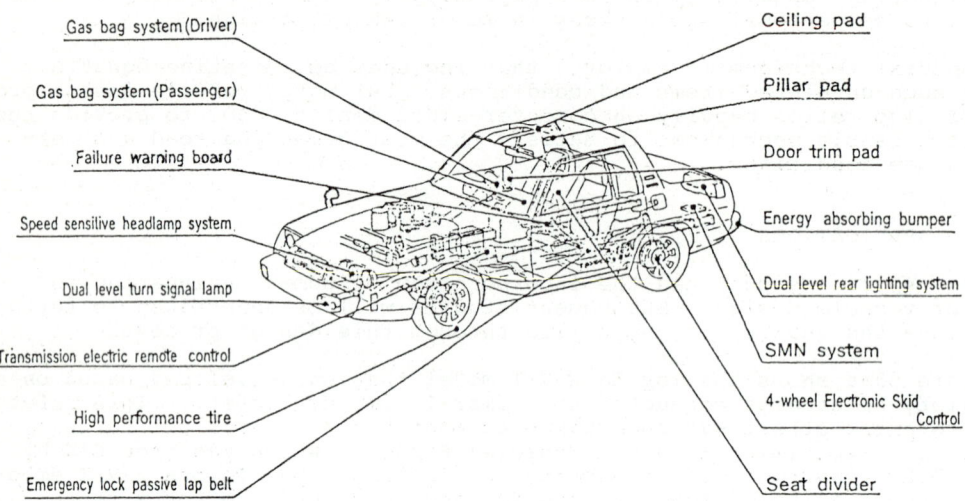

b

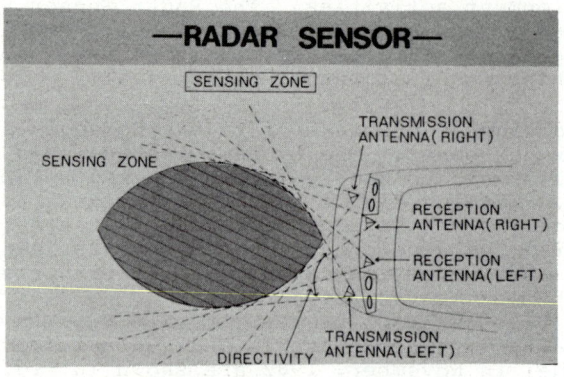

c

Fig. 35-a, b, c. TOYOTA ESV-I

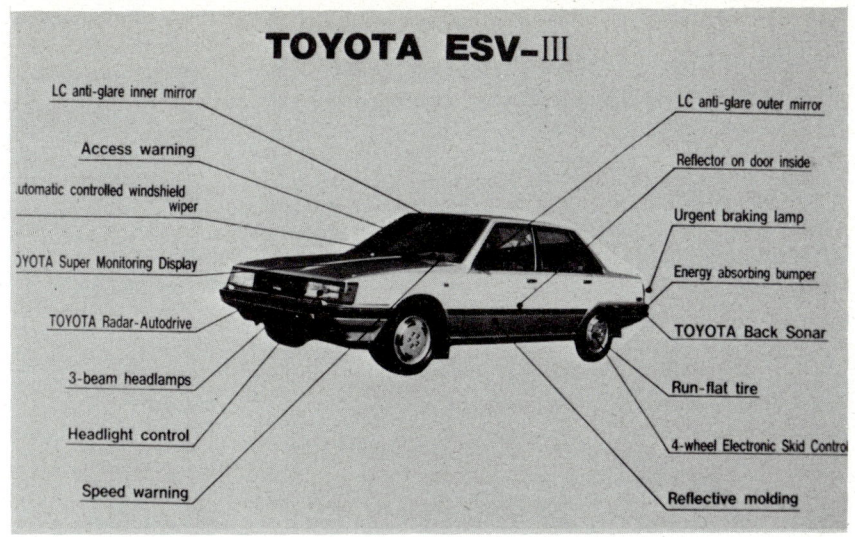

Fig. 36

advanced features as the previously described three-beam headlamp system, which is currently prohibited for production models by existing regulations governing headlamps.

Highly advanced design technologies continue to be realized through the use of computers, new materials, and the development of new sensors and actuators.

8. CONCLUSION

This report has outlined Toyota Motor Corporation's efforts to incorporate human engineering in motor vehicle design. Although many automobile-related technologies are not directly related to human engineering, such as steel plating, welding, and painting, it is hoped that this report served to clarify the importance of technology in improving the quality of human life.

We are deeply committed to the concept of designing motor vehicles for people; for those who drive and ride in our vehicles, and to meet the most-demanding environmental conditions.

The focus of our commitment now turns toward electronics. We believe that the development of electronics will lead the way into the future. In line with this belief, our efforts will be directed toward the development of new system configurations, control methods, and improved motor vehicle reliability.

CAD/CAM/CAE
Development Systems

From Solid Modeling to Finite Element Analysis

David A. Field
General Motors Research Laboratories

Abstract

Solid modeling, finite element mesh generation and analysis of finite element solutions are obviously tightly connected in the design redesign cycle. However, practical realizations of each of these aspects of mathematical analysis of solid objects requires a significant amount of internal independence. Each aspect has to be justified on its own merits and their actual development reflects this independence. Together they form a powerful system and separately they perform useful functions which can interface with other computer codes and systems.

1. INTRODUCTION

Given the enormity of the subject of computer aided design I have chosen to address only three aspects. Although each aspect was developed independently of one another, I will stress their integration into computer aided design and mathemetical analyses of solid objects. I will begin by briefly describing early and recent solid modeling efforts at General Motors Research Laboratories. After focusing on solid modeling and its implementation in the second section of this paper, a detailed examination of applying solid modeling to finite element mesh generation will be given in the third section. The next and last section will discuss SPIFFE [11], a code which plots finite element solutions and provides assistance in deciding upon acceptance or rejection of the solid models.

2. SOLID MODELING

Early computer aided design efforts at General Motors resulted in CADANCE a computer graphics system which modeled scuptured surfaces. With the help of William Gordon a member of the Mathematics Department in General Motors Research Laboratories, the CADANCE system was built up from transfinite interpolation functions called Gordon surfaces, [19]. Given a network of space curves $X(s,t_i)=(x(s,t_i),y(s,t_i),z(s,t_i))$, $i=1,2,\ldots,N$, $0 \le s \le 1$, and $X(s_j,t)=(x(s_j,t), y(s_j,t), z(s_j,t))$, $j=1,2,\ldots,M$, $0 \le t \le 1$, the formula

$$F(s,t) = \sum_{i=1}^{N} X(s,t_i)\varphi_i(t) + \sum_{j=1}^{M} X(s_j,t)\psi_j(s) - \sum_{i=1}^{N}\sum_{j=1}^{M} X(s_j,t_i)\varphi_i(s)\psi_j(t).$$

where

$$\varphi_i(t_k) = \begin{cases} 0, & i \ne k \\ 1, & i = k \end{cases} \quad \text{and} \quad \psi_j(s_k) = \begin{cases} 0, & j \ne k \\ 1, & j = k. \end{cases}$$

defines a surface over the parameter domain $0 \le s \le 1$, $0 \le t \le 1$ which interpolates the network of space curves. That is, $F(s,t_i)=X(s,t_i)$ and $F(s_j,t)=X(s_j,t)$ for $i=1,\ldots,N$ and $j=1,\ldots,M$.

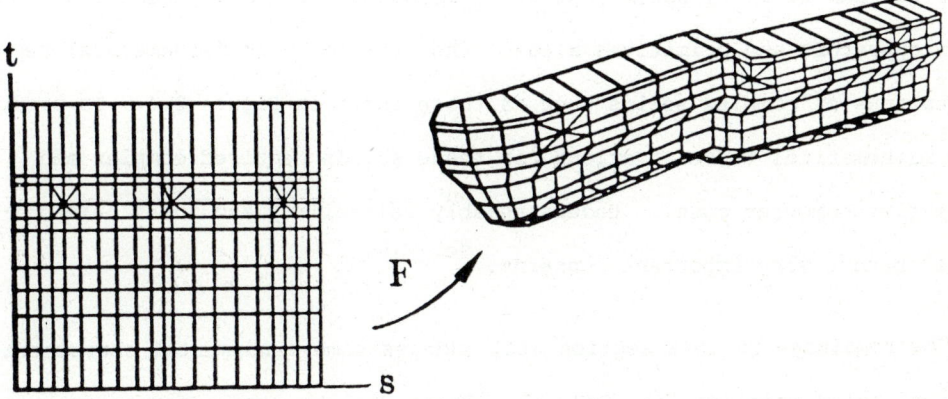

Figure 2.1 Gordon surface and interpolation network.

CADANCE has evolved to include special purpose forms of transfinite interpolation for three dimensional finite element mesh generation much like the relatively recent PATRAN-G commercial code [30]. However, from the inception of computer aided design efforts, problems like interference checking of automotive components were especially difficult to solve. For example, even though a complete geometric description of a new trunk was already available in CADANCE, it was much easier to build a wooden trunk and manually pack into the trunk as much luggage as possible. An easier method for creating many and more complicated solid objects was required.

A different approach to computer aided design started when the Computer Science Department at General Motors Research Laboratories collaborated with the Production Automation Project at the University of Rochester where the Part and Assembly Description Language (PADL) was developed [37]. In order to interface with the existing CADANCE graphics system and because a more general modeling capability was required, the solid modeler GMSOLID was created by extending ideas from the original PADL system [3,4].

Both PADL and GMSOLID were initially based upon union, intersection and difference Boolean operators with the fixed set of primitives shown in Figures 2.2 and 2.3. By specifying size, location and other parameters, Boolean operators and primitives allowed the user to build mathematical representations of complex solids such as those shown in Figure 2.4. To produce the mathematical representations for these solids required complex and extensive computer codes. Understandably reliability and "user friendliness" became very important concerns.

The remainder of this section will address some fundamental characteristics of solid modelers like GMSOLID. There will be some comments on human factors critical to the acceptance of solid modelers by the engineering

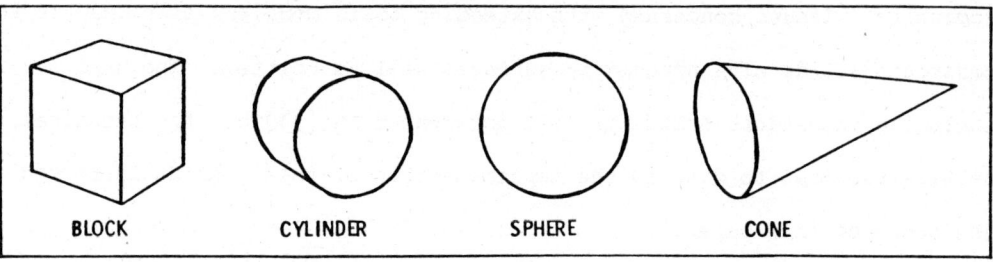

Figure 2.2 Primitive solids.

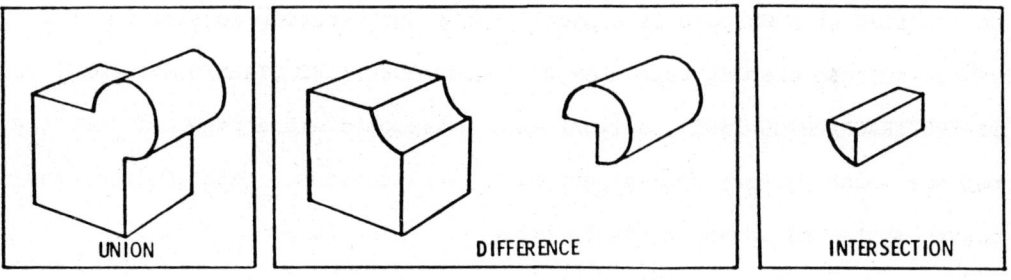

Figure 2.3 Boolean operations with block and cylinder.

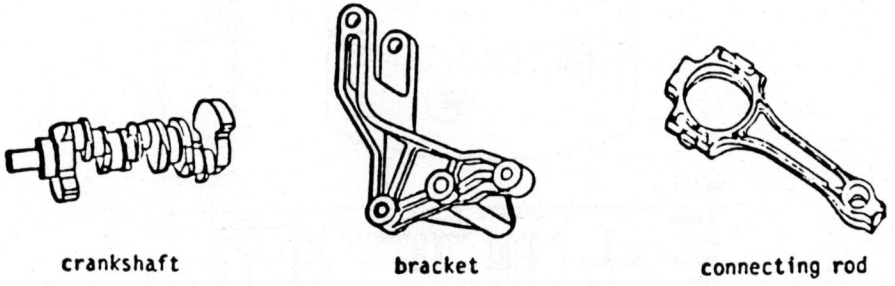

Figure 2.4 Solids designed using GMSOLID.

community. Issues concerned with extending solid modelers like GMSOLID to construct solids with nonquadric surfaces will be omitted. The need for including nonquadric solids is well documented in [33] but the technical details required to discuss the implimentation of these solids is beyond the scope of this paper.

Human factors are critical to an engineer's acceptance of a solid modeler. For most users of interactive graphics, graphic visualizations, the response time and an easy interaction with software are key issues. GMSOLID is currently implemented on an IBM 4084 and the user's physical interaction with GMSOLID is typically at a work station consisting of a vector refresh display tube (IBM 2250/3250/5080, DEC GT481/GTG2, ADAGE 4250 or SPECTRAGRAPHICS 1500), a light pen, a keyboard and a panel of function buttons which perform special graphical applications. GMSOLID is a menu driven system as shown in Figure 2.5.

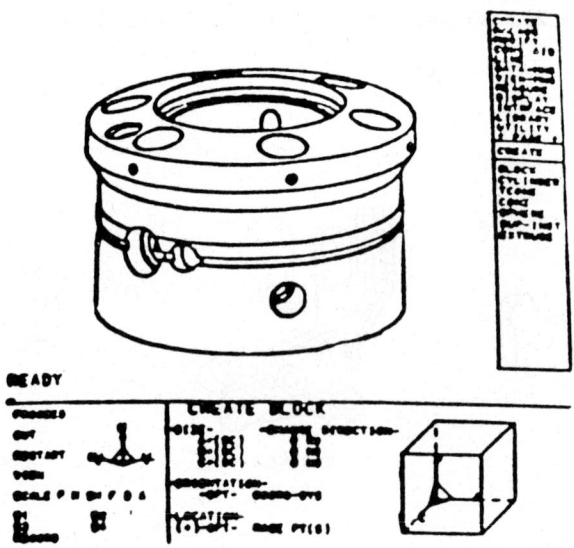

Figure 2.5 A GMSOLID menu.

The availability and response time of menu items is significantly impacted by the type of representation and processing of geometric data. Since GMSOLID manipulates true solids rather than points and curves defining solids as in surface and wire frame modelers, during the creation of a solid, GMSOLID encodes the solid's definition into a constructive solid geometry (CSG) tree as shown in Figure 2.6. Each node of the tree represents a solid with a corresponding Boolean operation. Associated with each node is also a matrix representing the transformation of a generic mathematical definition of a primitive solid to a desired scale and position in space. With such a binary tree structure complex solids have a well defined mathematical structure so that realistic graphical displays can be constructed along with a myriad of tools to manipulate these displays.

In addition to a constructive representation embodied in a CSG tree, a solid can also be represented through its outer boundary, namely its faces, edges and vertices. Whereas its CSG tree is a very compact representation and can be generated quickly, its boundary representation has significant storage and computational requirements. For example, a Boolean operation with a primitive solid and the current solid model creates a new solid model whose entire boundary has to be reclassified into various types of vertices, edges and faces. The union of two blocks shown in Figure 2.7 exhibits the classification of points, edges and faces. The classification process become more complex when nonplanar curves of intersection are created. Boundary representations require an extensive and very flexible data base [2].

Numerical stability of computational algorithms is a major issue with boundary representations. Furthermore, numerical proceedures for calculating intersection curves etc. can become quite cumbersome. Many special

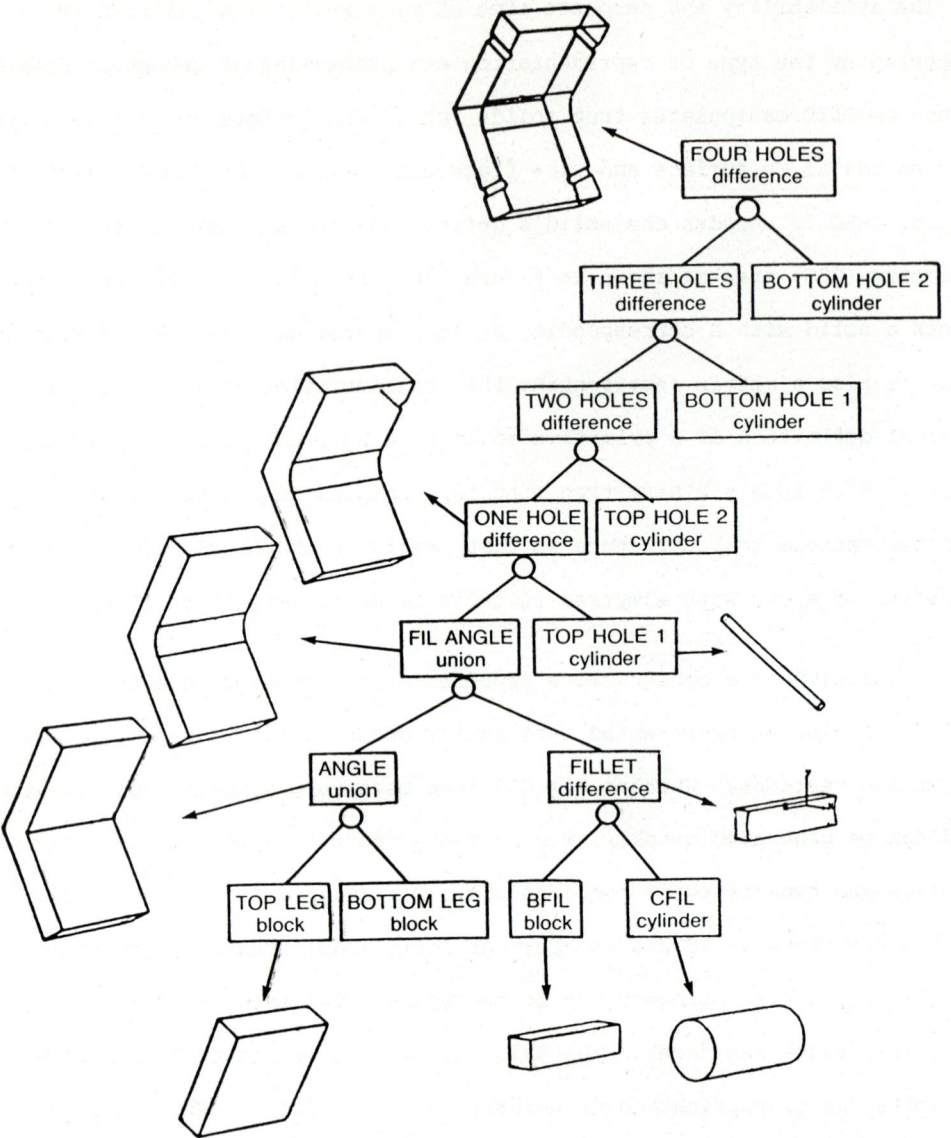

Figure 2.6 A constructive solid geometry tree.

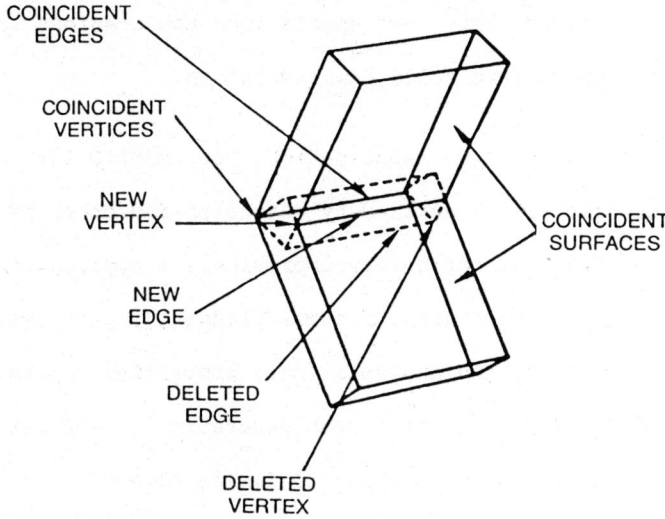

Figure 2.7 Some boundary representation reclassifications resulting from the union of two blocks.

cases, e.g. the intersection of a plane and a quadric surface, are developed to accelerate and control the reliability of numerical computations. Faster and more accurate algorithms are always prized, especially quick test algorithms which can avoid unnecessary calculations, see [12]. Sophisticated algorithms such as singular value decomposition [14,32] and Zangwill and Garcia's algorithm for solving systems of polynomial equations [17,26] are often called from scientific software libraries [8,22] or written as special purpose codes.

Whether CSG trees or boundary representations are prefered can best be answered in the context of specific applications and current state of the art algorithms. GMSOLID maintains both representations [2]. With both representations available, graphical tools for "building" complex solids becomes more flexible, communication with codes outside GMSOLID becomes less restrictive, and higher level and specialized applications can be based upon the most appropriate solid representation.

New directions and higher level applications for GMSOLID include solid modeling with nonquadric solids defined by extrusion and other parametric representations. GMSOLID is continually improving its applications in manufacturing (die design, NC machining, process planning), part design (initial design, part revision), and analysis (mass properties, packaging and clearances, kinematics, finite element mesh generation). The next section will focus on the applying solid modeling to finite element mesh generation.

3. AUTOMATIC MESH GENERATION

Because of the enormous size and complexity of solid modeling systems such as GMSOLID, there were advantages for developing a mesh generation code to function independently from GMSOLID. The major advantage was that code development could proceed independently of the solid modeler as long as the interface with GMSOLID was kept very simple. Thus an interface between GMSOLID and a mesh generation code were clearly specified at the outset. That this interface remained simple is one of the reasons that the mesh generation algorithm proposed here was sucessful in bridging the gap between GMSOLID and finite element analysis. However, the major success of the algorithm is that any solid geometry created by any solid modeler aan be decomposed into a finite element mesh. This accomplishment is remarkable because mathematical representations of solids for finite element analysis is extremely different from representations used by solid modelers.

The finite element method is an algorithm for numerically solving partial differential equations such as those arising in structural analysis of solid automotive components. A finite element solution is given by a linear combination of piecewise polynomials where each polynomial is defined on its own distinct subregion of the solid's geometry. A particular difficulty with the finite element method is that the solid must be decomposed into a union (mesh) of relatively small subregions called finite elements. Three dimensional geometries are usually decomposed into three element types, tetrahedra, pentahedra and hexahedra, see Figure 3.1. These elements are defined by assigning three dimensional coordinates to the nodes shown in Figure 3.1. Simpler versions of these elements having no midside nodes and straight edges are often used.

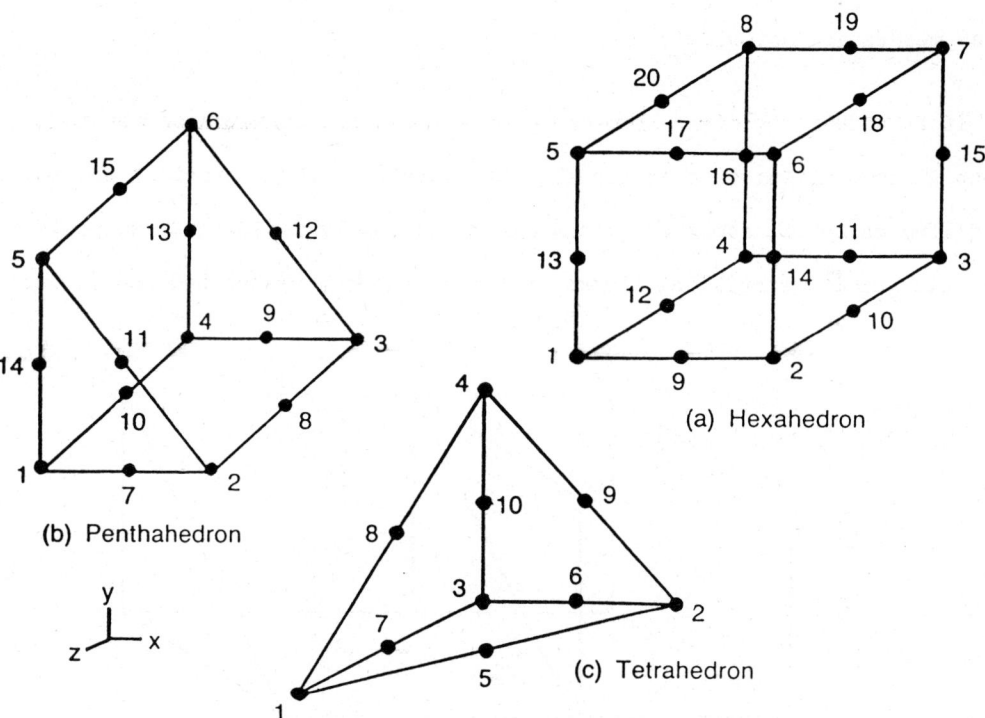

Figure 3.1 The quadratic isoparametric elements.

Decompositions of three dimensional solids into finite elements are generally constrained so that the edge of an element cannot lie on the interior of a face of another element, see Figure 3.2. Hexahedra are often used for their mathematical properties and for their relative ease in creating geometries. But even by allowing curvature of hexahedral edges, some geometries cannot be adequately decomposed with hexahedra. Pentahedra are then used to cope with difficult geometries but additional problems arise because the triangular faces of pentahedra are not compatible with the rectangular faces of hexahedra. Graphic displays to guide the construction of meshes are often unintelligable because of the concentration of element edges. This is especially true of tetrahedral meshes. This difficuly in visualizing the mesh during its construction is a major reason why tetrahedral meshes are not widely used.

It is abundantly clear from the preceeding brief comments on finite element decompositions that finite element geometric representations and solid modeling geometric representations can be extremely different. Not only are the primitive building blocks of finite element meshes and GMSOLID dif-

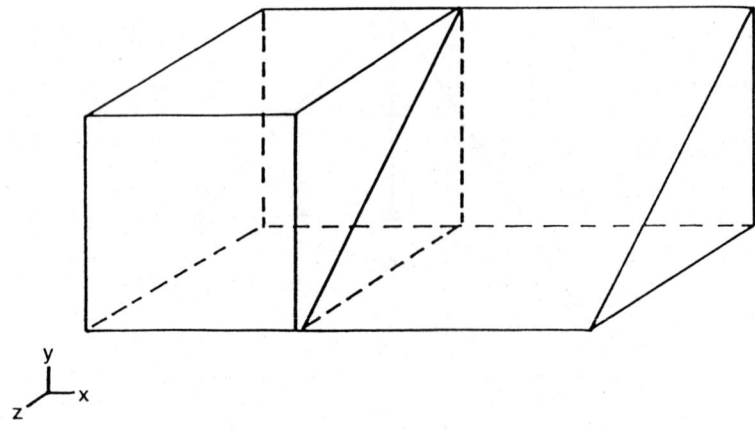

Figure 3.2 This union of a hexahedron and a pentahedron is not allowed.

ferent, but the geometric and analytic constraints of the finite element method make the transformation from solid modeling geometric contructions to finite element meshes a substantial obstacle. At this point it is worth commenting on mesh generators that have appeared in research journals or are commercially available. Typical mesh generators are based on piecewise polynomial interpolants much like the transfinite interpolation functions used in CADANCE, [20,30,36,39]. These generators require solids to be partitioned in simpler subblocks whose boundaries can be described with a network of curves. Through the transfinite interpolation functions a natural coordinatization compatible with finite element decompositions can be induced. When the geomerty of a solid is given by a solid modeler like GMSOLID, the solid must be completly rebuilt by the mesh generator. However, many solids are difficult and practically impossible to decompose into convenient subblocks. Indeed, solid modeling systems such as GMSOLID were developed because of this inadequacy.

Another approach to mesh generation is to extend two dimensional triangulation algorithms to construct three dimensional tetrahedra. There are many two dimensional triangulation algorithms, see [6,15,24,35] and the references contained therein. However, very little on three dimensional triangulations has been reported in the literature at this time except by Watson [38] and its dual version in [1,5,13,31]. None of these references specifically address the possible application to finite element mesh generation. Recently Nguyen-Van-Phai [29] reported a tetrahedral mesh generation strategy. During the mesh generation process extensive procedures are followed to ensure that new tetrahedra are compatible with those already created. The process appears to demand extensive user interaction and appears to be computationally expensive to triangulate a large number of nodes.

The mesh generation algorithm [7] developed in the Mathematics Department at General Motors Research Laboratories is essentially an application of Watson's algorithm [38]. Watson's algorithm is based upon the following theoretical characteristics of Delaunay triangulations [31].

> Given N random points in three dimensions there exists a unique decomposition of their convex hull into tetrahedra with the property that the 4 points defining the corners of a tetrahedron also define a circumsphere whose surface and interior contain none of the remaining N-4 points.

The unique decomposition described above is called a Delaunay triangulaton and the tetrahedra of the decomposition are called Delaunay tetrahedra. Because the construction of finite element meshes is the goal, extensions to Watson's algorithm are necessary. The finite element nodes used as vertices of Delaunay tetrahedra come from well defined and hardly random solids created by solid modelers. These solids dictate allowable point locations producing a nonrandomness of points which influences the uniqueness of the decomposition and can cause algorithmic difficulties. Watson's algorithm also creates a convex hull of tetrahedra so that some tetrahedra have to be removed.

An important characteristic of the mesh generation algorithm proposed here is the separation of two closely connected aspects of mesh generation, node definition and three dimensional triangulation.

> 1. <u>Node Definition</u>: Within and on the surface of solid coordinates of the finite element nodes are defined by the solid modeler. By controlling the distribution of nodes, important fea-

tures of solids can be emphasized and nodal density of finite elements can be predetermined before mesh generation begins.

2. **Three Dimensional Triangulation**: Nodes are automatically connected to form a mesh of well proportioned tetrahedral elements.

Although solid modelers can easily define nodes on and inside solids, how to define the "best" location of these nodes is a complex and important unresolved area of active research. On the other hand it can be shown that regular tetrahedra cannot triangulate three dimensions as equilateral triangles triangulate two dimensions. An approach to node definition can be found in [7].

Illustrations will be very helpful in understanding the geometric constructions of the three dimensional mesh generation algorithm. To clarify the illustrations and improve understanding, the algorithm and the illustrations will be given in two dimensional form while the three dimensional version, requiring only a few key word changes, will be given in parenthetical word substitutions. The algorithm's statement uses only one uncommon mathematical term. The <u>circumdisk (circumball)</u> of a triangle (tetrahedron) is the interior and boundary of the unique circle (sphere) defined by the three (four) vertices of the triangle (tetrahedron).

MESH GENERATION ALGORITHM

STEP1: Create a triangle (tetrahedron) which encloses all the given nodes, see Figure 3.3.

STEP2: Construct the Delaunay triangulation of the nodes of the enclosing triangle (tetrahedron) and one of the given nodes by connecting the node to each vertex of the enclosing triangle (tetrahedron), see Figure 3.4.

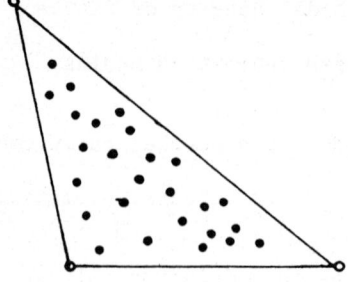
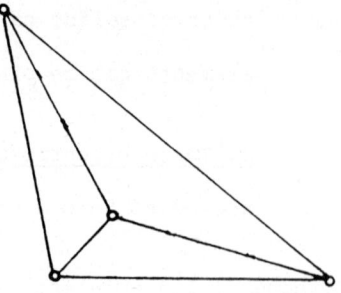

Figure 3.3 Nodes and enclosing triangle.

Figure 3.4 Delaunay triangulation with the first node.

Maintain a master list of the triangles (tetrahedra) and their associated circumdisks (circumballs).

STEP3: Insert the other given nodes into the triangulation one at a time as follows.

a) Determine which existing circumdisks (circumballs) contain the given node, see Figure 3.5.

b) Create a list of the triangles (tetrahedra) associated with the circumdisks (circumballs) determined in STEP3a. The union of these triangles (tetrahedra) is called an insertion polygon (polyhedron), see Figure 3.6.

c) Create a list of boundary edges (triangles) of the insertion polygon (polyhedron).

d) Create new triangles (tetrahedra) filling the insertion polygon (polyhedron) by connectimg the new node to the edges (triangles) of the insertion polygon's (polyhedron's) boundary, see Figure 3.7.

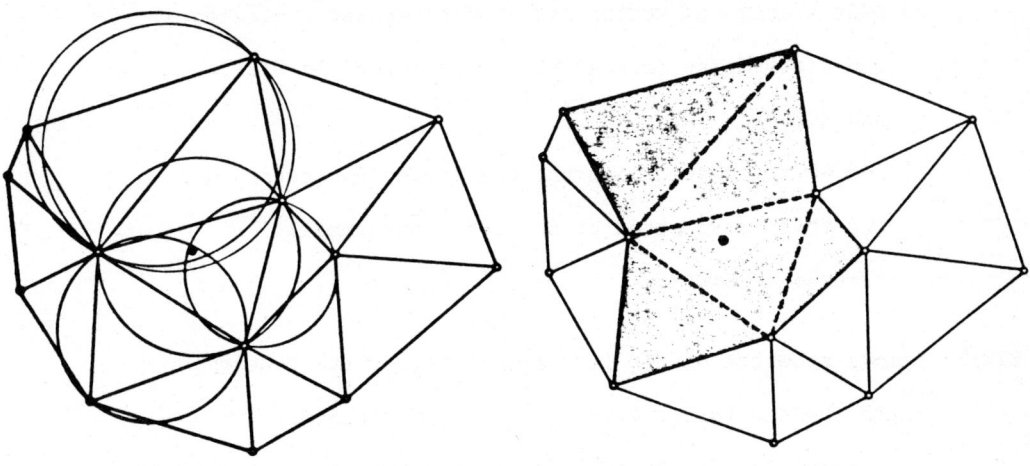

Figure 3.5 A new node and nonempty circumdisks.

Figure 3.6 The insertion polygon.

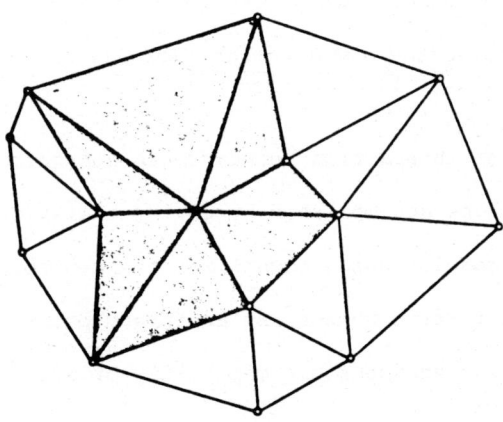

Figure 3.7 Triangulation of the insertion polygon.

e) Delete from the master list the triangles (tetrahedra) and circumdisks (circumballs) determined in STEP3a and STEP3b.

f) Add to the master list the new triangles (tetrahedra) and their associated circumdisks (circumballs) created in STEP3d.

STEP5: Remove from the master list all triangles (tetrahedra) which share a vertex with the enclosing triangle (tetrahedron). The union of remaining triangles (tetrahedra) in the master list is a Delaunay triangulation of the given nodes.

STEP6: Since the Delaunay triangulation in STEP5 is a convex hull, remove from the Delaunay triangulation those triangles (tetrahedra) whose centroid is not contained inside the given lamina (solid).

STEP6 indicates an interaction between mesh generator and solid modeler. This interaction can be accomplished by giving a list of centroid coordinates to the solid modeler which identifies the centroids lying in or on the solid. Quickly testing whether an arbitrary point lies inside, on or outside a solid is an indispensible capability of all solid modelers.

STEP3a can be troublesome when a node is computed (possibly incorrectly due to roundoff error) to lie on an existing circumball. Called an insertion degeneracy, this situation implies that Delaunay tetrahedra need not be produced during STEP3. Experience has shown that assigning the nodes causing an insertion degeneracy to be consistently inside or consistently outside the circumball will not resolve the difficulty. Perturbing the troublesome node off the circumsphere can lead to tetrahedra called "slivers", [7]. Slivers occur when four vertices of a tetrahedra are nearly coplanar and identified as tetrahedra having an extremely low ratio of radii of inscribed to circumscribed spheres. Almost all slivers can be avoided by judiciously defining node locations and the order of node insertion. Slivers can always be eliminated at the expense of not creating Delaunay tetrahedra. As described in [7] a limited amount of postprocessing can preserve Delaunay triangulations.

The one at a time insertion strategy of the mesh generation algorithm offers a unique opportunity for mesh refinement. Based upon a finite element solution, locations of new nodes can be determined. By saving the Delaunay triangulation before STEP5 and STEP6 are executed, the new nodes can be inserted as if they were part of the original mesh. In contrast to some mesh refinement techniques, see [18,23], special consideration for resolving nonconforming elements are not necessary. Unlike mesh generations based upon transfinite interpolation functions, the refinement is strictly local.

The tetrahedra created by the mesh generation algorithm tend to be well proportioned. This statement follows from the proven fact that for given node distributions in two dimensions Delaunay triangulations will produce as nearly equilateral triangles as possible [34]. Most of the details of the two dimensional proof carries over into three dimensions.

4. PROCESSING FINITE ELEMENT ANALYSES

Three dimensional finite element decompositions with thousands of elements and nodes are no longer unusual, see Figure 4.1. Whether finite ele-

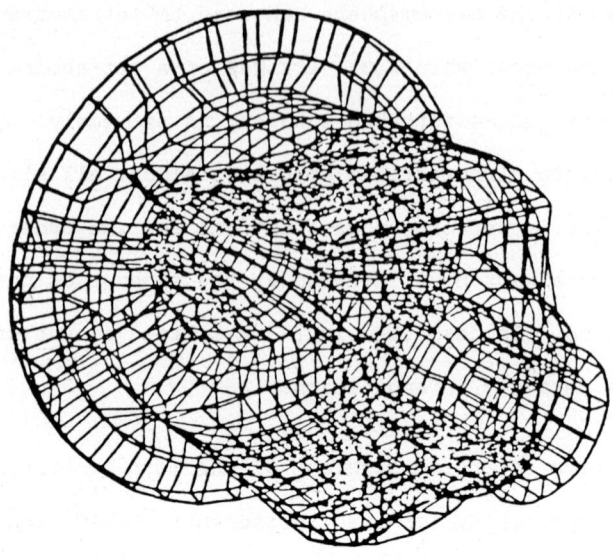

Figure 4.1 A finite element mesh with over 1000 elements.

ment decompositions are created automatically by a mesh generator or are laboriously "built" with finite elements, finite element solutions produce enormous amounts of data. Solutions are reported as numerical data associated with each node and each finite element so that increasing the number of finite elements to obtain accurate results increases the effort to distill a sensible interpretation of the numerical data. This section will be devoted to describing SPIFFE [11], an interactive computer graphics code which enables the user to display contours of actual finite element solutions on arbitrary cross sections of three dimensional finite element decompositions composed of hexahedra, pentahedra and tetrahedra. Thus SPIFFE contributes to the solid modeling design cycle by providing quick and accurate displays of mathematical analyses which can determine whether solid models are accepted, reanalized or redesigned.

The foundations of computing mathematically defined cross sections were published in [16]. The computer code reported in [16] was extended in [21] to plot stress contours on cross sections of hexahedral finite element meshes. Plotting the contours required the user to manually create plotting grids on cross sections and required solving systems of polynomial equations to compute stresses at the lattice points of the plotting grids.

SPIFFE retains the ability to mathematically define planar cross sections and adds the option of graphically choosing cross sections. This option is nearly indispensible for a user since poking at finite element displays via cross-hairs or some other positioning device is a great time saver and is often the only practical way to generate cross sections. Other significant improvements in SPIFFE over the codes reported in [16,21] are the inclusion of pentahedra and tetrahedra finite elements and that plotting grids on cross sections are automatically provided. This last improvement eliminates the tedious manual definiton of a plotting grid as well as the need to solve systems of polynomial equations. Although SPIFFE runs independently of commercial finite element software, SPIFFE's data structure is compatible with MSC/NASTRAN [28], a widely used and commercially available finite element code.

The remainder of this section will primarily focus on two topics. The more extensively discussed topic will be the plotting of contours with emphasis on minimal user effort and quick response time. The second topic, a natural development from the first, concerns geometric and analytic integrity of finite element decompositions.

The first step in plotting finite element solutions is to choose a surface upon which to display the solutions. To plot on exterior surfaces of finite element meshes, general codes such as MOVIE-BYU [27] are slow but more than adequate. However, plotting finite element solutions on planar cross sections must account for the mathematical formulation of the finite element method.

By emphasizing response time and user effort, SPIFFE has incorporated many tools for quickly choosing cross sections. For precise locations of cross sections the user is allowed to define planes mathematically by supplying a) three points on the intersecting plane, b) a point (or an node identification number) and a normal to the plane, or 3) the equation of the plane. Many graphical aids such as subbsetting finite element decompositions, specialized labeling routines, rotations, perspective and parrallel projections help the user to quickly choose a cross section. The algorithms associated with these features are similar in flavor to those solid modelers use to graphically display and manupilate solid models. Figures 4.2-4.4 illustrate typical displays produced by SPIFFE.

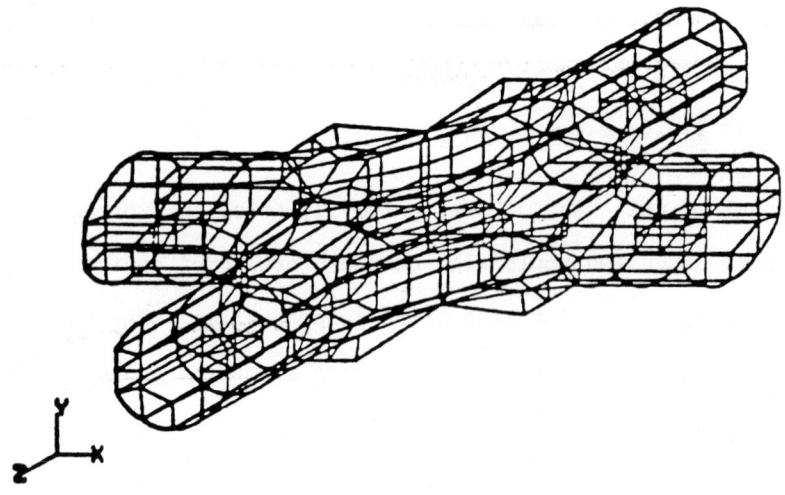

Figure 4.2 Oblique projection (cabinet).

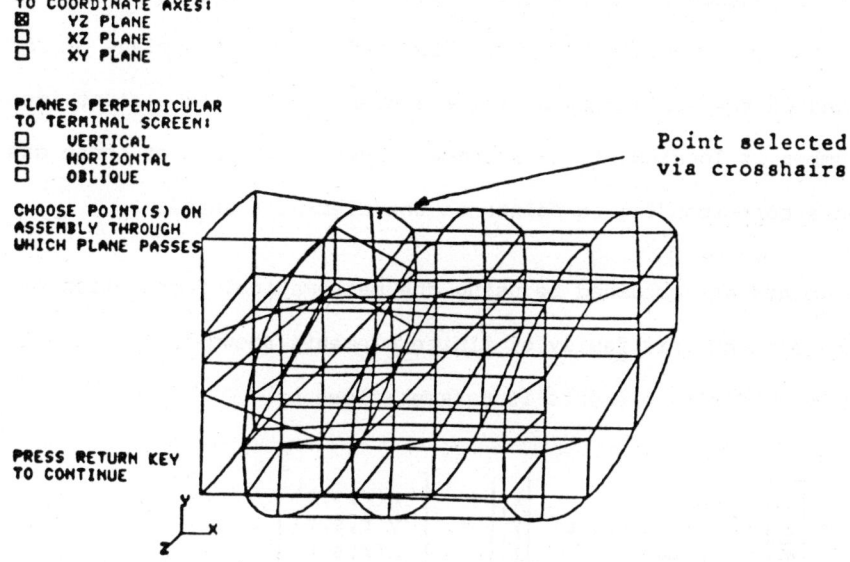

Figure 4.3 Selecting a cross section parallel to yz-coordinate plane.

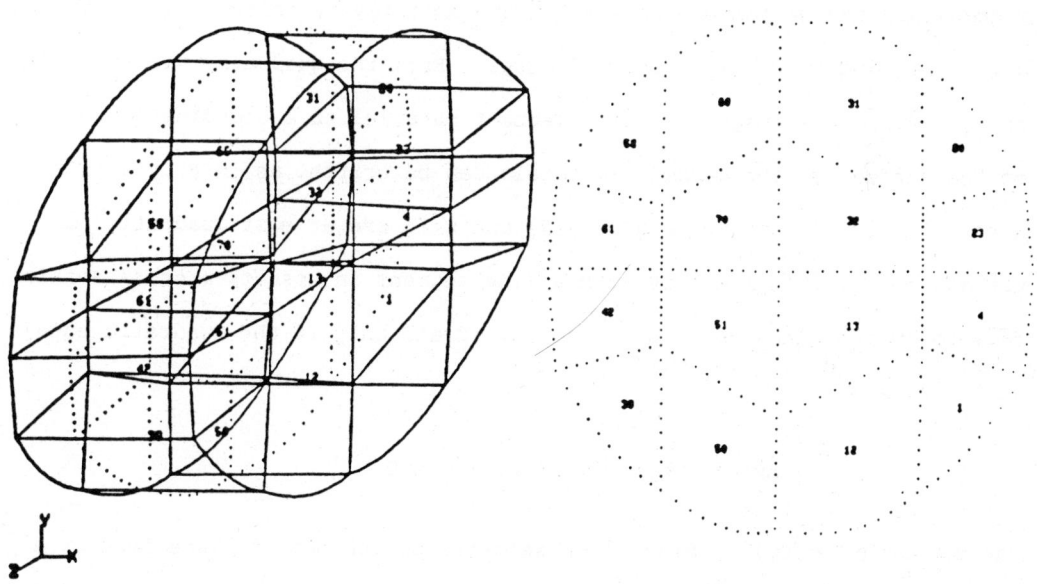

(a) Mesh with intersection.   (b) Normal view of intersection.

Figure 4.4 Two views of a cross section.

The cross section in Figure 4.4 is distinguished by dots and element numbers. Each dot represents a point located on both the plane of intersection and on the surface of a finite element. A finite element identification number is located at the average screen coordinate of those displayed dots corresponding to points on the element's surface.

These points are obtained be using the mathematical formulation of the quadratic isoparametric family of finite elements shown in Figure 3.1. This family of finite elements is parameterized by

$$T \begin{bmatrix} r \\ s \\ t \end{bmatrix} = \sum_{i=1}^{N} \phi_i(r,s,t) \begin{bmatrix} x_i \\ y_i \\ z_i \end{bmatrix} = \begin{bmatrix} x(r,s,t) \\ y(r,s,t) \\ z(r,s,t) \end{bmatrix}. \qquad (4.1)$$

where N is the number of nodes of the finite element, ϕ_i are the well known quadratic Lagrange shape functions, (x_i, y_i, z_i) are the coordinates of the N nodes and the parameters r, s and t are suitably restricted within the parametric domain $0 \leq r \leq 1$, $0 \leq s \leq 1$ and $0 \leq t \leq 1$. From the quadratic variation of ϕ_i in Eq. (4.1), the image of a line segment parallel to a coordinate axis and on the surface of the parametric domain can be written as $(x(\theta), y(\theta), z(\theta))$ where $\theta \in [0,1]$ and where $x(\theta), y(\theta)$ and $z(\theta)$ are at most quadratic polynomials in . The image of such a line segment intersects a cross section defined by the plane, $ax + by + cz = d$, if and only if the quadratic equation

$$ax(\theta) + by(\theta) + cz(\theta) + d = 0$$

has a root $\theta^* \in [0,1]$. Since line segments on the out of plane face of a tetrahedron (face 1,5,2,9,4,8 in Figure 3.1(c)) cannot be parallel to any of the coordinate axes, this out of plane face is mapped to one of the other tetrahedron faces and the composite transformation is used to determine $(x(\theta^*), y(\theta^*), z(\theta^*))$ on the out-of-plane face. A composite transformation is also used for the out of plane face (face 2,8,3,12,6,14,5,11) of the pentahedron in Figure 3.1(b).

To plot contours of finite element solutions on planar cross sections, a plotting grid of rectangles and triangles is constructed on the cross section, see Figure 4.5. The finite element solution must then be evaluated at the vertices of the rectangles and triangles. In contrast to reference [21], SPIFFE eliminates the tedium of selecting and connecting coordinates for the plotting grid and eliminates the need for solving a system of polynomial equations to evaluate the finite element solution at the vertices of the plotting grid.

Connecting the dots on the cross section in Figure 4.4 will create unsuitable triangles and rectangles for plotting contours. However, SPIFFE

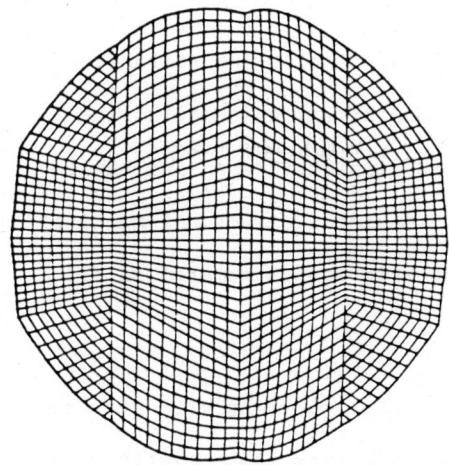

Figure 4.5 A cross section's plotting grid.

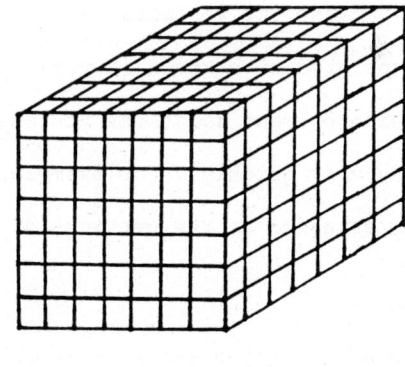

Figure 4.6 Decomposition of the parametric domain of a hexahedron into smaller blocks.

capitalizes on the fact that the (x,y,z) coordinates of these dots are already known and more importantly that their corresponding parametric (r,s,t) coordinates are also known. By extending Eq. (4.1) to include segments parallel to the coordinate axes and inside the parametric domain of the finite element, many (x,y,z) coordinates can be produced along with their corresponding (r,s,t) parametric coordinates. Noting that the additional parallel line segments decompose the parametric domain into small blocks, see figure 4.6, the (r,s,t) coordinates in each small block are appropriately connected in a connectivity table. The (r,s,t) coordinates are then mapped into the cross section by the finite element's isoparametric transformation and the connectivity of each small block is used to form a polygon on the cross section. The union of these polygons formed the plotting grid displayed in Figure 4.5. Experience has shown that the image of these connections have always been convex polygons and almost exclusively been triangles and rectangles. SPIFFE automatically triangulates the higher order convex polygons.

Because the vertices of the plotting grid's triangles and rectangles are associated with known (x,y,z) and (r,s,t) coordinates, calculating the finite element solutions at these vertices is straightforward and can be done very quickly. For example, in three dimensional structural mechanics the stresses at the vertices of the plotting grid are calculated as follows.

Let an (x^*,y^*,z^*) coordinate of a vertex of the plotting grid and its corresponding (r^*,s^*,t^*) coordinate belong to a finite element with displacements (u, v, w), $1 \leq i \leq n$. SPIFFE calculates the partial derivatives $\frac{\partial u}{\partial x}, \frac{\partial u}{\partial y}, \frac{\partial u}{\partial z}, \frac{\partial v}{\partial x}, \frac{\partial v}{\partial y}, \frac{\partial v}{\partial z}, \frac{\partial w}{\partial x}, \frac{\partial w}{\partial y}$ and $\frac{\partial w}{\partial z}$ and uses the chain rule to evaluate expressions such as.

$$\frac{\partial u}{\partial x} = \frac{\partial u}{\partial r}\frac{\partial r}{\partial x} + \frac{\partial u}{\partial s}\frac{\partial s}{\partial x} + \frac{\partial u}{\partial t}\frac{\partial t}{\partial x}$$

All the terms in the right hand side are computable from Eq. (4.1) and.

$$T \begin{bmatrix} r^* \\ s^* \\ t^* \end{bmatrix} = \sum_{i=1}^{N} \phi_i(r^*,s^*,t^*) \begin{bmatrix} u_i \\ v_i \\ w_i \end{bmatrix} = \begin{bmatrix} u(r^*,s^*,t^*) \\ v(r^*,s^*,t^*) \\ w(r^*,s^*,t^*) \end{bmatrix}.$$

Since stress is not continuous across finite element boundaries, smooth contours can be plotted in SPIFFE whenever the user supplies consistent stresses at the nodes of the finite element decomposition. An approach similar to Meek and Beer's method [25] was used to assign consistent stress values across element boundaries in Figure 4.7(b).

Displays of cross sections as shown in Figure 4.4(b) present an opportunity to interrogate a finite element decomposition for its geometric consistency. Nonconforming boundaries of adjacent elements causing overlaps

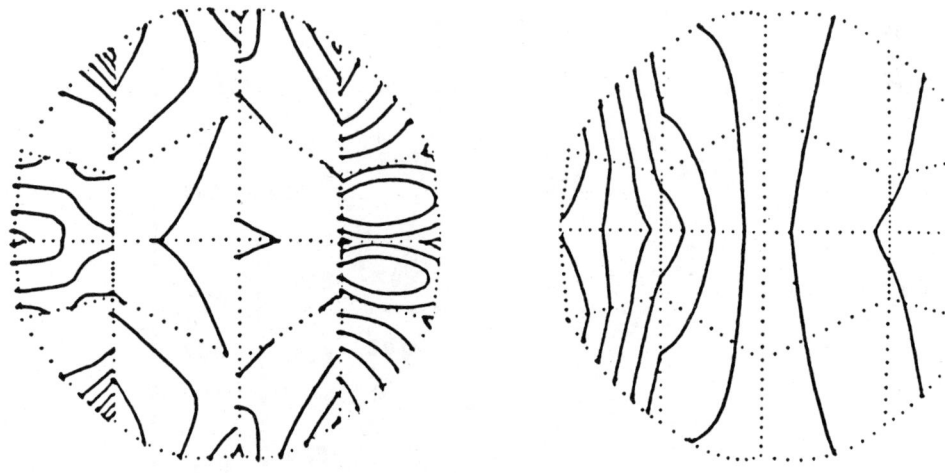

(a) actual solution   (b) smoothed solution

Figure 4.7 Stress contours.

or gaps between elements, missing elements or duplicate elements can be discovered by observing whether empty areas among the dots outlining cross sections of elemental surfaces are labeled. Figure 4.8 illustrates typical inconsistencies which may occur.

Experience has shown that a misplaced node can cause highly distorted surfaces of elements and can cause unlabeled regions of its own cross section. This latter kind of geometric inconsistency is directly related to the analytic formulation of the finite element. A basic assumption of the finite element method is that the isoparametric transformations defining finite elements must be invertible. Determining the invertibility of these transformations is a difficult and important problem. In recent articles by the author [9,10] theorems and algorithms are given for determining the invertibility of many two and three dimensional quadratic isoparametric finite element transformations.

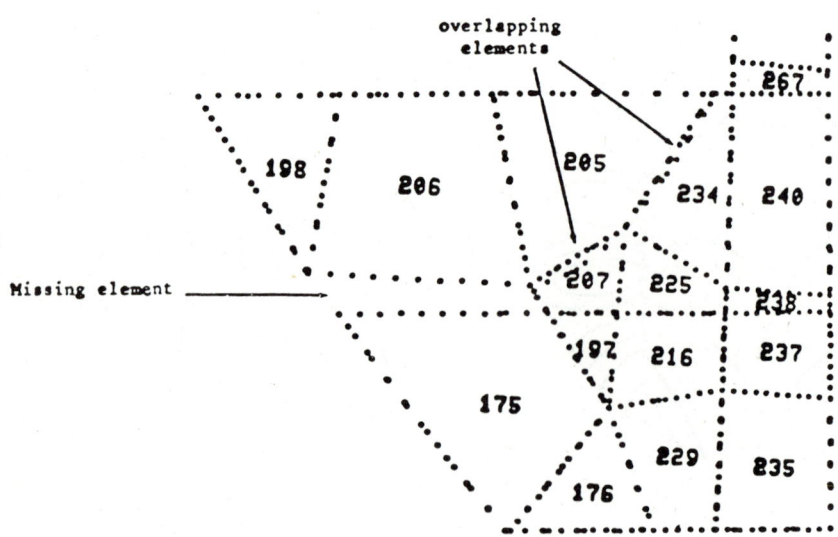

Figure 4.8 Typical inconsistencies.

REFERENCES

1. Bowyer, A., "Computing Dirichlet Tesselations", The Computer Journal, Vol.24, 162-166 (1981).

2. Boyse, J.W., "Data structure for a solid modeller", General Motors Research Publication, GMR-2768 (1979).

3. Boyse, J.W. and Gilchrist, J.E.,"GMSOLID-Interactive modeling for design and analysis of solids", IEEE Computer Graphics and Applications, Vol.2, pp.27-40 (1982).

4. Boyse, J.W. and Rosen, J.M.,"GMSOLID-A System for interactive design and analysis of solids", Society of Automotive Engineers Technical Paper Series, No.810196 (1981).

5. Brostow W. and Dussault J.P, "Construction of Voronoi polyhedra", Journal of Computational Physics, Vol.29, pp. 81-92 (1978).

6. Cavendish, J.C., "Automatic triangulation of arbitrary planar domains for the finite element method", International Journal for Numerical Methods in Engineering, Vol.8, pp.679-696 (1974).

7. Cavendish, J.C., Field, D.A. and Frey, W.H., "An approach to automatic three-dimensional finite element mesh generation", General Motors Research Publication, GMR-4533 (1983).

8. Dongarra, J.J., Moler, C.B., Bunch, J.R. and Stewart, G.W., LINPACK Users' Guide, Society for Industrial and Applied Mathematics, Philadelphia, PA (1979).

9. Field, D.A., "An algorithm for determining invertible quadratic isoparametric finite element transformations", Mathematics of Computation, Vol.37, pp.347-360 (1981).

10. Field, D.A., "Algorithms for determining invertible two and three dimensional quadratic isoparametric finite element transformations", International Journal for Numerical Methods in Engineering, Vol.19, pp.789-802 (1983).

11. Field, D.A. and Hatton, M.B.,"Plotting three dimensional finite element solutions on planar cross sections", General Motors Research Publication, GMR-4608 (1984).

12. Field, D.A. and Morgan, A.P.,"A quick method for determining whether a second degree polynomial has a solution in a given box", IEEE Computer Graphics and Applications, Vol.2, pp.65-68 (1982).

13. Finney, J.L., "A procedure for the construction of Voronoi polyhedra", Journal of Computational Physics, Vol.32, pp. 137-143 (1979).

14. Forsythe, G. and Moler, C.B., Computer Solution of Linear Algebraic Systems, Prentice-Hall (1967).

15. Frederick, C.O., Wong, Y.C. and Edge, F.W., "Two-dimensional automatic mesh generation for structural analysis", International Journal for Numerical Methods in Engineering, Vol.2, pp.133-144 (1970).

16. Frey, A.E., Hall, C.A. and Porsching, T.A., "An application of computer graphics to three dimensional finite element analyses", Computers and Structures, Vol.10, pp.149-154 (1979).

17. Garcia, C.B. and Zangwill, W.I., "Finding all solutions to polynomial systems and other systems of equations", Mathematical Programming, Vol.16, pp.159-176 (1979).

18. Gago, J.P. de S.R., Kelly, D.W., Zienkiewicz, O.C., and Babuska, I., "A posteriori error analysis and adaptive processes in the finite element method: part I - adaptive mesh refinement", International Journal for Numerical Methods in Engineering, Vol. 19, pp.1621 1656 (1983).

19. Gordon, W.J., "Spline-blended surface interpolation through curve networks", J. Math. Mech., Vol.10, pp.931-952 (1968).

20. Gordon, W.J. and Hall C.A., "Construction of curvilinear coordinate systems and applications to mesh generation", International Journal for Numerical Methods in Engineering, Vol.7, pp.461-477 (1973).

21. Hall, C.A., Porsching T.A. and Sledge, F., "STRSIT: contour plotting of stresses on planes of intersection", Computers and Structures, Vol. 12, pp.221-224 (1980).

22. IMSL Library Reference Manual, 7500 Bellaire Boulevard, Houston, Texas.

23. Kelly, D.W., Gago, J.P. de S.R., Zienkiewicz, O.C., and Babuska, I., "A posteriori error analysis and adaptive processes in the finite element method: part II - error analysis", International Journal for Numerical Methods in Engineering, Vol.19,pp.1593-1619 (1983).

24. Lawson, C.L., "Software for C surface interpolation" Software III, Academic Press, New York, pp.159-192 (1977).

25. Meek, J.L. and Beer, G., "Contour plotting of data using isoparametric element representation", International Journal for Numerical Methods in Engineering, Vol.10, pp.954-957 (1976).

26. Morgan, A.P., "A method for computing all solutions to systems of polynomial equations", ACM Transctions on Mathematical Software, Vol.9, pp.1-17 (1983).

27. MOVIE.BYU, contact Hank Christiansen, Civil Engineering, 368 CB BYU, Provo, Utah 85602.

28. MSC/NASTRAN, The MacNeal-Schwindler Corporation, 815 Colorado Boulevard, Los Angeles, California 90041.

29. Nguyen-Van-Phai, "Automatic mesh generation with tetrahedron elements", International Journal for Numerical Methods in Engineering, Vol.18, pp.273-289 (1982).

30. PDA: PATRAN User's Guide, PDA Engeneering, 1560 Brookhollow Drive, Santa Anna, California, 92705.

31. Rogers, C.A., Packing and Covering, Cambridge Mathematical Tracts, No.54, Cambridge university Press.

32. Sarraga, R.F., "Algebraic methods for intersections of quadric surfaces in GMSOLID", Computer Vision, Graphics, and Image Processing, Vol.22, pp.222-238 (1983).

33. Sarraga, R.F.and Waters, W.C., "Free-form surfaces in GMSOLID: goals and issues", General Motors Research Publication, GMR-4481 (1983).

34. Sibson, R., "Locally equitriangular triangulations", The Computer Journal, Vol.21, pp.243-245 (1978)

35. Suhara, J. and Fukuda J., "Automatic mesh generation for finite element analysis", in Advances in Computational Methods in Structural Mechanics and Design, University of Alabama Press, pp.607-624 (1974).

36. SUPERTAB interactive finite element model generation system, Structural Dynamics Research Corporation, Cincinnati, Ohio.

37. Voelker, H.B. and Requicha, A.A.G., "Geometric Modelling of Mechanical Parts and processes", Computer, Vol.10, pp.48-57 (1977).

38. Watson, D. F., "Computing the n-dimensional Delaunay tesselation with applications to Voronoi polytopes", The Computer Journal, Vol. 24, No. 2, pp.167-172 (1981).

39. Zienkiewicz O.C. and Phillips D.V., "An Automatic mesh generation scheme for plane and curved surfaces by isoparametric co-ordinates", International Journal for Numerical Methods in Engineering, Vol.3, pp.519-528 (1971).

PC Based Customized VLSI

Terumoto Nonaka
Nippon Gakki Co., Ltd.

§1 Introduction

During the 1970's, the microprocessor became an important device for the system designers which provided a software solution for certain design problems. More recently, in order to optimize hardware for the end application, electronics manufacturers are also moving to integrate their systems or subsystems onto their own customized chips. This custom VLSIs have proven themselves the optimum answer for circuit implementation in a wide range of applications. As a sequence, increasing use of customized VLSI is rapidly changing the electronics market. System designers require more specialized chips as application requirements grow in complexity. Market projections estimate that by the late 1980s, more than half of the industry's semiconductor sales will come from custom chips designed specifically for a single user, or from semi-custom chips that can be tailored to meet unique requirements.(1) More significantly, it predicts that nearly 40% of these will be designed by their end users.(2)

Custom circuits make special user-designed hardware possible and provide several well-known advantages over equivalent designs implemented with standard circuits : lower power dissipation and minimal thermal radiation, improved reliability, lower product assembly and test costs, improved performance and additional functions, space savings, light-weight and design security.

Compared with gate array LSIs, the design of full custom or standard cell LSIs can be incredibly detailed, achieving reduced chip size, the realization of highly complex functions, and higher operating speed. Additionally, there is a wide acceptable range in the electrical characteristics and there is no limit on the number of gates used.

On the other hand, the shortage of skilled VLSI designers has been critical in the semiconductor industry. To meet a requirement of a rapid increase in the number of new customized VLSI designs, performance and productivity of chip designs must be increased.

§2 PC based design system

With today's increasing complexity of VLSI circuitry, an environment must be provided to the designers. This requirement can be met by offering computer aided engineering (CAE) system that matches the existing design tools. CAE is also becoming the solution to overcome the bottleneck of LSI design capability by making the system designer to design his own customized chips. The rapid growth of the semicustom business means its success in developing method which assists the system designers in designing their own customized VLSIs.

CAE workstations provide us a highly productive path. Table 1 shows the change of EWS (Engineering Workstation) in LSI designs. The advancing technology in micro electronics is continually making more computing power available at lower costs and the recent personal computer (PC) technology has created inexpensive CAE tools. As shown in Table 1, the recent development has focused on the personal workstation which operates as the local and low cost schematic capture.

Table 1 EWS in LSI designs

~1981	• Handcrafted drawing with templates and rulers
1981~1983	• Schematic capture
1983~present	• Simulator • Special purpose hardware • Low cost schematic capture

We have developed the schematic capture system called "YIS-LOGiC" employing the personal computer. The basic functions of YIS LOGiC are :

1) logic capture and retrieval

   The user can use the tablet and the pen interactively to design the logic circuit diagram on the color display.

2) netlist extraction

   The circuit coding data (netlist) is automatically compiled from the circuit diagram created.

3) error check capability

   This incorporates the automatic error check functions, minimizing human errors encountered when designing the circuit. These are open of input terminal, output short circuit, incorrect wired logic and excess of fan out.

4) plotter and printer interface

The corresponding system configuration is shown in Fig. 1 and Table 2.

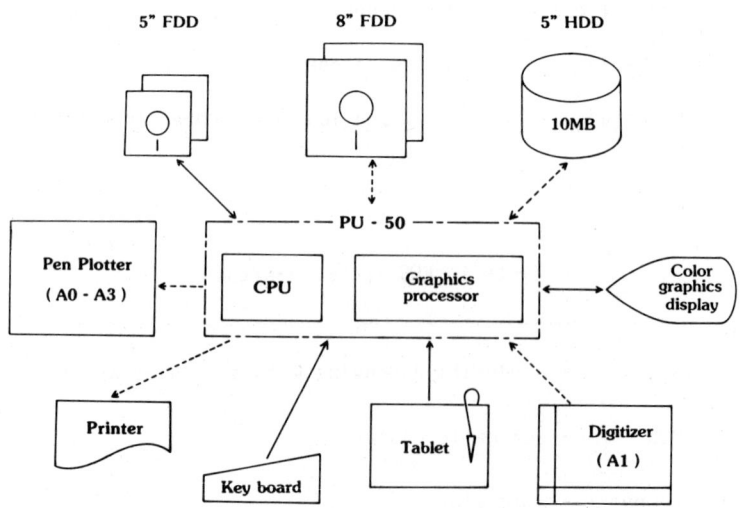

Figure 1   Hardware configuration

Table 2 System configuration

Processor and Memory
• CPU   6502 • RAM   256 KB + 256 KB for graphics • ROM   4 KB (Boot program + other subroutines) "kanji" ROM is standard. • DISK 5·1/4 inches Floppy Disk Drive × 2 (Totally 656 KB = 328 KB × 2)   DMA (Direct Memory Access) is supported.
Color Graphics
• 14 inches color CRT • 512 × 384 pixels • RGB Separate-drive (Linear Input) • Simultaneous display up to 8 colors from a palette of 256 colors. • Max 25 lines of 80 characters • High-speed graphics board for Vector generation • Intelligent graphics board with 16 bit CPU (Z8001)

This personal computer is equipped with high speed graphic board with customized vector generation chip and intelligent graphic board with 16 bit CPU (Z8001).

The basic features of this personal CAD system are :

A) Low cost thus ensuring continuity for designer's exclusive use.
B) Commands are input from menus on a tablet. Instructions and its strings can be freely stored in a menu as macrofunctions, allowing quick and friendly editing and screen operations.
C) Hierachical design enables large scale integration to be designed quickly and efficiently.
D) Page mode and dual window capability.

Mask geometry editing softfware has also been developed which runs onto this personal computer and can be used as a sub-system connected to the turnkey CAD system. Mask level interface provides another new realm of design freedom for the engineer. This can handle versatile geometric editing for a layout of handcrafted LSI circuits. Expert engineer can design supermacros, super speed cells or complex function cells which performs custom chip with the freedom to build in major logic structure or with the capabilities which are not available in any other semicustom way.

Basic functions of this mask geometry editing system are :

A) Any kind of geometric layout structure can be created, including areas, rectangles, polygons, paths, arcs, circles and cells.

B) A wide assortment of editing commands (MOVE, ROTATE, COPY, MIRROR, SWAP, EXPAND, CUT, DELETE, ARRAY, etc) ensure easy creation and modification of complicated geometries.

C) A1 size digitizer is supported for speedy input of layout patterns.

D) Cell edit in place function allows easy editing when a specific cell in a chip or a block will be modified or replaced.

E) Data classification of 64 layers, various line types (solid, dash, dot etc) and filling pattern (dot, grid, solid etc).

The personal computer can be applied to various tasks, such as schematic capture and retrieval, netlist extraction and mask layout design. This personal CAD can be used as a stand-alone system, reducing the burden on a host computer and/or workstation, and eliminating influences to other users on response speed. It can be also connected to a host computer and/or workstations via network to form a distributed system as shown in Fig. 2.

This personal workstation is low cost, stand-alone with easy transportation and personal use system which results in a short design time and low design cost. Netlist can be extracted automatically and interfaces to various tasks, such as logic/timing simulation, layout software, rule check software and layout versus

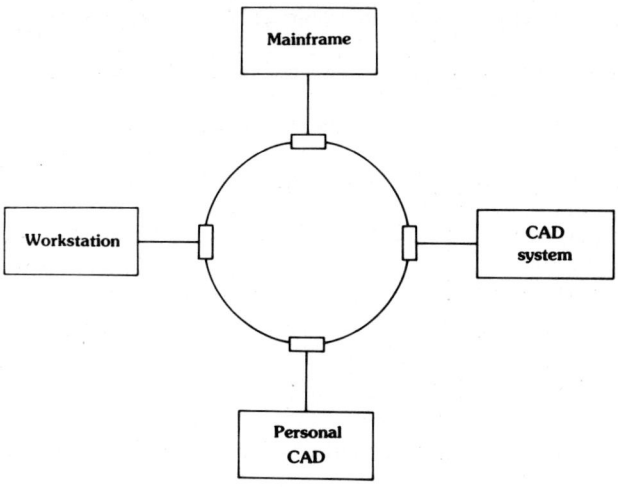

Figure 2   Function distributed system

network consistency check, which runs on a host computer or upper-level workstation.

A common trend through all the custom LSI approaches is flexibility and, previously, full custom LSI provided a solution. Semi-custom LSI will become the performance technology of the late '80s with the advantage of cost efficiency, such as low design cost and short development time. In order to enhance the design flexibility, however, composite structures in one chip can be another solution which is designed based on this personal CAD system. This is a mixture of several parts such as, standard cell part, memory blocks, array structure, PLA blocks and/or mask edit part. This composite structure results in use of silicon area more efficiently and higher operating speed.

§3   SITL technology

Advanced CMOS technology for customized VLSI applications is dividing the digital VLSI world into two fields. At one hand, CMOS will take over low-power medium-speed applications, while the other will be dominated by the high-speed ECL bipolar technology. However, ECL can operate only at higher power level and usually requires heat sink for high density chips. This means the ECL technology is not suitable for consumer VLSI applications with reasonable costs.

We have investigated various LSI techniques for improving switching speed with moderate power consumption. A unique logic circuit has been developed — Static Induction Transistor Logic (SITL) —utilizing the static induction transistors.(3) SITL can be another technology by making it about two to three times faster than CMOS circuits with similar $2\mu m$ technology, while CMOS has improved dramatically and will satisfy many VLSI applications.

The major mechanism of current transport in SIT is a majority carrier injection due to barrier height control at the intrinsic gate in the channel. Barrier height potential at the intrinsic gate is controlled through dielectric coupling both from a gate and a drain. Since the channel consists of material of very low impurity concentrations, junction capacitance becomes very small. In terms of the base series resistance, there is no limitation on the concentration of impurities in the gate diffusion region, meaning it has smaller value than the bipolar transistor. Accordingly, it shows very small power-delay product.

SIT can be designed to have either saturating or non-saturating current-voltage characteristics. SIT has been applied to I^2L circuit technology, where an inversely operating SIT serves as an output transistor. This SIT Logic (SITL) measures power-delay product as low as 0.2 fJ with a simple fabrication process, and has been put to practical use in LSIs.(4) The Schottky-type SITL has been reported for higher speed applications.(5)

Figure 3 shows a typical structure of a planar type junction gate SIT for integrated circuits. This SIT can be designed for a normally-off type where the channel is completely pinched off by gate to channel diffusion potential. Accordingly, there exists a potential barrier in the channel. A forward bias

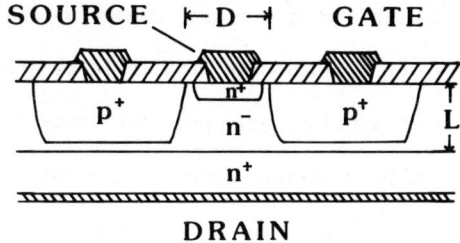

Figure 3 Structure of the planar type junction gate SIT

voltage applied to the gate reduces this barrier thus establishing a current flow from source to drain. This normally-off SIT exhibits saturating I-V characteristics.(6) However, the conventional SIT exhibits non-saturating I-V characteristics which is called the normally-on type. These are shown in Photo 1.

In the normally-off SIT, the channel dimension and the impurity concentration in the channel must be designed to build a potential barrier between the source and the drain without an applied bias voltage. Fig. 4 shows a potential distribution in the channel from the source to the drain which is simulated by a numerical computation.(7) When a gate forward voltage increases, a potential barrier becomes lower and its peak position moves toward the source. There then appears the virtual drift base region in the channel at a certain forward bias voltage in

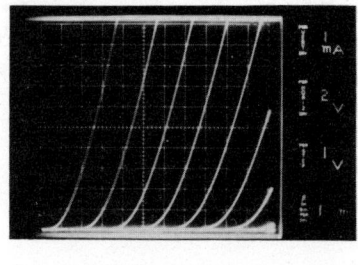

NON SATURATING I-V SIT

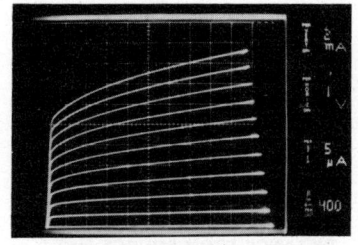

SATURATING I-V SIT

Photo 1 Current-voltage characteristics of the 2 type SITs

which there exist holes with almost the same number of electrons in an n channel SIT. This appearance of the virtual base results in saturation on I-V characteristics. Electrons are localized near the center of the channel at a small gate voltage, with a high bias voltage spread out widely between the gates flowing from the source to the drain. The impurity concentation is 1×10^{14} cm^{-3} in the channel in Fig. 4, whereas there exist about 1×10^{17} cm^{-3} electrons in the channel injected from the source. Therefore high current capability can be achieved at on state. In low current regions the main operating mechanism is the potential barrier control resulting in a positive temperature coeffcient. In high current regions, current is affected both by channel resistance and drain resistance exhibiting a negative coefficient. There is no current crowding and no thermal run-away.

Figure 5 shows an equivalent circuit of the normally-off SIT, which consists of a gate-source and a gate-drain diode, junction and diffusion capacitances, channel

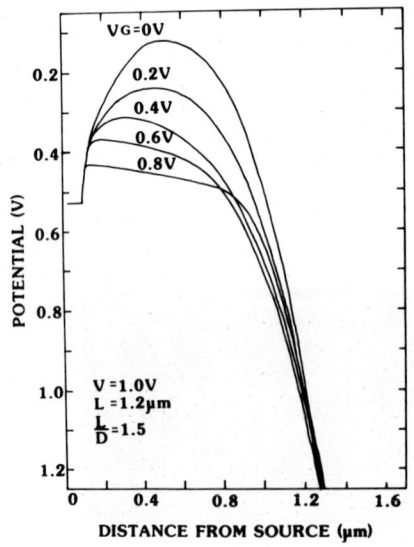

Figure 4 Potential distribution in the channel of the normally-off SIT

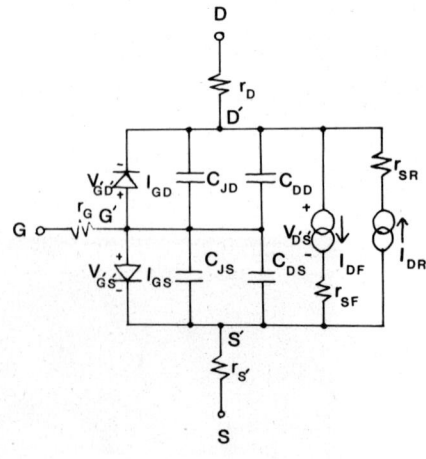

Figure 5 Non-linear large signal equivalent circuit of the normally-off SIT

current source, and resistances.(8) This is a nonlinear large signal model and is used for digital LSI design. I_{DF} is a channel current expressed by :

$$I_{DF} = I_{OF}\left\{\exp\left(\frac{qV_{GS}^*}{kT}\right) - 1\right\} \qquad [1]$$

$$V_{GS}^* = \eta_F\left(V_{GS} - \frac{1+\mu_F}{\mu_F} r_s I_D + \frac{1}{\mu_F} V_{DS}\right) \qquad [2]$$

where I_{OF} is constant, q a unit charge, k Boltzmann's constant, T temperature, η_F effective partition ratio of the gate voltage, μ_F voltage amplification factor, and r_S series resistance in the channel. I_{DR} is similar to I_{DF}, which is an inversely operating current. The three additional resistances, the gate series resistance, the source resistance, and the drain resistance, are modeled in Fig.5. These are the ohmic resistances modeled from the electrode to each active region to improve its accuracy. Channel resistance plays an important role in demonstrating the effect of negative feedback on a high-drain current region, which is expressed as in the Equation [2].

The Schottky SITL is generally DTL (Diode Transistor Logic) circuit implemented with a current supplying resistor, a normally-off type n-channel SIT and a set of Schottky barrier diodes. Basic concept of this circuit structure is similar to STL technology.(9)

This logic uses one respective SIT acting as an inverter whose source is connected to the grounded potential, whose gate forms the logic input and whose drain is equipped with several logic outputs in the form of Schottky diodes as shown in Fig. 6. The Schottky diode acts as output decoupling element. Another Schottky diode clamps the n-channel SIT and prevents the SIT from going too deep into saturation resulting in fast switching speed. A resistor is provided at the input for the current supply of the SIT. Each gate is powered by this resistor connected to an internal local power bus (VGG). This resistor also

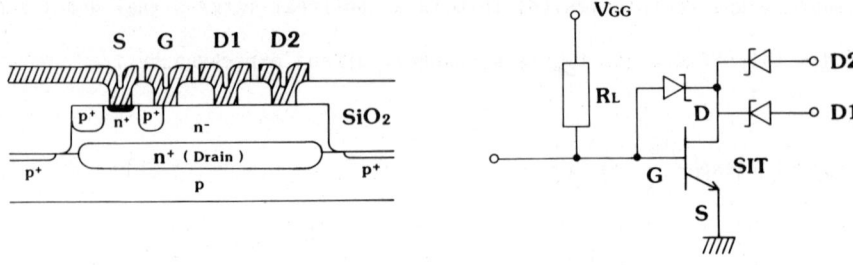

Figure 6   Structure and circuit of the Schottky SITL

acts as the output pull-up for any preceding gates attached to this input. This SIT itself is usually without a direct drain output terminal as the Schottky diodes are integrated in the drain zone. Therefor the SIT and Schottky diodes can be considered as a one integrated component.

The I/O circuits of the Schottky SITL can be designed to be fully compatible with standard LS or S-TTL environments. It can be designed with large current driving capability, and power is supplied by a VDD power supply (5.0 V), whereas power to the internal gate is supplied by a VGG (2.0 V). Logic is designed in TTL like form, therefore conversion to dynamic circuits such as the dynamic MOS circuit is not necessary.

This SITL technology can be suitably extended to the standard cell applications as following reasons :

- fast speed   (1nsec/gate)
- high density   (5k ~ 10k gate)
- low power   (200 μw/gate)
- gate structure (rectangular configuration)
- little dependency of speed on wiring length
- high speed I/O (TTL/ECL)

Figure 7 shows typical gate delay time of the 2 μm rule Schottky SITL versus power dissipation of 2NAND with 3mm loading condition. The gate structure of SITL is efficient and its simplicity allows a rectangular configuration which is compatible with automatic layout. Fig. 8 shows typical gate delay with length of metal interconnection. This shows little layout dependency on wiring length than CMOS technology because of small logic swing of approximately 0.3V, wired logic configuration and small field capacitance. At the same time, SITL provides more output drive capability than CMOS. Therefore, this process is well-suited to bus-oriented systems in which a larger interconnection length usually presents.

High performance applications for SIT LSIs have been implemented specifically in computer graphics and signal processing systems. We have developed a system which uses high speed SITLs to create a graphic displays with a high cost performance. In recently developed new porduct, YGT-1000, there are three customized chips, including a vector generator (5500 gates), video controller (3200 gates) and frame controller (3500 gates). The highest part of these devices is operated at 64 MHz of video frequency in speed to provide 60Hz non-interlaced refreshing of raster scan with 1024× 800 resolution. Photo 2 shows a photograph of the vector generator chip.

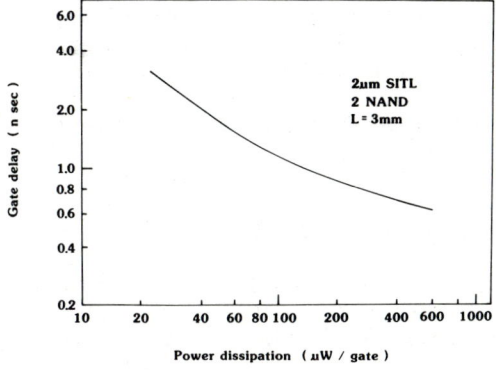

Figure 7 Typical gate delay versus power dissipation

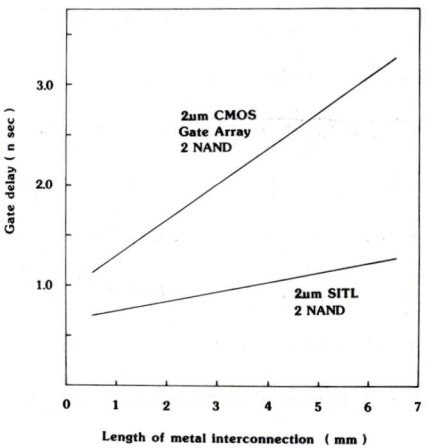

Figure 8 Typical gate delay dependency on metal wiring length

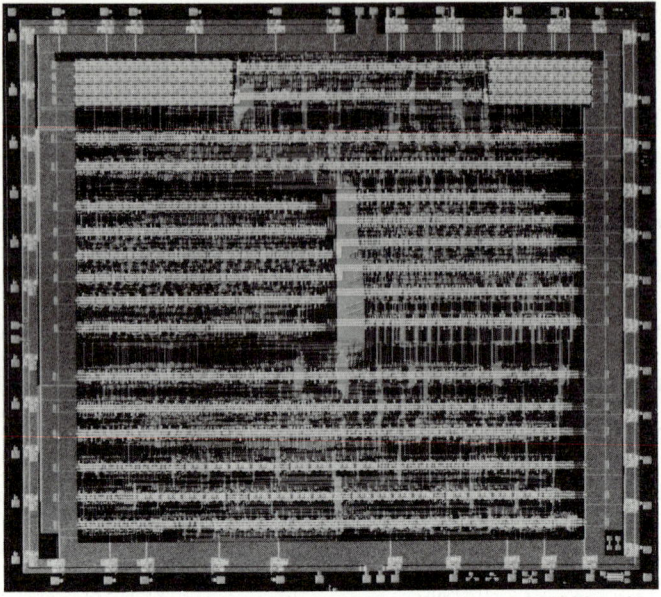

Photo 2 Chip photograph of vector generator made with SITL technology

The static characteristics of the SIT is determined mainly by the channel dimension, the channel length and the impurity concentration in the channel. In the normally-off SIT, the channel should be completely pinched off by the built-in voltage. Accordingly, the channel dimension and the channel length become smaller and the impurity concentration in the channel should be higher in the scaled down device. The speed-power performance and the packing density will also be improved by scaling the physical dimensions.(7)

Operating voltage is an important factor in all aspects of VLSI technology. Low voltage is desirable to minimize power dissipation on a chip and to avoid problems associated with high electric field. In order to operate at lower voltage with reasonable speed performance, a large transconductance is required with which a small voltage can control a large current. As shown in Equation [1], SIT has a high transconductance in the exponential form. This means that the drain current of the SIT can be controlled by one order of magnitude through use of only 60 mV of a gate voltage. Meanwhile the drain current of the MOS transistor is only proportional to $(V_G-V_T)^2$, where VG is a gate voltage and VT is a threshold voltage, because the current is controlled by the gate through a capacitive effect.

Next, the scale down properties of SIT are discussed and compared with the bipolar transistor. In the bipolar transistor the base width should become smaller to improve the frequency response. However, the base width is limited by punch through which is not scaled down. The series resistance of the base layer will increase by decreasing the base width. Nevertheless, impurity concentration in the base is limited by both punch through and injection efficiency. Accordingly, the base width and its impurity concentration is determined to have sufficient punch through voltage.(10) While in SIT, the intrinsic gate is controlled through dielectric coupling from the gate. The SIT seems to be the extreme device of the bipolar transistor in punch through condition, which is sometimes called the punching through device.(11) To reduce the channel length in SIT, the channel should be lightly doped, preferably with p-type material, in the n channel device, increasing the potential barrier height. The barrier height can be controlled by the channel dimension. Accordingly, the punch through voltage in the SIT can be controlled by the mask dimension of the gate diffusion.(7)

§4 Summary and trends

PC based VLSI design system has been developed. Recently, a long computing time required to simulate a complex VLSI is a major bottle-neck for design verification. Now, an advanced engineering workstation can deliver the performace of a simulation accelerator with its special purpose hardware. Workstation can also be linked to a real chips to simulate microprocessor-based systems in order to have a enough accuracy. However, a small personal computer can do only for small jobs. Therefore, it is considered that shared resource configuration in network with low cost personal workstations will be a future trend of CAE environment which is shown in Fig. 9.

Mask edit software on personal CAD provides a design flexibility for expert designers which enables the composite structure. A compact layout of a large complex cell also minimizes the silicon area penalty.

In the future, system designers will likely use silicon compiler to turn system specifications into VLSI chips automatically.

In terms of VLSI process technology, the SIT has been extensively implemented into customized VLSIs, while CMOS technology has dominated low power, high density and medium speed applications. SITL has been suitably applied to the PC based customized VLSI design system, because of its high speed, compatibility with automatic layout and high output driving capability.

SIT has short channel structure, a device which corresponds to the limiting case of the bipolar transistor with extremely thin base layer. The breakdown voltage is controlled by the static induction from the gate thus changing the potential barrier height in the channel.

GaAs SIT is even more attractive, especially in integrated circuit application, where two remarkable properties of GaAs such as high electron mobility and direct

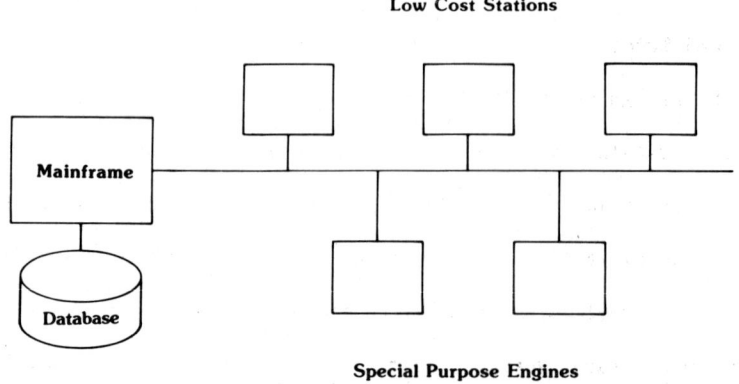

Figure 9 CAE environment

band structure increase the current density and decrease the minority carrier storage effect. Therefore, driving capability is improved while the switching time decreases. GaAs SIT will probably see the non-scattering or the ballistic transport when the source to drain distance becomes very small.(12)

§ Acknowledgement

The author would like to express sincere gratitude to Prof. Dr. Jun-ichi Nishizawa of Tohoku University for directing this research who invented the SIT. Also he wishes to thank Dr. Yasunori Mochida of Nippon Gakki Co., Ltd. for his direction to this work.

§ References

(1) W. Loesch : MIDCON Conf Rec, 6 (1982) 271

(2) S.C.Johnson : Electronics, May 3 (1984) 121

(3) J.Nishizawa, T.Terasaki, and J.Shibata : IEEE Trans. Electron Device, ED-22 (1975) 185

(4) J.Nishizawa, T.Nonaka, and Y.Mochida : Japan.J.Appl. Phys., 19 (1980) Suppl. 19-1, P. 279.

(5) J.Nishizawa, T.Nonaka, Y.Mochida, and T.Ohmi : IEEE J. Solid-State Circuits, SC-14 (1979) 873

(6) J.Nishizawa, T.Ohmi, Y.Mochida, T.Matsuyama, and S.Iida : Tech.Dig. IEDM, (1978) 676

(7) Y.Mochida and T.Nonaka : Semiconductor Technologies edited by J.Nishizawa, OHM North-Holland (1982) P 249

(8) T.Nonaka : Ph. D. Thesis, Tohoku University (1981) [in Japanese].

(9) H.H.Berger and S.K.Wiedmann : Digest of ISSCC 75, P 172

(10) B.Hoeneisen and C.A.Mead : Solid-State Electronics, 15 (1972) 891

(11) T.Ohmi : IEEE Trans. Electron Devices, ED-27 (1980) 536

(12) Nikkei Electronics, 1981.3.16, P. 108 [in Japanese].

Computer Aided Robotics

Hideo Matsuka
Taketoshi Yoshida

IBM Japan Ltd.

Abstract

This paper describes an interactive tool for assisting an operational plan or research activity for industrial robots. First we introduce the concept of a "role-your-own" system for developing a CAD/CAM application system. Second, we present a world model that is the basic and common component in the development tool. The world model represents objects such as a manipulator or a robotic working cell in a factory, and is composed of a geometric model, mechanism model and so on. Last, we show two application systems using the world model based on the concept of the role-your-own system. One is a task planning system for teaching and simulating robot motions. The other is a CAE system for studying advanced robotic systems. The importance of this model for application systems is then discussed.

1 Introduction

There are currently numerous research efforts directed at robotics systems. Among these, two areas receiving increasing attention are (1) a task planning system, and (2) a computer-aided engineering(CAE) system.

The first of these, the task planning system, determines the optimal process and schedule of a manufacturing cell, or program, using a computer for a series of robotic motions in work cells. When its only function is programming the robtic motions in an off-line mode, rather than directly teaching an actual robot, it is called an off-line programming system. Its greatest advantage is that it does not require stopping a production line to develop an application program for controlling a robot. For this reason, competition among researchers to develop a task planning system is keen (Derby 1982; Katajamaki 1984; Howie 1984). In order to realize this system, the precise representation of the status and the phenomena of the cell environment on a computer becomes a very important key(Matsuka 1985).

Research efforts on CAE systems often call, for example, for the study of an advanced robot. These require a designer to analyze and evaluate the performance of a designed robot in advance of producing the prototype. Design methods under which a new robot is experimentally developed through iteration between trial construction and testing are widely adopted, although these are costly and time consuming. If a CAE system has the capability, with various models, to simulate the physical phenomena of a robot, a designer will be able to accurately and promptly evaluate the performance of a designed robot without trial-manufacturing(Tsujido 1983; Chang 1985).

Thus, in order to realize a task planning system or a CAE system, the data models of a robot and its working cell becomes an important key. The purpose of this paper is to introduce data modelling of objects in a manufacturing cell, and to discuss the specific manipulation of the models under some application systems. In the first section, we introduce a "role your own" system, that is, a "hand-made" system for developing an application system. And then, we discuss the data model which mainly represents the shape and structure information of the objects, giving an example of a manipulator. In following sections, we explain how the data model is applied to a task planning and a CAE system for robotics systems.

2 A Tool for Application System Development

2.1 Utility Structure

In any application area of a task planning or CAE for robotics, numerous subsystems which assist in solving specific problems, for example, layout design, or kinematics/dynamics analyses, will be demanded. In order to build a hand-made or problem oriented subsystem, we will need application program interfaces composed of three functional groups of subroutines and utilities. One is a standard graphic management package called PHIGS (Bunshaft); a second is a data management handler that represents geometric data in a table form and stores them on a database (Matsuka 1983). Figure 2.1 shows a configuration of these two kinds of utilities as application interfaces. Application programs which mainly deal with the graphic output are easily developed by combining these two utilities. They are, however, insufficient for building a highly interactive application system. An interactive system needs a mechanism that allows the operator to enter parameters, and dynamically controls the execution of programs in response to an operator's request. That is, the execution of the programs must be controlled by the operator through menu/command instead of predefined sequence. This function, the third function, is called a dialog manager.

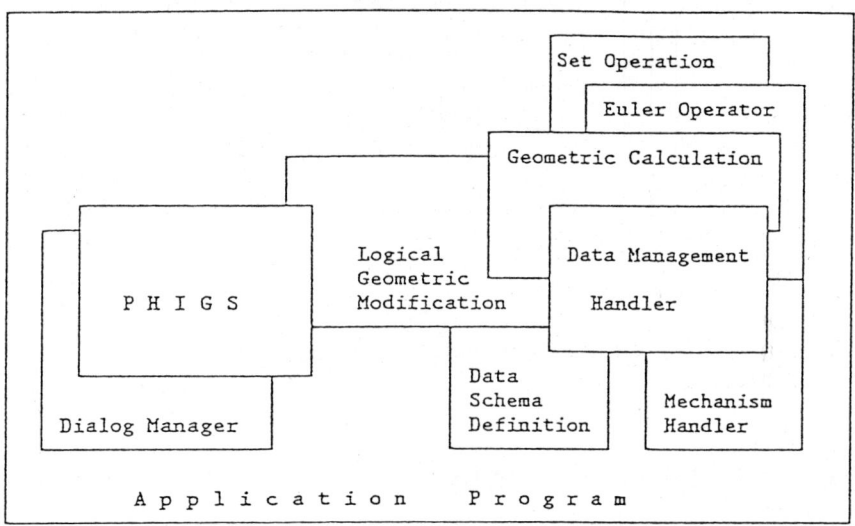

Fig. 2.1 Application Program Interface Layer

2.2 Dialog Manager

A dialog manager(DM) executes and manages, in the form of menu or commands, application programs on a graphic environment(IBM 5080 graphics system). It controls the execution of the application program by an operator's interaction, as well as manages a graphic display screen. A hierarchical menu or command system is registered for each application subsystem. A menu corresponds to the execution of a specific program, or moves the hierarchy level up and down. Besides such menus, it is possible to register a global menu that does not belong to any hierarchical level, and can be selected at any time. The program to be executed is not necessarily link-edited, that is, it is dynamically loaded if so-specified. In either case, the data entered by the operator are passed to the program after the data type conversion.

Figure 2.2 shows the display screen layout of the IBM 5080 graphic terminal supported by the DM. The DM has a function that displays status of the control environment to help the operation. A set of available menus varies from time to time, because the hierarchy level of menu moves up and down. Besides menu, the history of hierarchy movement, parameters and messages are simultaneously displayed on the screen. The region displaying a graphical shape can be arbitrarily defined by view-port and window specification from a user program.

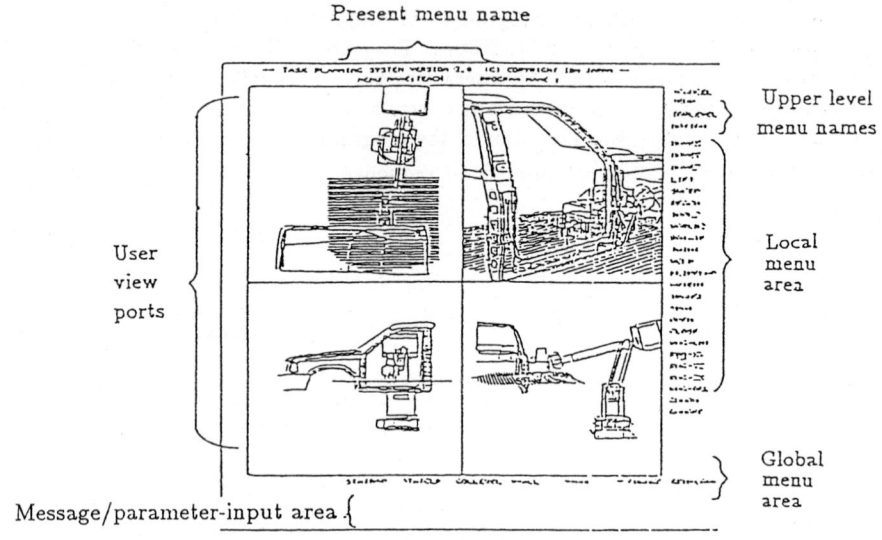

Fig. 2.2 Layout of IBM 5080 Graphic Screen

3 Data Model (World Model)

A data model of an object in the real world that is modeled by a user's intention and abstraction on a computer is called a world model. We offer, as a typical example of a world model, a model of the work-environment of a manufacturing cell in a production line. The world model consists of various object models such as manipulators, pieces of equipment, processed or assembled parts, and related conveyance systems. Figure 3.1 illustrates the structure of object models with various attributes such as geometry, structure/mechanism, input/output communication and sensors (Mayer 1981; Jayaraman 1984).

3.1 Geometric Model

A geometric model is the most fundamental model among various application models. It can be abstracted to three classes : wire-frame, surface, and solid models (Baer 1979). A solid model is especially effective for calculating collision detection, mass property, static stability and so on.

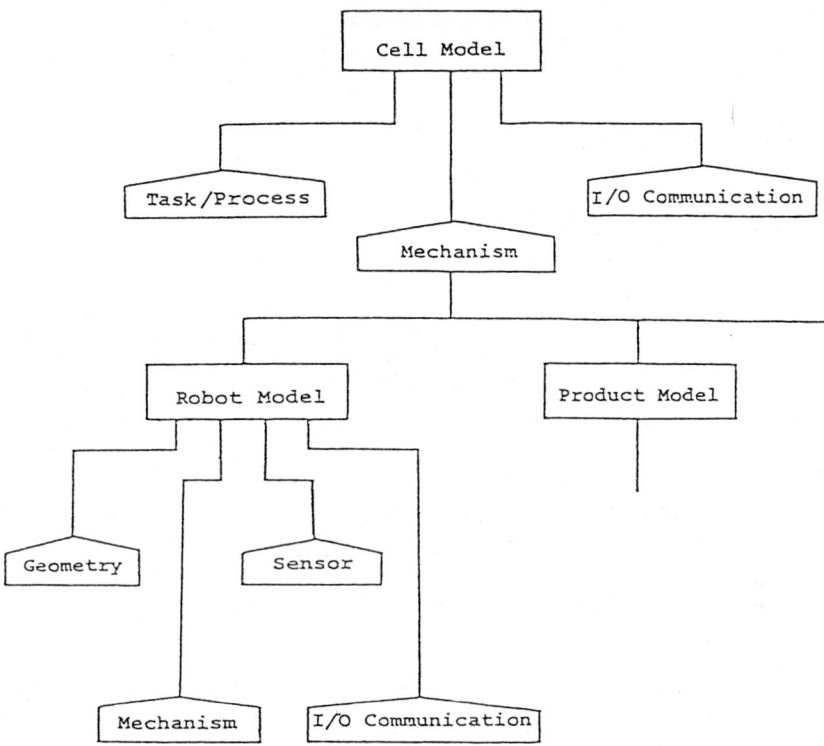

Fig. 3.1 World Model of Robotic Work Cell

The structure of the model is shown in Fig.3.2. The topological relation between face and edges is described by a boundary representation method called "winged edges". For the geometric model, its modelers are provided by utilizing the data management handler shown in Fig.2.1. The main functions of the modeler are as follows:

1. primitive volume generation such as cuboid and prism
2. transformation and rotation of a model
3. local topological modification
   Euler operator level macro operator level ,e.g. drill,lift,glue
4. logical geometrical modification
   translate an edge, a face rotate a face, a volume
5. set operation,e.g. union, intersection,and difference

An important feature of the modelling is that all the topological operations needed to deal with solid models consist of Euler operation (Mantyla 1982; Kawabe 1984).

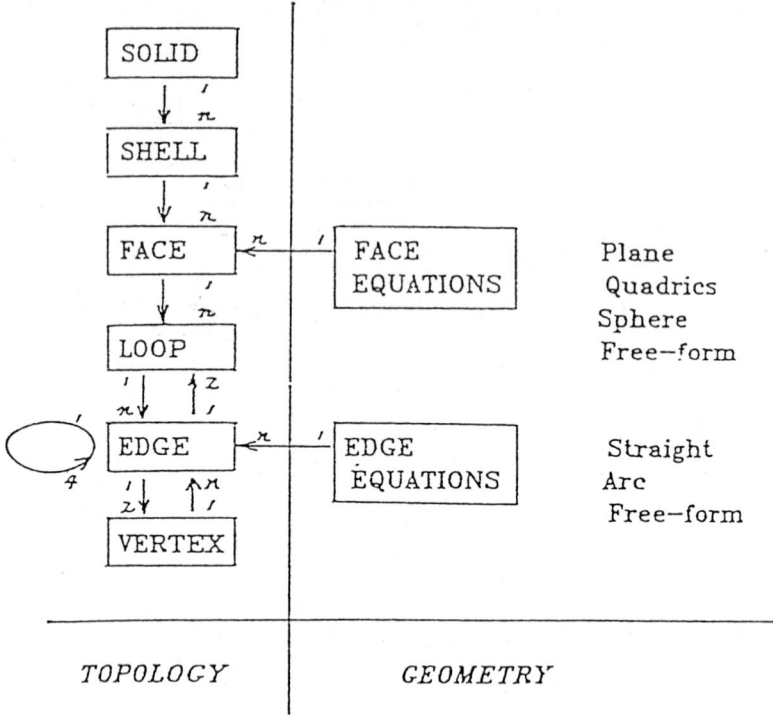

Fig. 3.2 Solid Model Structure

3.2 Mechanism Model

A mechanism model provides information on the kinematic relationships and constraints among constituent parts. Figure 3.3. shows an example of a manipulator mechanism model being developed in the task planning system and the CAE system.

The "MECHANISM" represents the configuration of the arm and the attached end effector. After the attachment of a tool is completed, the two coordinate frames are fixed. The linkage mechanism is described in form of connection of each link; L_0, L_1,..., by joint; J_1, J_2, ..., via "ROUTE", this model can represent the mechanism with a loop or a branch. The "LINK" table has the attributes such as weight and principal axis of a link. The 'JOINT" table has attributes such as joint type (prismatic, revolute) and constraint conditions of movable coverage (Kawabe 1984).

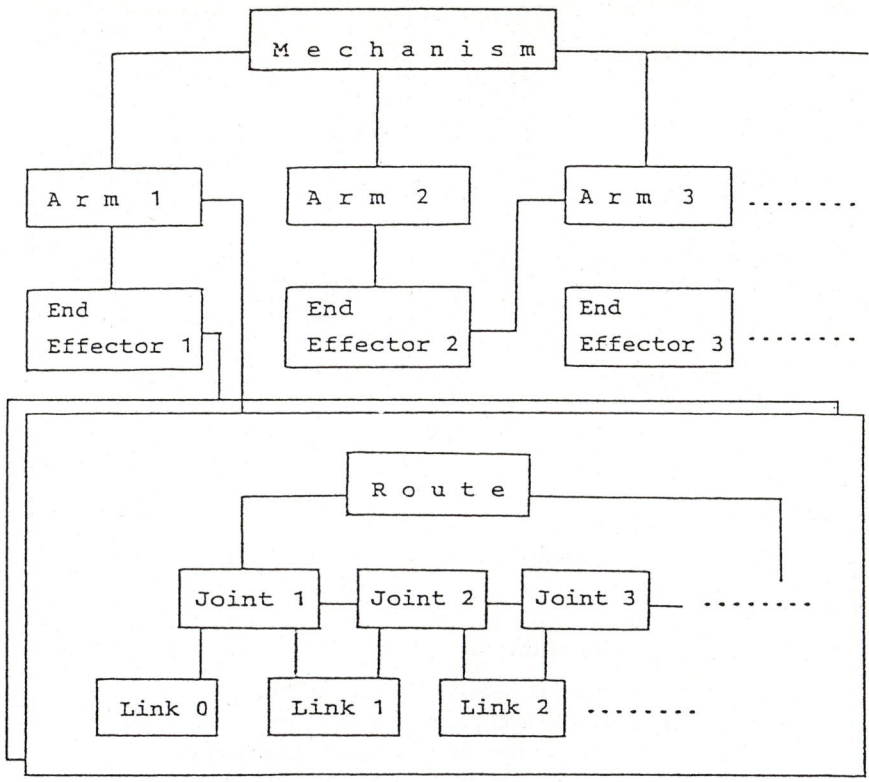

Fig. 3.3 Robotic Model structure

When the constraints for a manipulator are as follows, the relative position between adjoining link and link are expressed by four parameters representing by Denavit-Hartengerg shown in Fig.3.4 (Matsuka 1983a,1983b).

1. Simple chained opened link mechanism moves in 3-D space
2. Each joint connects two links and joint type is prismatic or revolute
3. Each link is a rigid body
4. The end of chained mechanism is fixed at the ground

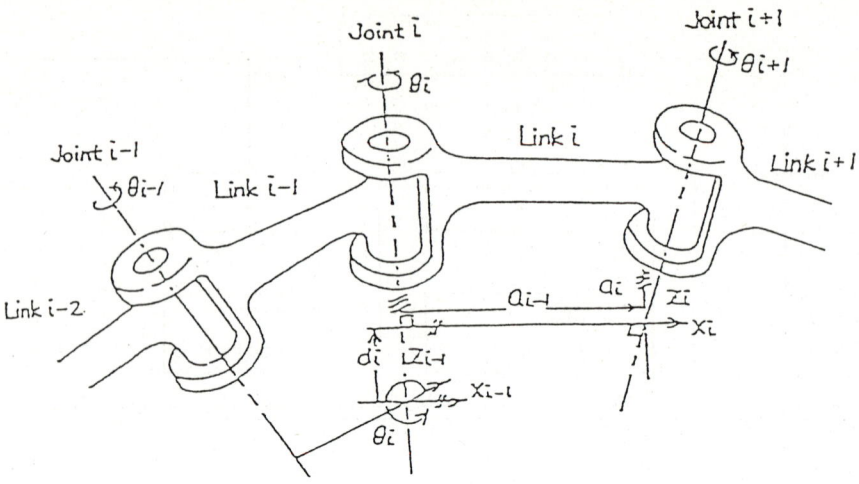

$link_0$: the reference frame

$\vdots$

$link_n$: the tip of the link

$joint_i$: connection between $link_{i-1}$ and $link_i$
z_i : the motion axis of joint
x_i : the axis along the common normal from z_{i-1} to z_i
y_i : the axis composed of the right hand coordinate
a_i : the distance between the origins of coordinate system (i-1) and i measured along x
d_i : the distance between x_{i-1} and x_i measured along z_{i-1}
α_i : the angle between the z_{i-1} and z_i axes measured in the right-hand sense about x_i ; and
θ_i : the angle between the x_{i-1} and x_i axes measured in the right-hand sense about z_{i-1}

Fig. 3.4 Denavit-Hartenberg Representation

4 Application System: TPS

A task planning system (TPS), in which a computer assists a planner in structuring precise plans for controlling the complex robot-motions, plays an important role in a *computer integrated manufacturing* (CIM), since frequent alteration of the product requires that the lines lend themselves to prompt and easy modification. Moreover, at the very time when more sophisticated and cooperative robots are required, and consequently, it is becoming increasingly difficult to specify accurate robot-motion through the teaching-by-showing method – teaching by moving a real robot through its steps – there has developed a severe scarcity of factory engineers possessing the required robot-training skills.

4.1 Ideal System

A robotic task includes the following three steps: task specification, layout design, and programming synthesis. In task-specification, the task and the processes to be performed in each manufacturing cell are defined and appropriate equipment is selected from not only the engineering data of a CAD system, but also equipment-specifications. At the layout-design step, the selected equipment is located by surveying the working-range of the equipment. Then the environmental model, which contains the necessary information for the next program-synthesis-step, is created. In the programming synthesis step, the robot motions are programmed. Through the iteration of programming and simulation, the precise motion processes for synchronization of various pieces of equipment are determined. The final results are then transferred to a real manipulator.

A desired configuration of the system is shown in Fig.4.1. The main feature of the system is a graphical off-line programming system based on a model which replaces the teaching-by-showing method. The off-line system is closely connected with a run-time system. A data model should maintain consistency with an everchanging environment of a manufacturing cell. This shows an intelligent system that the robot can judge its activity by itself and also that of an autonomous model building system by referring the changing environmental information via various sensors. To realize the system, we intend to develop a high performance industrial workstation to maintain the environment model(Mclean 1983).

4.2 Experimental System

At present, we have developed a task planning system for spot welding which clearly separates the off-line part from the run-time system (Kawabe 1985). The graphical motion teaching method is similar to the operations which are performed in teaching a real manipulator with a pendant. Each link of the arm is continuously moved around the joint by the valuator input device of the graphic terminal. The position and orientation of the hand is set with respect to a hand coordinate frame,or object coordinate frame or an object coordinates attached to a car body. After teaching, the system can, using animation, display the motion. The simulation mode has interactive collision detection capability between moving and stationary objects. One of the features of the method is that a moving object can stop before it hits a stationary object. On stopping, the distance from the stationary object can be specified by a programmer. This method can be utilized for the trace of the boundary of the object with a constant distance. Since the hitting direction is also obtained, the next motion to avoid the collision can be determined (Kawabe 1985). Cycle time spent to perform a certain motion is estimated. A programmer can stop an arm at the specified time on the graphic screen.

Through the iteration of teaching and simulation, precise plans for synchronization of various pieces of equipment, even for abnormal or emergency situations, can be edited. The final results are then automatically translated into an intermediate robot language. The intermediate language achieves robot-independence of the control language (Grossman 1985). Although each robot has both its own control language and coordinate system, the TPS generates a robot-independent program in an object-oriented language using an absolute coordinate system. This program is to be converted into an existing robot control language using a robot-dependent translator. We have already developed translators for the PUMA, the NACHI and IBM's SCARA-type robots.

All the 3-D object models representing a manipulator and a work cell are displayed on the screen,and taught motion commands effect to the models. Reactions of the models are immediately shown to the programmer through a graphic display. Some examples of a robotic working cell displayed on the IBM 5080 graphic terminal are shown in Fig.4.2.

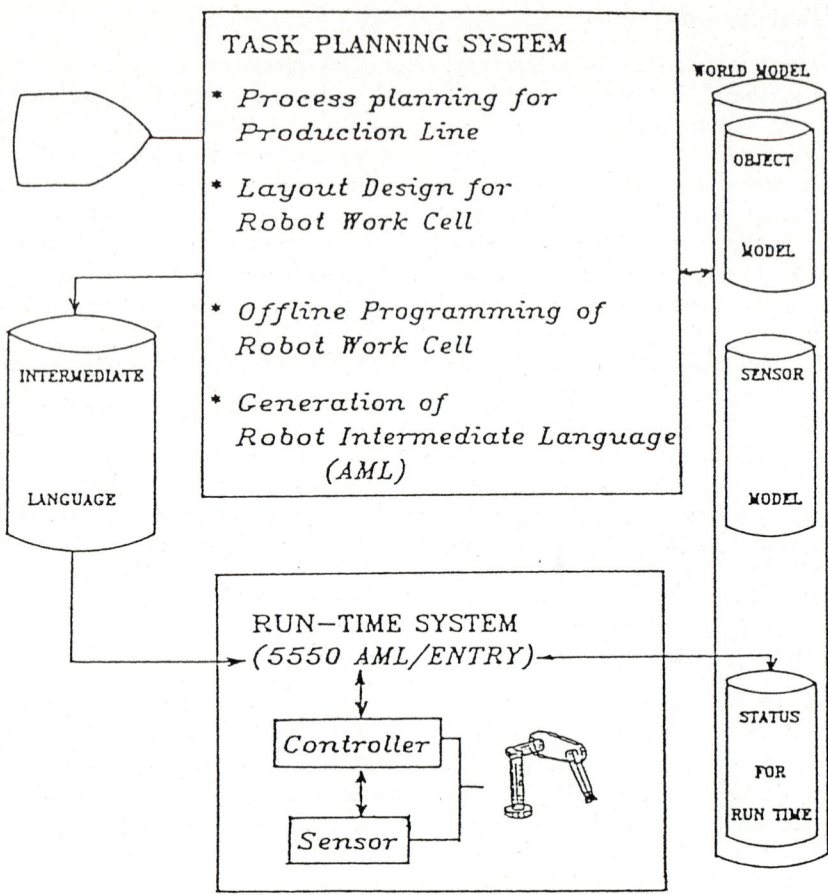

Fig. 4.1 Robotic Task Planning System

5 Application System: CAE

Most contemporary manufacturing equipment is composed of machatronic devices, the name given to the complex system formed by combining a mechanical structure and a microcomputer; mechanical functions are controlled by the computer, which in turn receives feedback-signal from sensors in the mechanism. A robot is one such instrument.

Despite the fair complexity inherent in this sort of mechanism, its design must, nonetheless, focus on ease-of-operation. A means for system tuning, that is, for tailoring the mechanism's functions to individual work cells must be incorporated as well. Furthermore, the effect directed to the prototype development should be kept to a minimum. In order to satisfy these requirements, a computer aided engineering(CAE) system must be supplied which allows the design of mechatronic equipment of high performance and reliability.

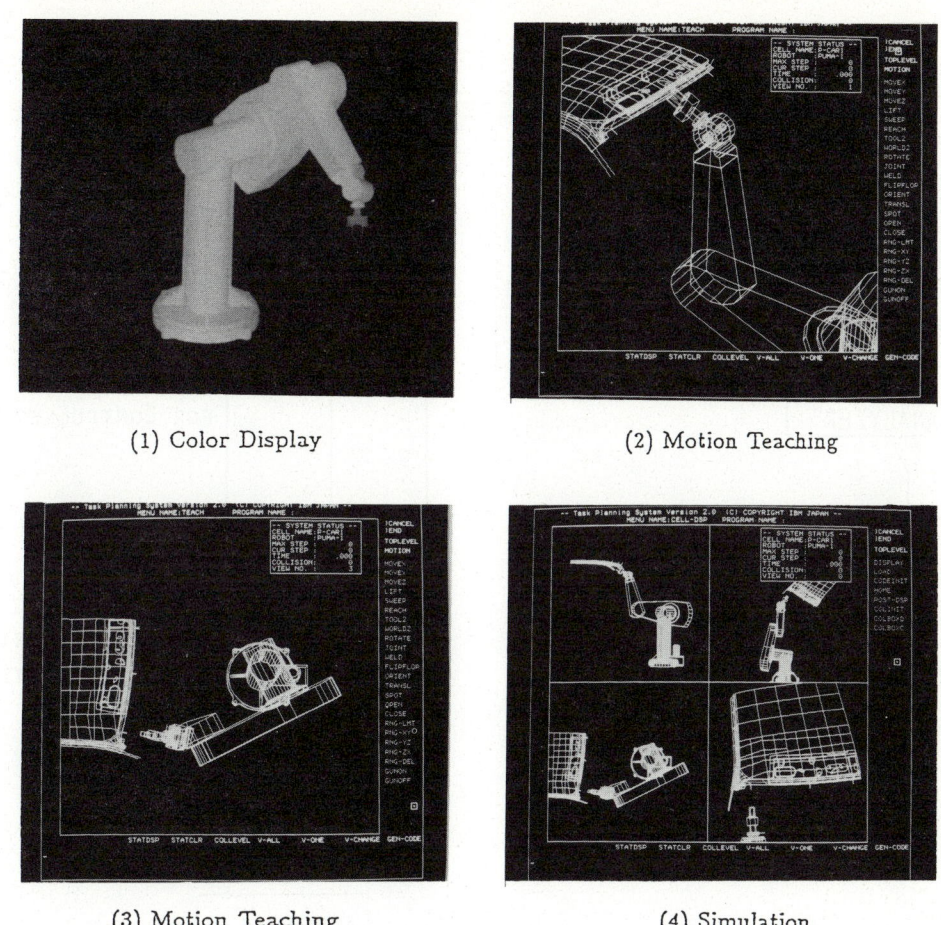

(1) Color Display   (2) Motion Teaching

(3) Motion Teaching   (4) Simulation

Fig. 4.2 Results of Spot Welding Plan

5.1 Ideal System

Figure 5.1 shows the coming CAE system. It will be composed of the data gathering/system identification part from the analog/digital sensors of the actual robot, the control synthesis part, simulation part and a micro codes generation/transmission part to the actual robot. These components are closely connected for real-time communication with each other. To realize the CAE system, many problems must be solved:

1. multi-task and real-time operating system for an engineering workstation must be created;
2. high speed arithmetic boards allowing simulation of the concurrent processes of the actual control system must be mounted;
3. communication function between the models and physical sensor signals is to be established.

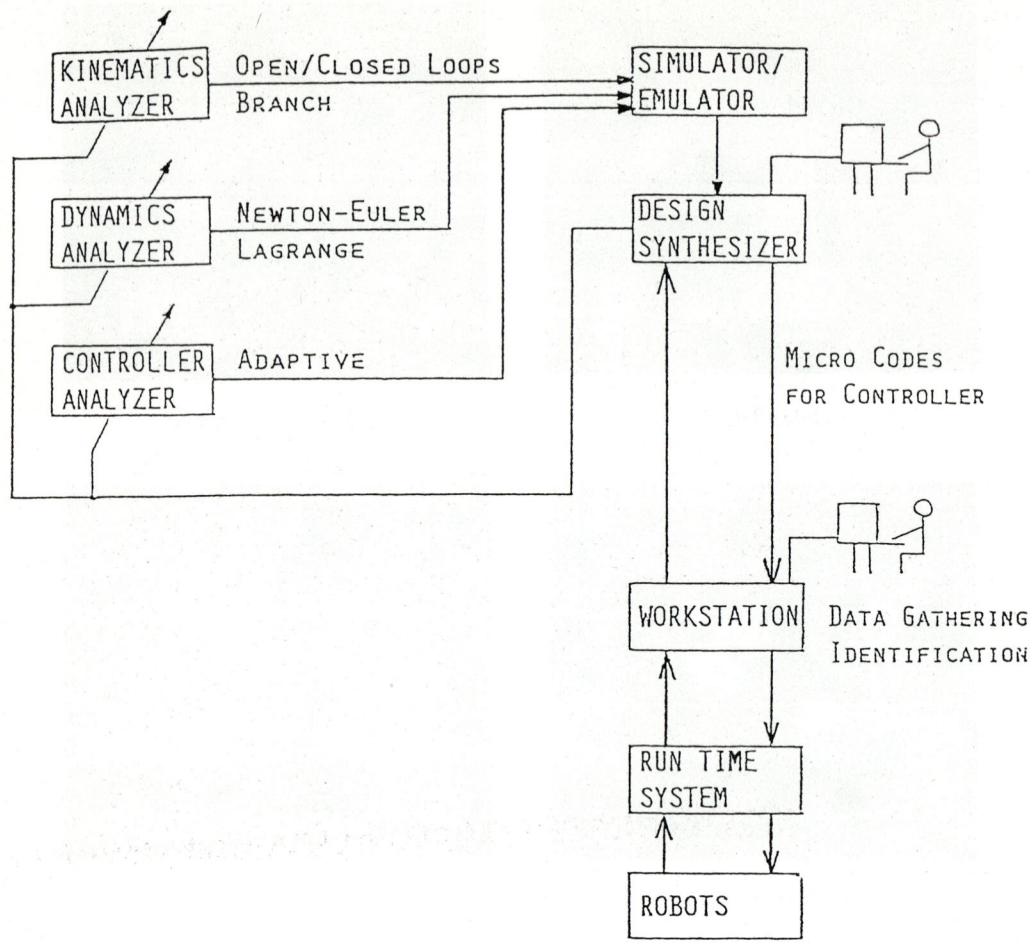

Fig. 5.1 Robotic CAE System

5.2 Experimental System

We are undertaking the effort to develop various scientific packages analyzing kinematics, dynamics, multi-variable control performance, and so on, for a robot. In this, we introduce kinematic and dynamic analysis using the robot model.

Kinematics

There are two kinds of kinematics for a manipulator. One is to calculate the position and direction of the end effector from each joint variable at arbitrary time. The other, what we call the inverse kinematics, is to get each joint variable from the given position and direction of the end effector.
The latter solution is obtained from the fact that the compound vector of the end effector (composing of positional vector and direction vector), u, equals to the vector $\xi(q)$ obtained from the

generalized coordinates of each joint:

$$\xi(q) - u = 0$$

Generalized coordinates for motions are as follows:

in case of a rotational joint : $q = \theta$
in case of a prismatic joint : $q = d$

A general solution method does not exist. However, if the Jacobian matrix is not singular, the solution can be obtained by using the Newton-Raphson method. Here the key idea is

$$du = \frac{\partial f(x)}{\partial q^T} dq$$

where J(q) is a Jacobian matrix given by

$$J = \begin{pmatrix} z_0 \bar{\sigma}_1 + (z_0 \times p_1^*)\sigma_1 & z_1 \bar{\sigma}_2 + (z_1 \times p_2^*)\sigma_2 & \cdots & z_5 \bar{\sigma}_6 + (z_5 \times p_6^*)\sigma_6 \\ z_0 \sigma_1 & z_1 \sigma_2 & \cdots & z_5 \sigma_6 \end{pmatrix}$$

where σ_i is the indicator of the joint type ($\sigma = 0$; prismatic ; $\sigma = 1$; revolute), $\bar{\sigma}_i = 1 - \sigma_i$, z_i the z axis of joint i, and p_i^* the positional vector of the ith coordinate system with respect to the reference coordinate system.

Some results of the analysis of the JPL manipulator are illustrated in Fig. 5.2 (1). When the position of the end-effector is

$$0 < t < 60/130; x = 1, y = 0, z = 1.8t^5 - 4.5t^4 + 3t^3$$

$$60/130 < t; x = -6t^5 + 15t^4 - 10t^3 + 1, y = 1 - x, z = 0.3$$

and the direction is $(-2/\sqrt{2}, 2/\sqrt{2}, 0)$, the general coordinates, speeds, accelerations and forces at each joint are shown in Fig. 5.2 (2 - 4). (Luh 1980)

Dynamics

In the dynamic analysis there are two kinds of analyses: the force analysis deriving general force from a given trajectory, and the trajectory analysis (motion analysis) calculating general position from general force.
The force analysis, what we call the inverse dynamics, is derived from the Newton-Euler equation shown by Luh (1980). The results are shown in Fig. 5.2 (3 - 4). In case of trajectory analysis, the motion equation is as follows:

$$\tau(q, \dot{q}, \ddot{q}) = H(q)\ddot{q} + C(q, \dot{q}) + G(q) + J^T(q)f_e + B\dot{q} + kq + f_e$$

where

τ : $n \times 1$ generalized force vector
H : $n \times n$ inertial matrix
C : $n \times 1$ centrifugal and Coriolis vector
G : $n \times 1$ gravity vector
J : $n \times n$ Jacobian matrix
f_e : $n \times 1$ external force vector
B : $n \times n$ damper coefficient vector
k : $n \times n$ spring coefficient vector

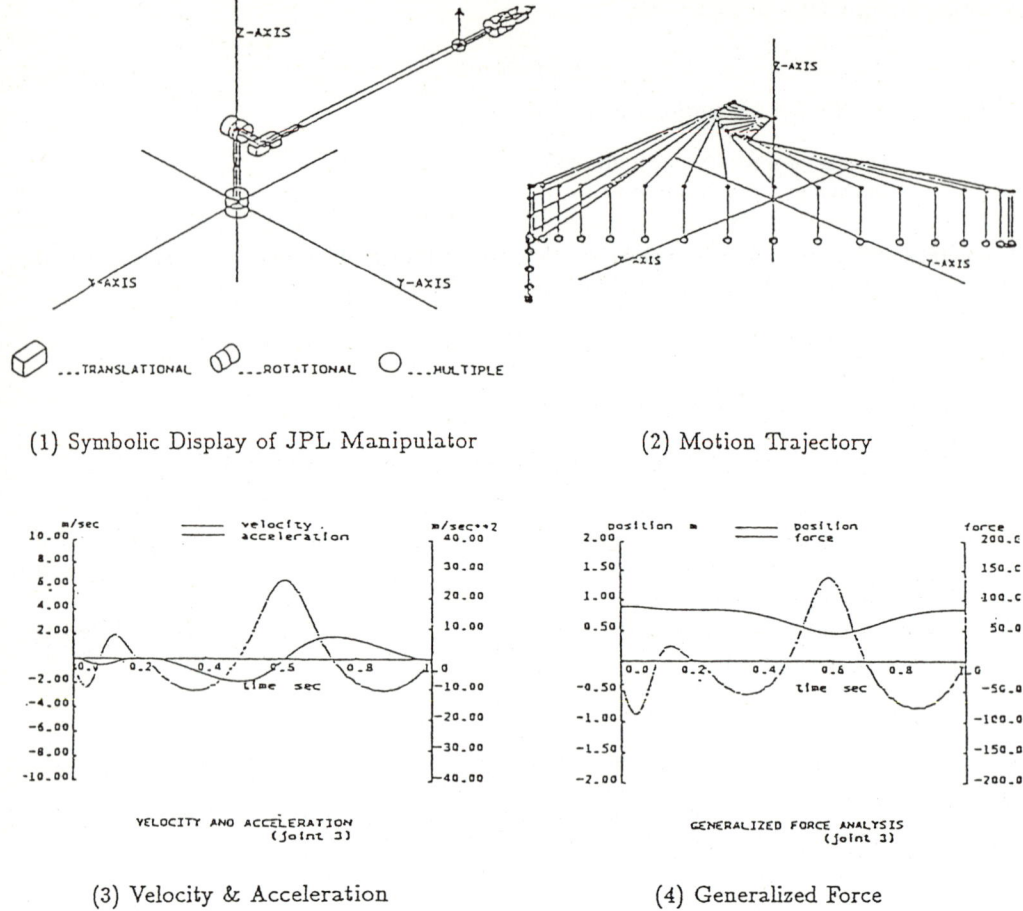

Fig. 5.2 Results of Kinematic Analysis

The solution is obtained by recursive integration by Hollerbach(1980). As q during sample period is constant in the trajectory analysis, the results of the integration becomes errors. In order to get a proper division during a sample period Δt, we analyzed the positional input to the force analysis and the position obtained from this trajectory analysis by divided the motion for one second in details. Figure 5.3 shows proper accuracy and calculation time for the sampling time against a high speed manipulator (Matsuka 1983; 1984).

The simulation incorporated with a control system is shown in the following. The block diagram and control strategy is shown in Fig. 5.4(1). The reference trajectory is given by

$$q_i = 3t^5 - 7.5t^4 + 5t^3; i = 1, 2, 3.$$

and the sampling frequency is 200 Hz. The results are illustrated in Figures 5.4(2-3), where Fig.5.4(2-3) show the results with no compensation and the ones with dynamics compensation.

(1) Error Analysis

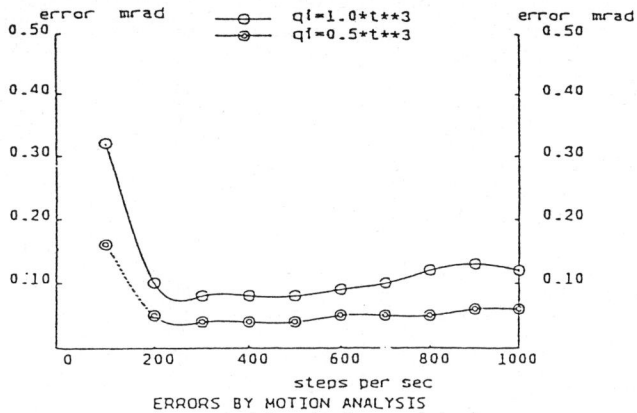

ERRORS BY MOTION ANALYSIS
(computed in single precision)

(2) Computation Time

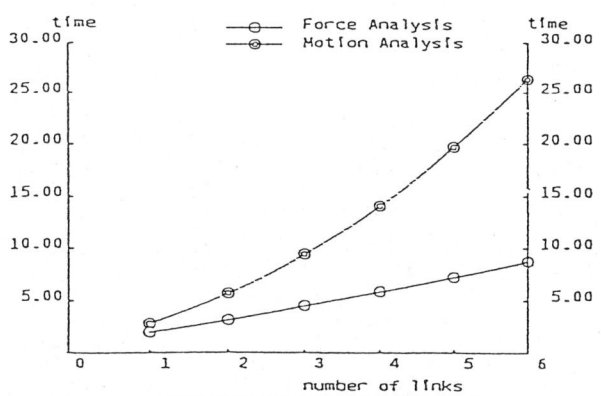

COMPUTATION TIME FOR DYNAMICS

Fig. 5.3 Dynamic Analysis

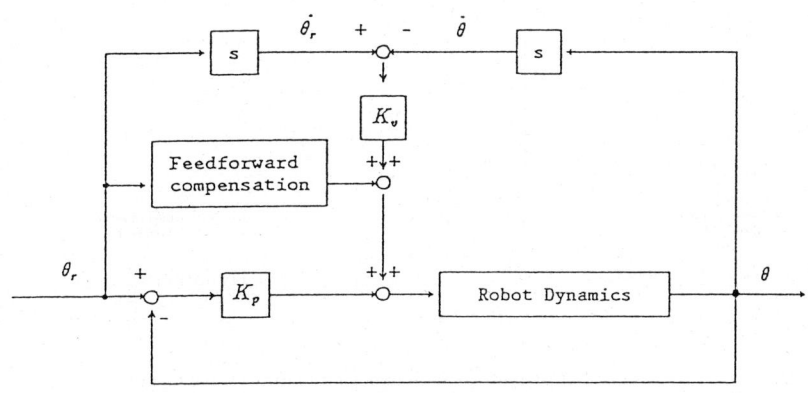

(1) Block Diagram of Control System

Fig. 5.4 Applied to Control Performance Analysis

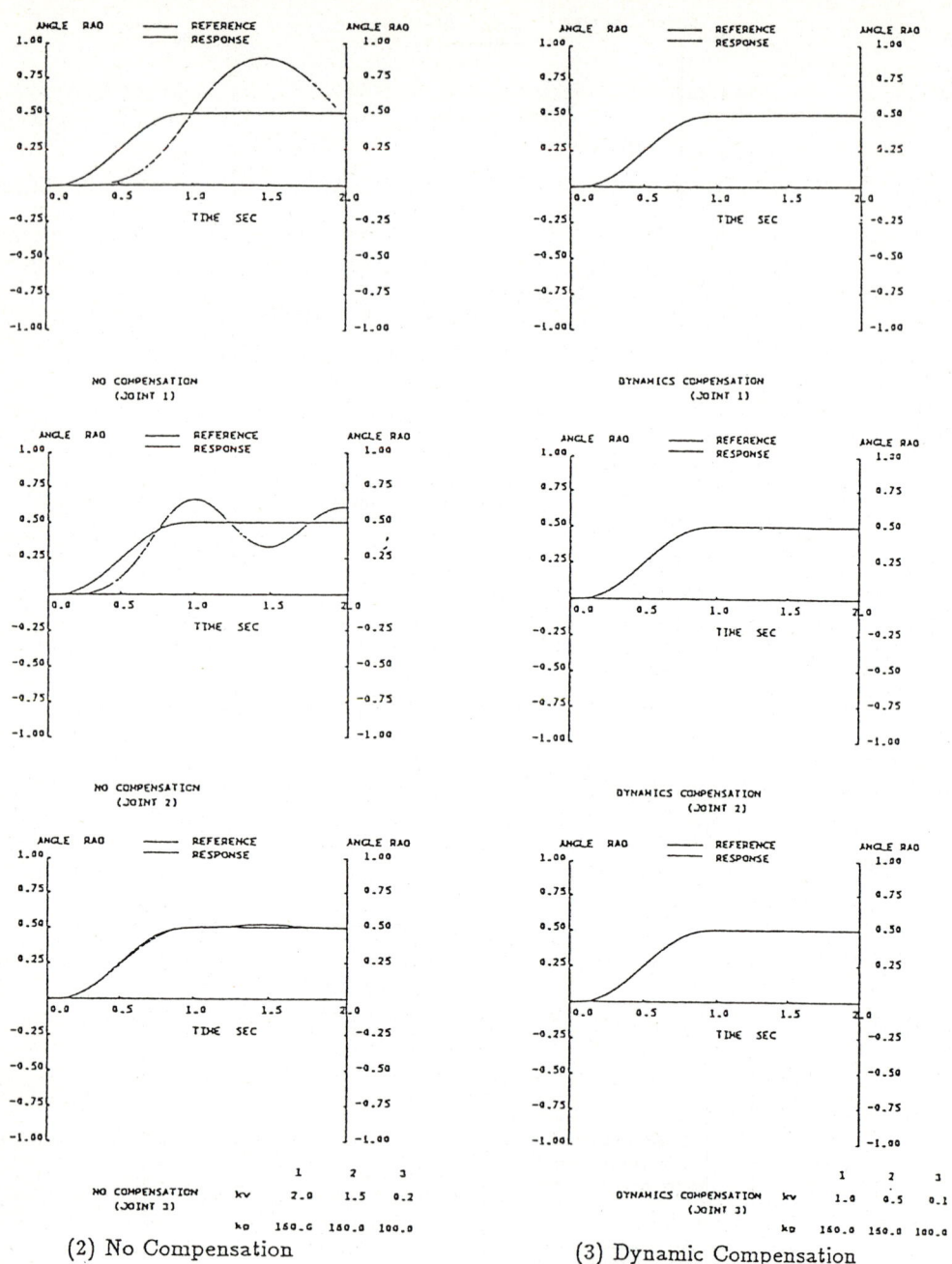

(2) No Compensation (3) Dynamic Compensation

Fig. 5.4 Applied to Control Performance Analysis

6 Conclusion

In this paper, we described the modelling on a manipulator, and some of its applications. To program robot-tasks within a working environment and to simulate the physical phenomena on a robot, we have developed world models which are composed of attribute data such as geometry and mechanism. Two examples of a robotic task planning system and a robotic CAE system are provided as useful illustrations of the model. There is no doubt that the world model is essential for robotic applications. We are applying and extending the world model to various application systems such as task planning/controlling/monitoring of working cells in the production line. The authors wish to express their appreciation to S.Kawabe, H.Okano who have developed the task planning system together and to the members of Kato Labo. of Waseda University for the CAE system.

REFERENCES

Baer,A.(1979) C.Eastman and M.Henrion, Geometric Modelling: a Survey. Computer Aided Design, Vol.11,No.5: 253-272
Bunshaft,A.J.,PHIGS Implementation at RPI, X3H31/83-50, Rensselaer Polytechnic Institute.
Chang,K.H.,H.Funakubo,et al.(1985) Programming for a System to Assist in the Control of Robots. Proc. of the 15th ISIR,pp767-774,Tokyo
Derby,S.J.(1982) General Robot Arm Simulation, Part1 & Part2. ASME Comp.Eng.Conf.
Grossman,D.and W.M.Short (1985) AML - Much More Than a Robot Language. Robot 9: 18.14-18.22
Howie,P.(1984) Graphic Simulation for Offline Robot Programming. Graphic Simulation: 63-66
Jayaraman,R.(1984) Simulation of Robotic Applications. IBM Research Report, RC10714, Sep.
Katajamaki,M.and J.Kanerva (1984) CAD/CAM-revoutionizing Robot Applications Design. Proc. of 14th ISIR: pp691-700
Kawabe,S.and K.Kajioka (1984) Solid Modelling Based on Euler Operation. Proc. of the 29th Convention of IPSJ: 1757-1758
Kawabe,S.and H.Matsuka (1984) One Method of Object Modelling. Proc. of 59th Fall Convention of Precision Machinery: 273-274
Kawabe,S.,H.Ishikawa,A.Okano and H.Matsuka (1985) Interactive Graphic Programming for Industrial Robots. 15th ISIR: 699-706
Kawabe,S.,A.Okano and T.Yoshida (1985) Robot Task Planning System Based on Product Modelling. COMPINT85, IEEE: 699-706
Mantyla,M.and R.Sulonen (1982) A Solid Modeler with Euler Operators. IEEE CG&A, Vol.2, No.7: 17-31
Matsuka,H.,M.Noguchi, et al.(1983) Study on Computer Aided Robotic Design System. Proc. of the 58th Fall Convention of IPSJ: 1573-1575
Matsuka,H.,S.Uno and K.Sugimoto (1983) Concept and Tools for Interactive Graphics – A-IDAS –. Computer & Graphics, Vol.7,No.3-4, Pergamon Press Ltd.: 215-224
Matsuka,H.(1983a) The State-of-the-arts about Dynamic Analysis for 3-D Link Mechanisms. J.of Japan Society of Precision Engineering, Vol.49,No.12: 108-115
Matsuka,H.,M.Noguchi,et al.(1983b) Development of Computer Aided
Robotics Design System. Proc. of Annual Convention, Robotics Society of Japan: 223-224

Matsuka,H.,K.Taniguchi (1984) Dynamic Analysis System for Multi Link Mechanism. Proc. of the 59th Fall Convention of IPSJ: 1769-1770

Matuska,H.and S.Kawabe (1985) Robotic Task Planning System. J.of Robotics Society of Japan, Vol.3,No.2: 40-45

Mayer,J.(1981) An Emulation System for Programmable Sensory Robot. IBM J.of R & D, Vol.25-6: 955-961

Mclean,C.,M.Mitchell,et al.(1983) A Computer Architecture for Small-batch Manufacturing, IEEE Spectrum, Vol.20,No.5: 59-64

Tsujido,Y.and M.Oshima (1983) Realtime Motion Simulation of Robotics. Proc.of ICAR,Tokyo

Walker,M.W.and D.E.Orin (1982) Efficient Dynamic Computer Simulation of Robotic Mechanism. J.of Dynamic System,Measurement and Control,ASME Trans.: 205-211

Computer Aided Engineering
Will There Be Any...Ever?

Ralph E. Miller, Jr.
Boeing Commercial Airplane Company

R. Peter Dube
Boeing Computer Services Company

ABSTRACT

An embryo CAE system is sketched which reflects both the computer-based tools and the humans that are to use these tools. Key to this description is the inclusion of not just technical aspects but social and management aspects as well. A methodology, based on the IPAD work, is suggested which can help one to identify critical CAE system aspects, such as the activities, the data requirements, and their interactions, both for technical and management processes.

An ideal CAE system is described which is heterogeneous in both hardware and software. The IPAD prototype software work for data management, executive, and network communications are described.

Various marketing approaches, as well as some of the economic factors which appear to influence the potential emergence of the needed CAE system, are evaluated. It is concluded that market forces, not technology, are constraining the emergence of a fully integrated CAE system, effective for engineering data management and task and data communication in a heterogeneous computer environment.

AN OVERVIEW OF THE TASK

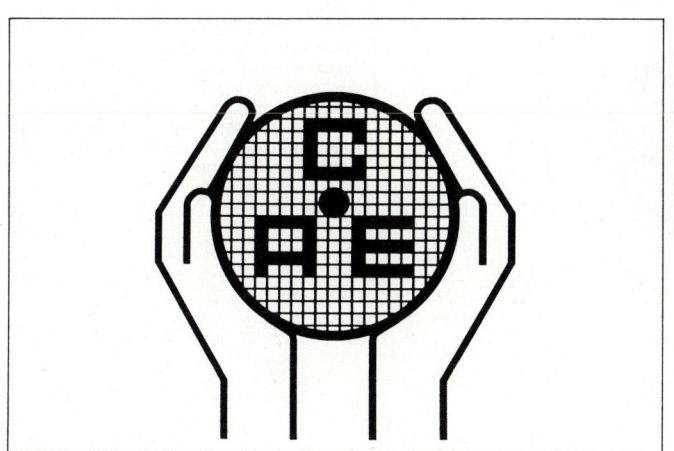

Certainly, a lot of time, money, and hard work have been spent over the past 10 years trying to create a computer aided engineering (CAE) tool. To date, however, a satisfactory, fully developed form of computer aided engineering does not exist. In fact, CAE exists only in an embryo state, leaving one to wonder whether in any realistic sense there will ever be any CAE system. In order to understand the question posed, we need to first examine the essential nature of CAE.

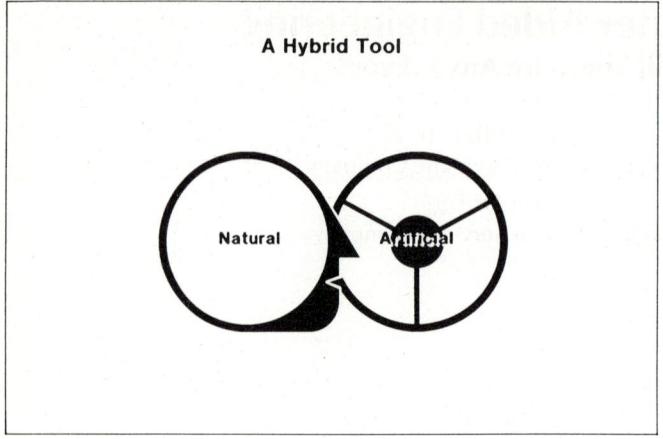

CAE is obviously a very high technology. It is a tool that can greatly increase our effectiveness. Of course, it is a hybrid tool, a combination of the natural and the artificial.

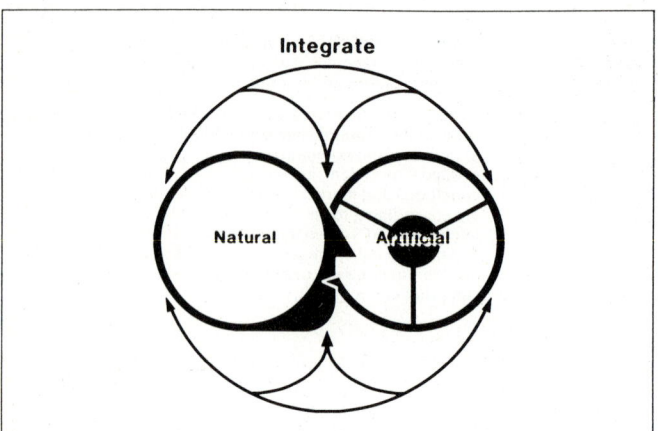

The natural aspect of CAE is due to the involvement of human beings. CAE is a particular subcategory of human problem solving (ref. 1). More precisely one could call it group or distributed problem solving. The artificial aspects are the tools the humans use, the computers and associated software. If we are to succeed in producing CAE we must successfully integrate the natural with the artificial. We need to recognize that even though there is no explicit description of the functioning of the natural elements (the humans) in their role as tools within the system, there is always an explicit description of the functioning of the artificial (computer tool) processes. But, in fact, we need explicit descriptions not only of the artificial processes, but of the natural, largely social or managerial processes as well.

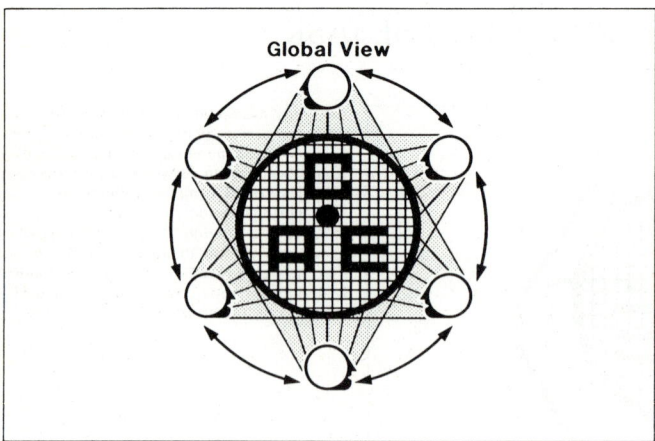

Because of the hybrid nature of engineering as a problem-solving activity, and more particularly of CAE, one of our most pressing tasks is to analyze our work situations on the widest and most inclusive basis. So far, the absence of such a global view has blocked all our attempts to design systems, to use existing tools in a group situation, or to reorganize groups of people into effective computer-based organizations. What is missing with CAE is not merely an overview, the "Big Picture" in a static sense, limited to large-scale elements.

What we need is a concept of the way the "Big Picture" and all the component details mesh together in action. This concept of the vital whole in action, including not only the technical but the social and management aspects as well, will provide us with a synergistic whole. Only when we have achieved such a synergistic whole will we have developed a problem-solving process which is greater than a simple sum of the component parts.

Daunting as this task may be, at least we don't have to invent many of these processes out of thin air since in fact we are solving engineering problems and using computers every day. So, naturally, we already have an intuitive grasp of CAE. What we need is a formal description of these processes if we are to build on or improve them.

APPROACHING THE WHOLE - SELECTING THE SIGNIFICANT

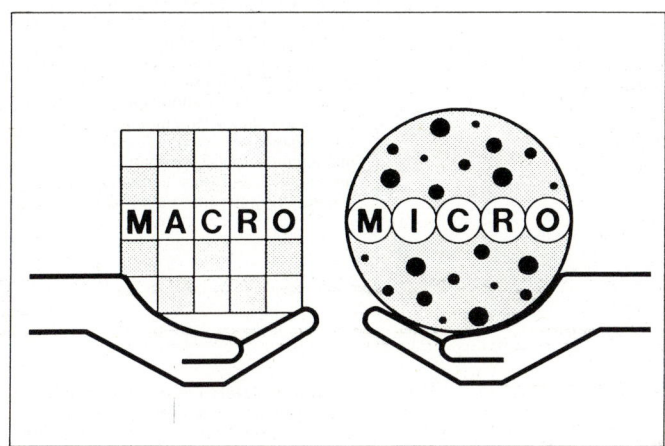

CAE, the whole, could be considered as being composed of macroaspects and microaspects.

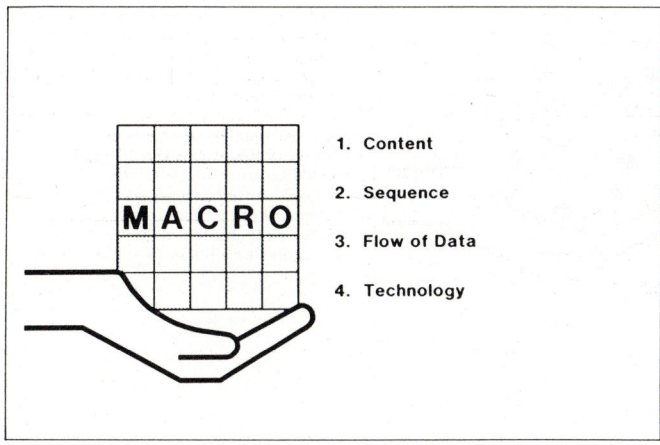

1. Content
2. Sequence
3. Flow of Data
4. Technology

The macroaspects are concerned with the human activities during problem solving. We could examine these macroaspects through an analysis of problem-solving case studies. Here we would 1) enumerate and describe the content of problem-solving activities, 2) describe the sequence of these activities, 3) describe the flow of data within and between these activities, and 4) identify the particular technology most appropriate to each activity.

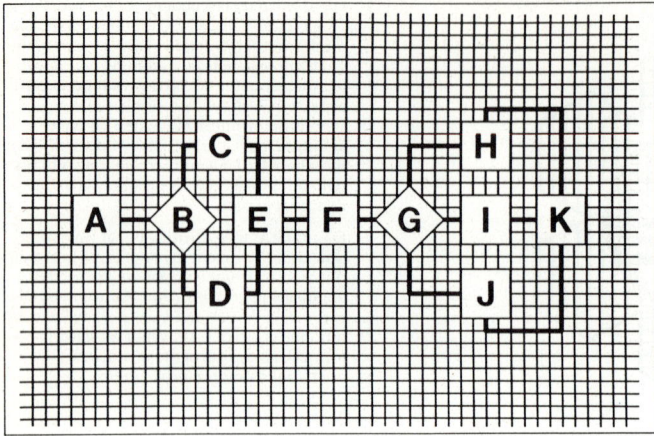

Past studies (ref. 2, 3) have shown that a flowchart-like methodology is effective for such descriptions.

Management Processes

1. Interaction
2. Strategies
3. Responsibility
4. Tools

Our examination of the macroaspects must be broader than merely technical. We could also apply an approach similar to that above to the management processes which are associated with the technical processes (ref. 4). Here we would 1) identify management interaction during problem solution, 2) identify optimal organization of CAE strategies, 3) establish work breakdown structures and allocation of responsibility to the organizational structure, and 4) match the management activities to the best management tools for resource allocation and monitoring, task assignment and scheduling, and status reporting.

- Bits and Pieces
- Rapid Change
- Different Problems, Organizations, and Applications

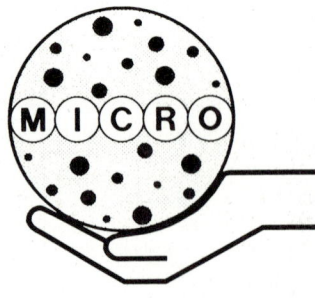

At the microlevel we need a change of approach. While our macrostudies would be concerned with the architecture and the whole of the process, for the micro we must be concerned with the small bits and pieces. The microaspect, the level where the humans and the tools meet, is undergoing rapid change. At the level of the microspect, the CAE process is something that will be applied in many different ways on many different problems to many different organizations. Thus we should focus not necessarily on what we observe today but on what we would like to find, the ideal form. A knowledge of what we want in the ideal will illuminate both the constraints to growth and the inevitable tension between the natural and the artificial elements of CAE.

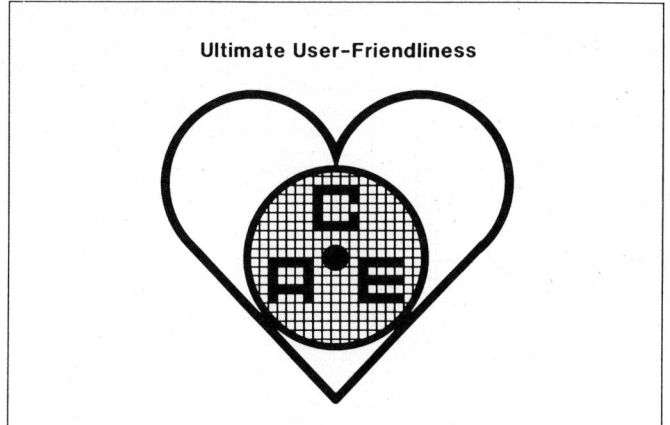

The Ideal CAE Environment

Ideally the CAE environment would be the ultimate in user-friendliness. It would be completely accessible to each user, require minimum on-the-spot programming, and respond to requests for information in an appropriately timely manner. Necessarily, in a project involving more than one user in the problem-solving process, all data and all users' discoveries should be available to all other users. The computer system should not only furnish this data with a minimum of effort on the part of the requestor, it should also organize the answers and correct mistakes in the style of request.

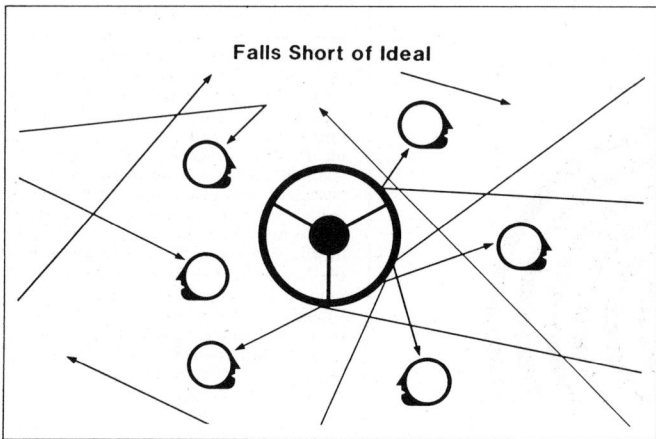

Obviously, today's computing environment falls far short of this ideal. But in fact, not all of the necessary areas of growth are technological; of the four problem areas I will discuss next, only two are problems with the technology; the other two comprise problems that have arisen naturally from the social organization of work.

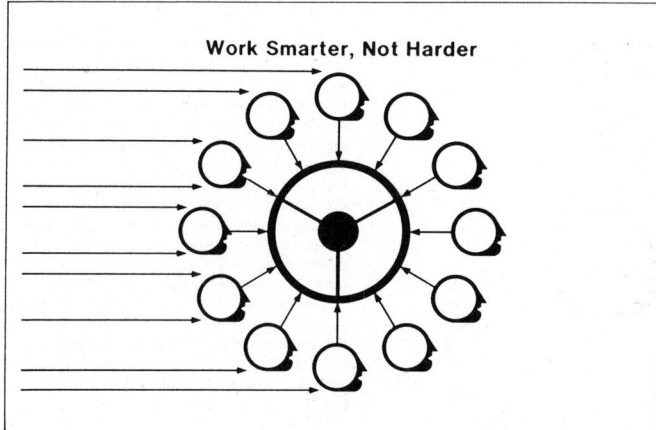

Planning for CAE

The first of the natural or human problems of today's computing environment is a legacy of the precomputer workplace: simply, poor organization. It is an age-old problem that the application of computers greatly magnifies. The change wrought by computers has been a change of degree: in the computer age, more than ever, we must "work smarter, not harder." The fast pace of the computer age illuminates the shortfalls of precomputer organization: planning becomes crucial.

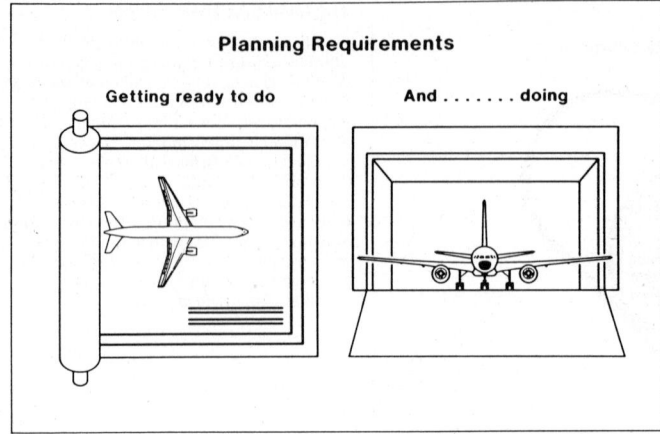

All human projects require the two steps, both the "doing" and the "getting ready to do." Engineering is the kind of endeavor that benefits from a tidy separation of the "doing" and the "getting ready," and planning embodies this separation. When we worked at a human pace and identified problems in the midst of our design process, we could get along all right by correcting those problems ad hoc.

But when we work at the pace of computers—as we do now—the computer with its great speed absorbs most of the incremental, formerly human steps. A single error can require multiple corrections: we must redo the design, rectifying the original mistake, as well as adjusting later steps in the design sequence. But, of course, the solution is not to slow down the speed of the computer—there would then be no benefit in CAE. The fact is there is no practical limit to how fast we should want our tools to go. Ideally, the only constraining factor in speed of computing should be the human's limitations. Computing power and capacity should not even be noticed by the user. To put it simply, planning has always been advisable and good practice, and in the computer age it has become an indispensable, constant requirement.

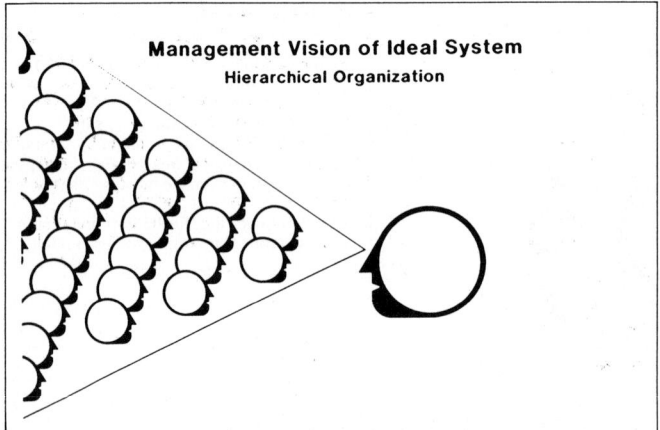

Access to Data

Access to data is the second problem area of the microaspect where again the human or social factor is the impediment to growth. Because the roles of managers and executives on the one hand, and engineers and technical people on the other, are different, their current views concerning an ideal CAE are different. Roughly speaking, while engineers may be concerned only with the technology of a project at hand, managers want to know what—and how much—each engineer is doing; they want to control—or at least monitor—the exchange of information between users. As it happens, the managers' current attitude is also at odds with the imperative need for integration at every level of CAE—within the computer system, within groups of problem solvers, even within the individual.

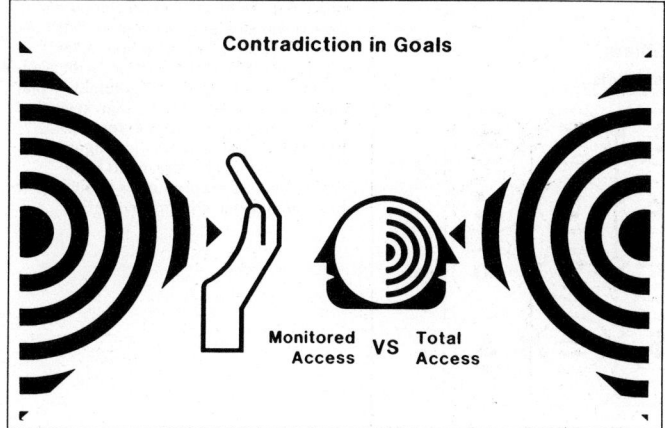

Glossing over this contradiction between the goals of managers and the needs of engineers will not make the problem go away, but in time the managers' viewpoint will certainly yield—yield to irresistible demands for increased employee effectiveness, product quality, and overall productivity in problem solving for the organization.

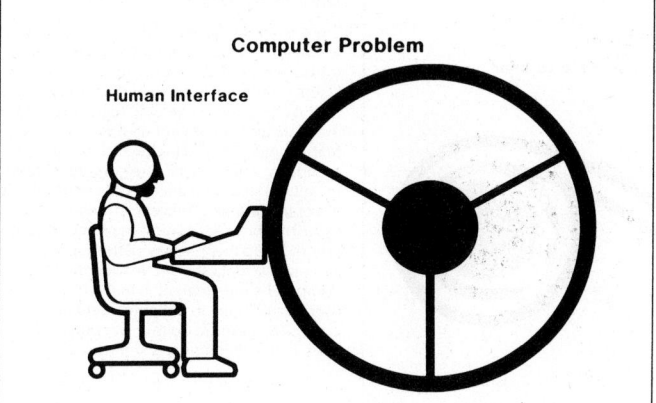

User Interface

In addition to these natural human factors of the workplace, two technical problem areas, the user interface and data management, will require improvements which constitute adaptation to human traits and needs. The computer's interface to the human is currently both too complex and too unnatural. It lacks interfacing features that are complementary to the human body. Even though we have keyboards and sketchbooks for the fingers and light boards for the eyes, computers still lack features natural to the human—the most important of all, of course, being language.

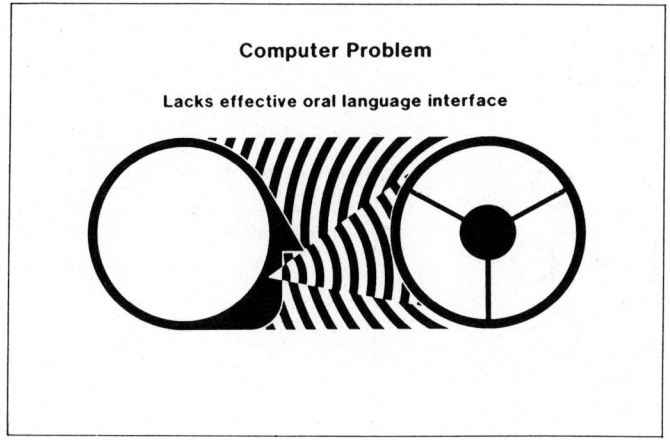

We need an effective language interface that will allow the human to use his or her native language. Ideally, this would be done orally. Even though researchers into artificial intelligence have applied themselves for a number of years to this problem, they are still struggling for effective solutions.

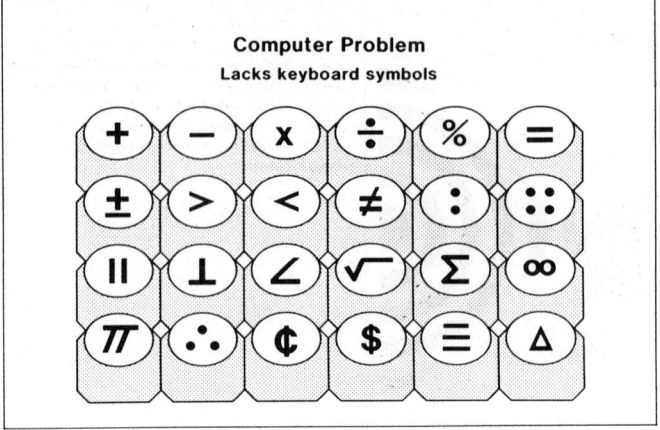

Meantime, we need to equip computers with a keyboard language interface that uses symbols. LISA™ (Apple Computer, Inc.), enables users to select functions with symbols such as "word processor," "graphics," "plan," and so forth—much simpler than keying in 14 characters. More symbols for CAE applications and mathematical signs for "divide," "add," et cetera, need to be incorporated. The computer's equivalent to the scratch pad needs improving.

Data Management

The second of these problem technical areas is data management, including the transparency of the data within the computer. Users should not have to know where the data is; they should be able to simply ask for that data and get it. Similarly, data management should be the task of the computer, not the user. The computer should be able to store and access data without the aid of the human. The user should not have to worry about versioning or configuration control. And the integrity of the separate packages of data should be the job of the computer's data manager. In every case, the user should have to know less and the computer more. To do this, of course, we'll have to make computers smarter.

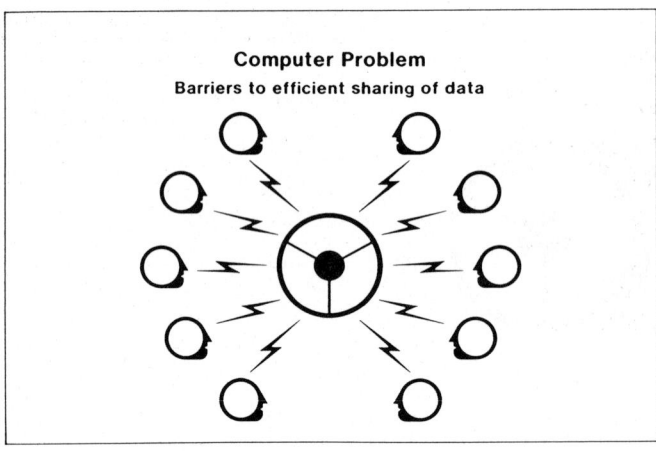

We should remember that the whole advantage in computers consists of increasing access to information. As we have seen, some of the barriers are technical and can be found within the machines; other barriers are social and exist between users, outside the machine. Because we want to promote sharing of data between the user and the machine, between the different tasks any one user is working on, and between individuals and groups of users, we need to eliminate both kinds of barriers, social and technical.

IPAD - THE IDEAL

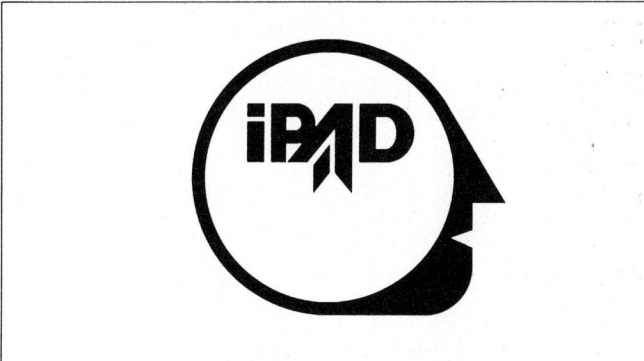

Integrated Programs for Aerospace-Vehicle Design

Let's turn now from what have been fairly general considerations to a concrete example: IPAD. IPAD's full name is Integrated Programs for Aerospace-Vehicle Design; it is a Boeing contract, sponsored by the National Aeronautics and Space Administration (NASA). IPAD is two things: a plan for an ideal future CAE system and an existing, operational prototype of some key elements of a CAE system. We will first discuss the origin of IPAD, then visualize it as a complete, ideal system, and later describe what the current IPAD system can do.

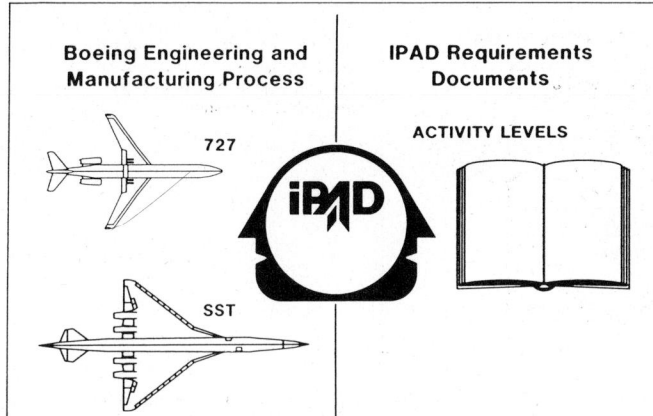

IPAD, although not a system designed to answer any specific engineering need, does have a specific, practical origin. The IPAD Requirements Documents (ref. 5, 6) were evolved from a long, in-depth study of the interaction of design and manufacturing in an actual engineering process: the design of two Boeing commercial transport airplanes, the supersonic transport and the 727 subsonic. This process—beginning with the initial research level, through preliminary design, and into production—was broken down into activity levels that defined the process. Each of these activities was further broken down to identify, in detail, the activities and flow of the design process for the two airplanes. IPAD grew out of the needs of a specific environment but is being developed toward a flexible form which integrates the many diverse applications and processes of engineering.

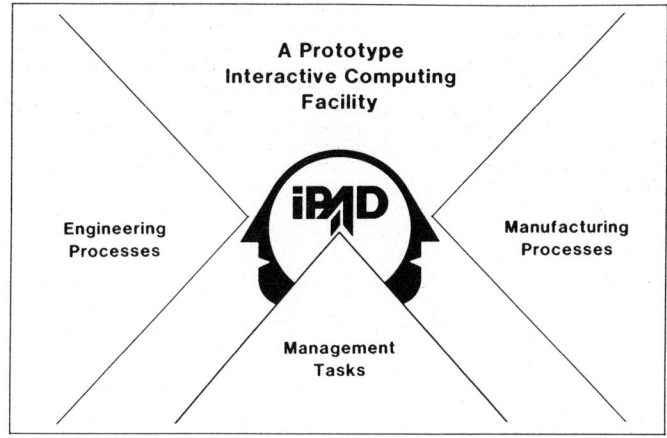

IPAD is conceived as a prototype, interactive, computer aided engineering and design facility, a facility to support engineering and manufacturing processes and associated management tasks. Begun in 1976, IPAD is an ongoing project to identify and develop prototypes of efficient future computing environments. The IPAD software system is designed for distributed computing and will reside on hardware from a variety of vendors, with a mix of mainframe and mini- and microcomputers (ref. 7, 8).

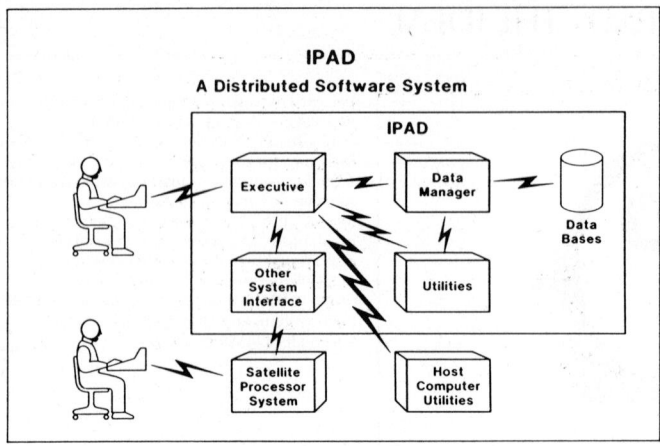

Its three major software components will be its executive, called IPEX, its data manager, called IPIP, and its interfaces to other computer systems. In addition, it will of course have its user interface and necessary system functions.

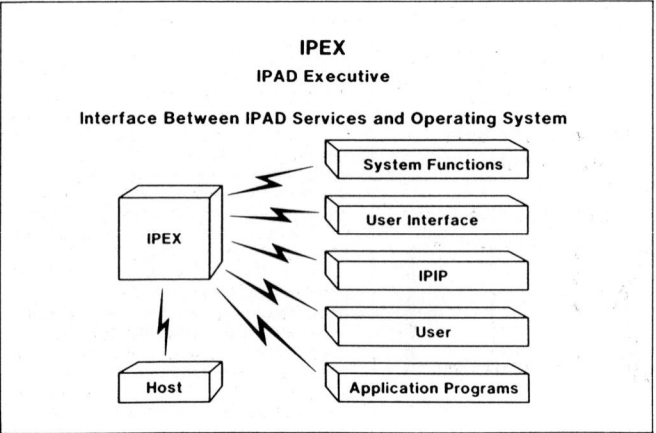

The executive, IPEX, will be the interface between all IPAD services and the operating system. IPEX will provide both the communications interface across the heterogeneous machines and a uniform user interface. It will be, in essence, a traffic director. The most critical function of IPEX will be to support the distribution of data and software functions to the computers and their operating systems.

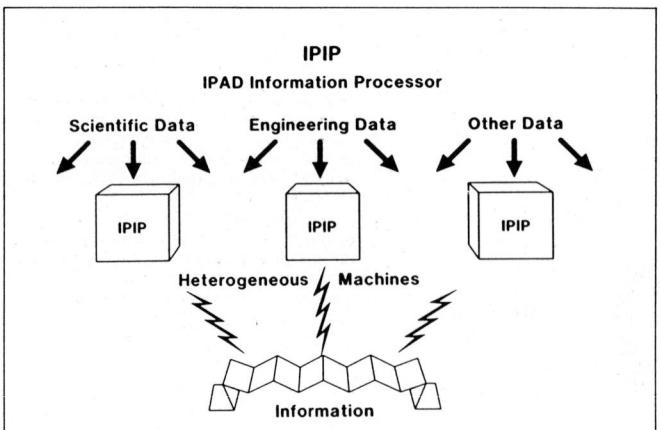

But IPIP, the IPAD Information Processor (ref. 9, 10), will be the core of the IPAD system. IPIP will manage data distributed across the heterogeneous machines. The underlying requirements of engineering and manufacturing have made unique demands on IPIP. It is designed particularly for science and engineering and their many data models, types of data, and types of data organization. Especially important are configuration control and versioning of data, and management of geometric information.

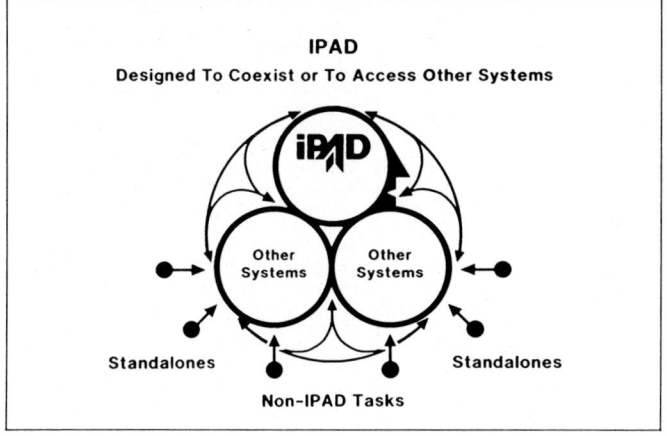

IPAD is designed to exist simultaneously on multiple computers and to access still others. This will allow our users access to the many effective standalone systems that have been developed over the years in the pursuit of automated engineering and manufacturing. These will enter the system as non-IPAD tasks.

All together, the large integrated software system that is IPAD in the ideal will consist of the IPEX system as executive, IPIP as our information processor, and the access to the standalone systems.

THE IPAD REALITY

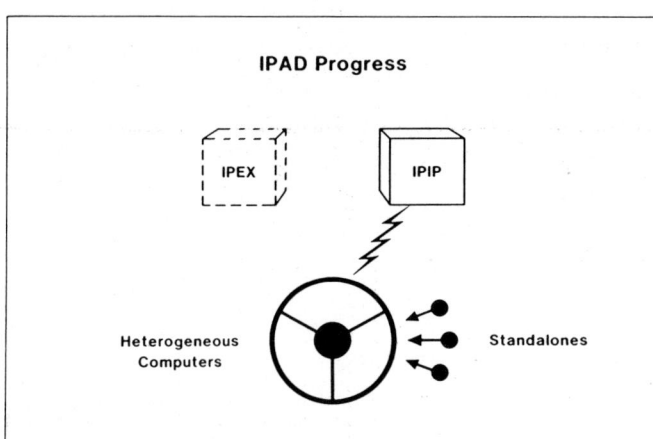

IPAD does not yet include all the previously mentioned features and capabilities. But perhaps the foregoing discussion of what is needed and of the IPAD plans will provide a perspective for looking at progress on IPAD to date. To summarize this progress very briefly: a fully developed IPEX does not exist, IPIP does exist, and communication access to heterogeneous computers and standalones is partial but growing.

Work on an IPAD executive has only been underway since last October with the start of fiscal year 1984. But although the executive, the powerful IPEX traffic director and communications interface, is still missing, significant progress has been made toward integrating different vendor computers in the area of information processing.

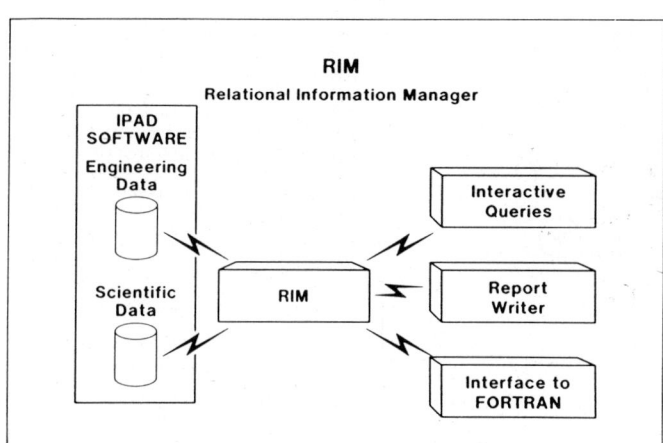

Considerable progress has been made in engineering data management. In fact, RIM (Relational Information Manager, ref. 11), a forerunner of IPIP, has emerged as a commercial product. RIM is a relational data manager that supports relational models of engineering and scientific data, such as floating point representations of vectors and matrices. In addition, RIM has a report writer, an interface to FORTRAN, and a very powerful processor for interactive queries. However, although a RIM data base may be accessed in read mode by multiple users, access for update is restricted to a single user.

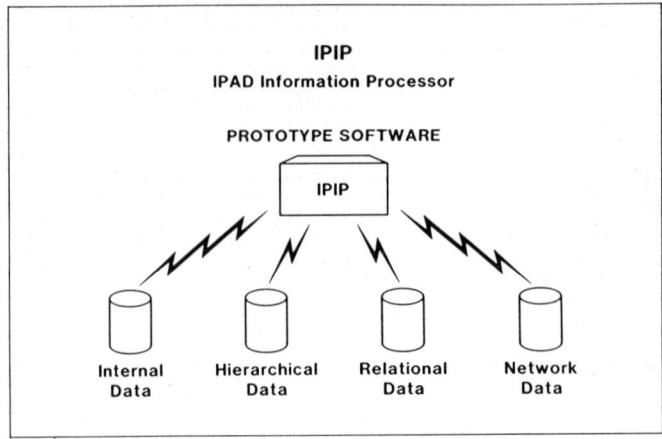

IPIP, on the other hand, is a much more sophisticated prototype. It is a multiuser, multischema, multimodel system: the user can represent hierarchical, relational, and network data models. IPIP manages an internal data inventory and supports a variety of engineering data. And IPIP is unique in its embedded ability to manage geometric information (ref. 12-14)—a significant feature, since geometric data constitute the central thread throughout the entire engineering and manufacturing process. To be successful, a data manager must be complex enough to handle not only ordinary text and numerical data but the data characteristics unique to geometry as well. Version 5.0 of IPIP is available from the IPAD Project as prototype software for U.S. industry, and IPIP was made available in April 1984 as a product by Control Data Corporation.

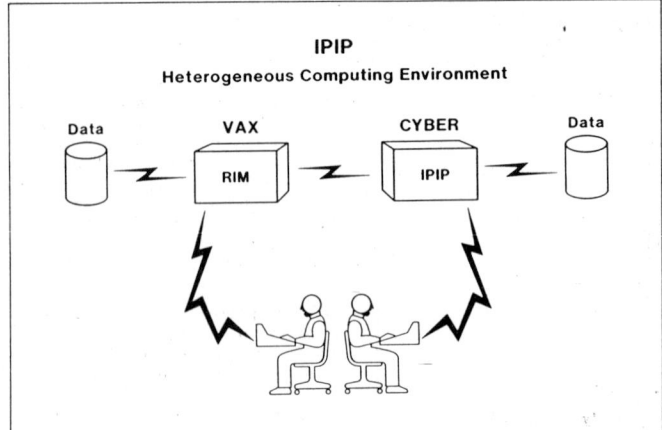

To date, IPIP, as installed on the CYBER, can provide data management for tasks being processed on the CYBER. Through the high speed bus network, it can also provide data management for tasks being processed on the VAX-780. Thus IPAD is now able to communicate tasks in real time between processes executing on heterogeneous vendor computers (CYBER and VAX). All data translations, including floating point, are available on IPAD, which, as a network software system, was built to adhere to the International Standards Organization model for categories and subcategories of data.

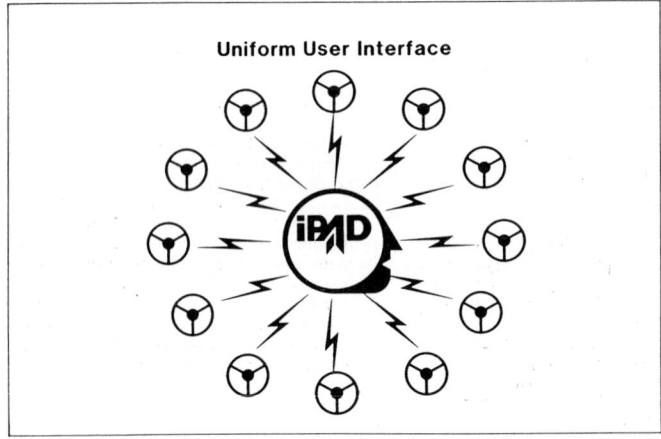

IPAD has developed substantially but still has a long way to go. Today it includes a limited ability to manage data for tasks being processed on heterogeneous machines, but this capability needs to be expanded. Moreover, an executive that can handle traffic between many kinds of machines is needed. And very importantly, a uniform user interface—uniform and more natural for humans—is still lacking. The lack of a more natural user interface steals precious time from engineers, who could be engaging more directly in productive work, rather than struggling with computers.

CAE AS COMMUNICATION - BETWEEN PEOPLE

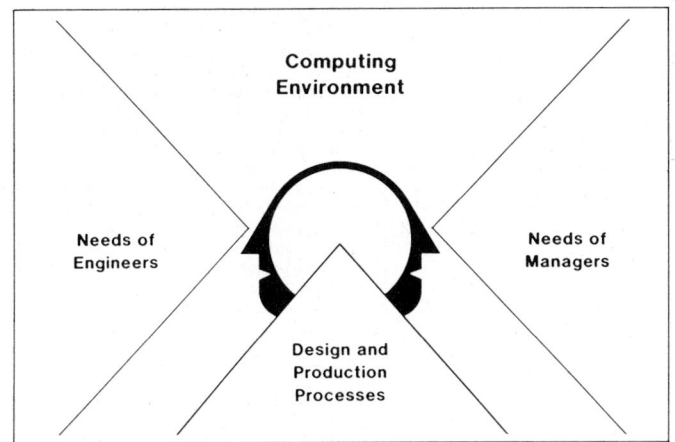

Let us look further now at the current limitations of the computing environment, at the needs of the potential human users of CAE, and at the way those needs can be met. I mentioned earlier the conflict between the needs of engineers and managers; let's look more closely at those needs, with a glance first at the design and production processes that bring engineers and managers together.

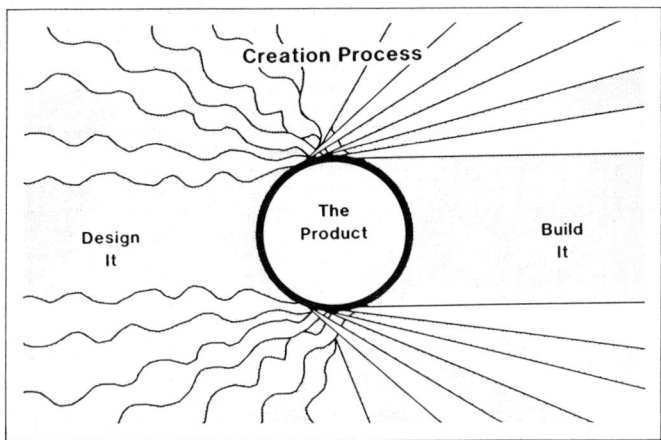

The creation of products is composed of two processes: a "design it" process and a "build it" process.

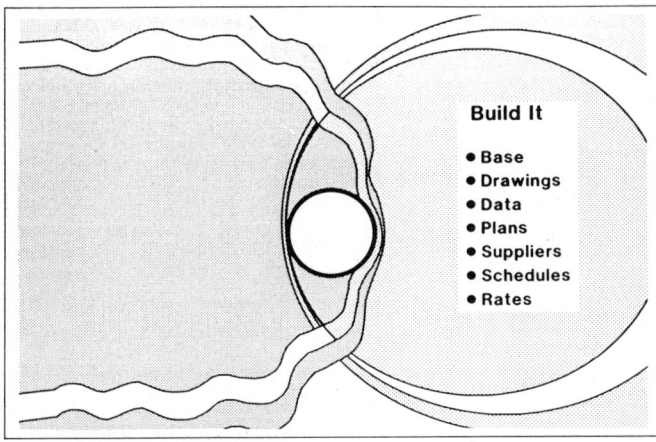

The build-it process is quite well defined: in the build-it process there is a known customer base, drawings or data sets exist, tool and production plans exist, and suppliers and schedule and production rates are committed. Firmness and structure characterize the build-it process.

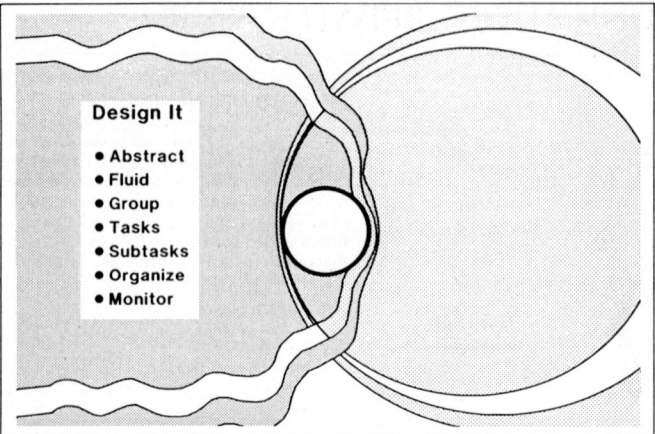

In contrast, the design-it process, though it overlaps the build-it process to some degree, is abstract and dynamically fluid, as are the data which represent the product up to the point of engineering release to manufacturing. The design-it process is essentially a group problem-solving process composed of many individual tasks which are aggregated into group tasks and focused by management into a design. All these tasks combine in a linked, iterative manner to produce the product design. This combining of tasks is often defined by a work breakdown structure which identifies subtasks performed by a single employee and tasks performed by groups of employees, a grouping of subtask and task which permits management to organize and monitor the progress of the project.

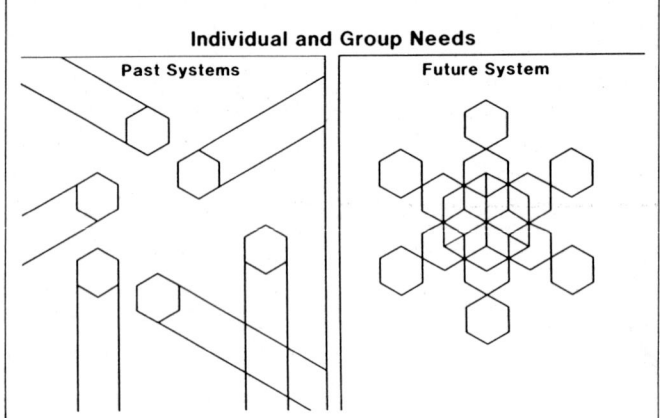

These group problem-solving processes define different needs. In the past, researchers in the computer field have focused on the needs of individuals. Thus, we have arranged it so that an individual user can create subtasks for computer execution and create and modify the contents of a private data base over many sessions. We have also made it possible to safeguard the privacy of these data until the user-owner releases them for general use. In the future, however, we will need to provide a system that maintains and manages the many private data bases and recognizes that they are part of an overall information bank. In order to design a CAE system, we need to focus on the group as well as on the individual.

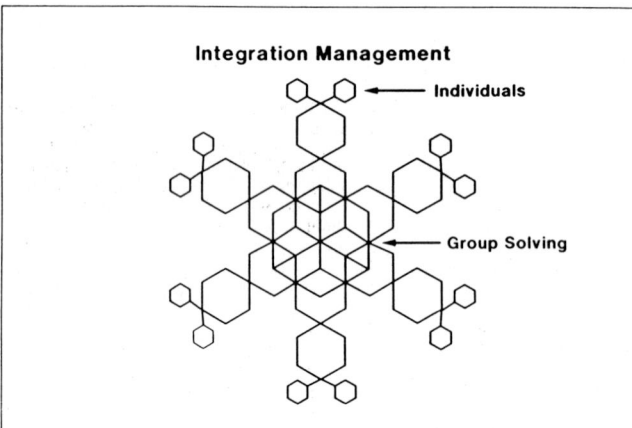

At the group level, subtasks are combined into tasks, the aggregate of which forms the total group problem-solving effort. Those elements of a system that support integration also support group problem solving, and the CAE system needs to help manage this integration of effort. Such a system should provide the means to associate subtasks, tasks, resource budgets and actuals, and to schedule plans and to report actuals; and management needs to be able to coordinate exceptions to the work plans at the various subtask and task levels through the data base system.

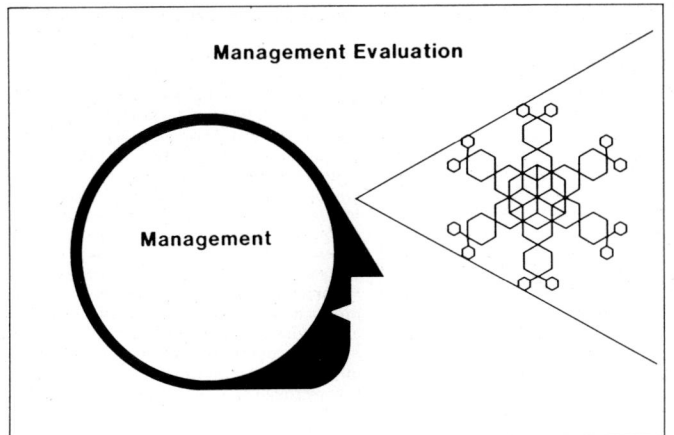

Through these associations, management will be able to evaluate and report on progress during the design process. As work progresses, the data base system needs to be able to advise the workforce of the availability and currency of data. The computer system needs to maintain an awareness of the status of the work throughout the group problem-solving process.

CAE AS COMMUNICATION - BETWEEN SYSTEMS AND TASKS

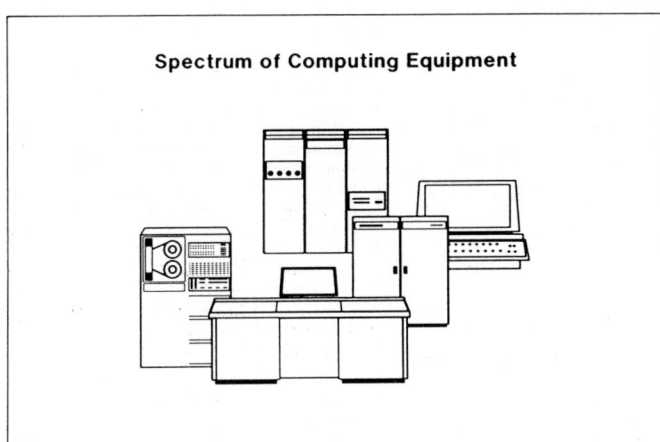

Note that, in discussing how to cope with the needs of engineers and managers in the build-it and design-it processes, for simplicity's sake we ignored the fact that all the engineers do not use the same kind of computing equipment. Even though some organizations have attempted to rely on a single computer vendor, all but the smallest of organizations already utilize multiple computer vendors and have an installed base of various computers that match specific computing capabilities to specific tasks.

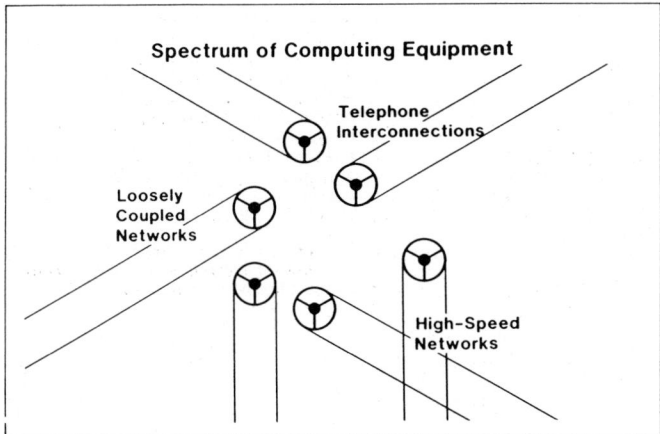

This spectrum of computing capability is interconnected variously through communication facilities such as standard telephone interconnections, loosely coupled networks such as Ethernet™ (Xerox, Inc.), and high-speed networks such as Hyperchannel™ (Network Systems, Inc.). Magnifying this variety and complexity is a layer of software heterogeneity superimposed on top of the heterogeneity of computers. This complexity is an even greater problem in the use of data management systems than it is in the technical applications. Unless we can impose some kind of unity on this random evolution of heterogeneous systems, the natural consequences will be confusion and unreliability.

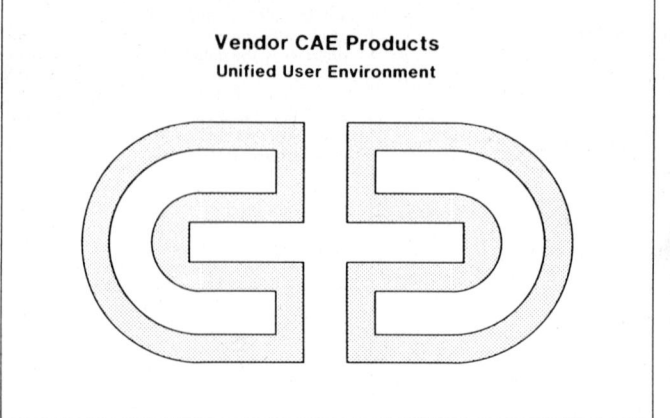

There have been attempts to provide a unified user environment; some vendors are beginning to address the problems of integration. The Control Data Corporation system ICEM™ (Integrated Computer-Aided Engineering and Manufacturing) is typical of the integration of a particular set of applications with a data base and operating system capability. ICEM integrates engineering and manufacturing functions including geometric modeling, design, drafting, analysis, schematics, and numerical control, providing common access to the same version of all data and drawings from the inception of design work to the production process. But the ICEM capabilities are, naturally, CDC CYBER-based, and ICEM's communication capability does not allow for integration of geographically dispersed users through multiple, dispersed CYBERs.

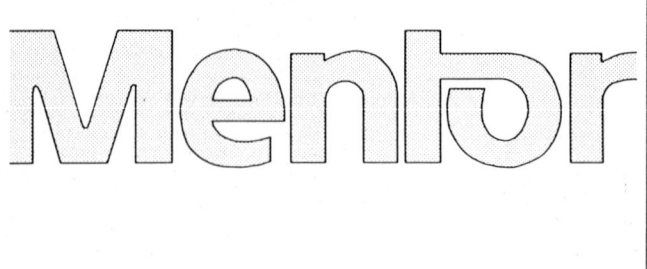

The IDEA™ system from Mentor Graphics integrates design and simulation for digital logic component design. IDEA does offer, through the Apollo Ring system, a distributed problem-solving environment. Two important features are distributed in the Mentor Graphics IDEA system: data at the file level and processing power. System growth thus becomes more flexible because overall computing power is augmented as each workstation is added to the system. This feature prevents degradation of performance due to computational overload, a problem which can seriously limit the growth of systems dependent on a single CPU. An additional strength of IDEA is the integration into the system of management communication and documentation tools with complete access to the design files. However, the data file is the only distributed data element; the data base at the record level is not distributed.

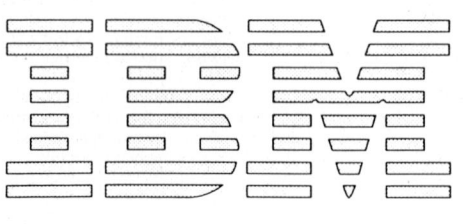

IBM's introduction of CADAM™ (a development by Lockheed) made an early contribution to CAD. In addition, IBM's new graphics device, the 5080, and a relational extension to their IMS™ data base management system have recently been introduced. The 5080 provides full 3D and graphics transformations internal to the device. This will provide geometry creation at the user terminal. The new DB2™ relational data management capability, because it is coupled to IMS, permits both relational and hierarchical data models and also utilizes the extensive production capabilities, such as backup and recovery, which have long been present in IMS.

Even though these developments represent steps toward integrating the homogeneous computing environment, none address the problem of heterogeneity.

PRESSURES ON CAE - THE HUMAN FORCES

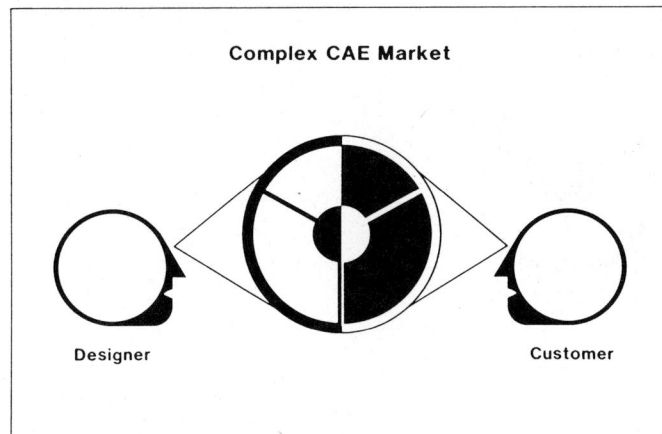

One important reason products have not progressed beyond the CDC, Mentor Graphics, and IBM examples is that the market for more advanced products, presuming it does actually exist, has been very difficult to find. Perhaps we can account for this difficulty if we examine some of the factors that affect design and, more particularly, marketing. The CAE product required is necessarily very complex. As discussed earlier, complexity naturally arises from the unavoidable requirements of group problem solving. But it is when we approach the marketing area that we are forced to acknowledge certain complications produced by cultural norms and assumptions: computer systems do not exist in a value-free context any more than any other objects for sale. One result is that what would appear logical to the designer of computer systems may not seem desirable or logical to the customer.

Integration by Customer

Two approaches to marketing are possible: separate sale of components or sale of integrated systems. Since most goods on the mass market are sold as separate components, most consumers—even engineers and managers—will assume that computer systems are also sold as separate components. This approach has been followed successfully for those consumer goods that require only a very loosely coupled interaction between individuals for a high cost-benefit ratio, such as cameras, automobiles, televisions, computer hardware and so forth.

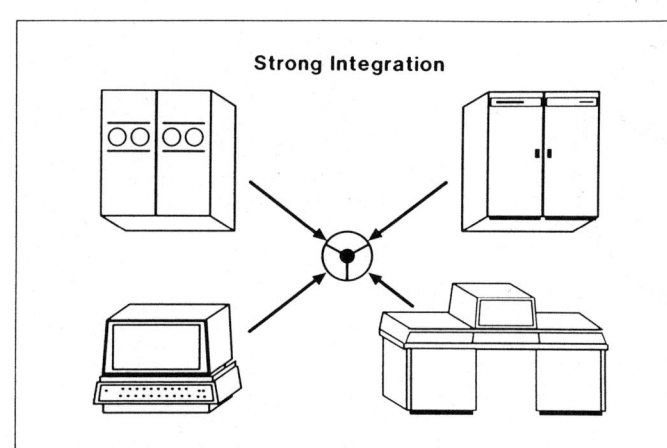

However, this approach has not proved as successful for CAD systems, and software systems in general. Their cost-benefit ratios have suffered from a lack of rationalization. This is partly because computer aided engineering systems generally require a high degree of cooperation and integration of activities and data among large numbers of people. Also it is naturally very difficult to design software components for a system in which the integration of those components is left to the vagaries of the marketplace. If we continue to avoid a confrontation with cultural expectations by letting the consumer be responsible for system integration, we are merely begging the question.

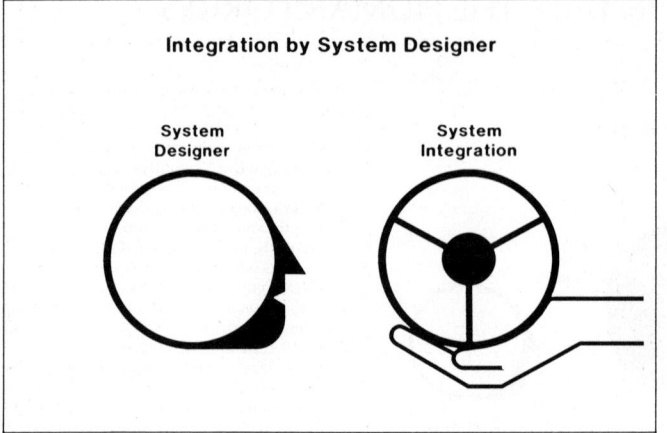

Integration by System Designer

Until vendors face these issues, the customers themselves will continue to have to perform ad hoc system design and marketing functions inside their companies. If we are to make progress, we must finally give the responsibility for system integration back to the designer.

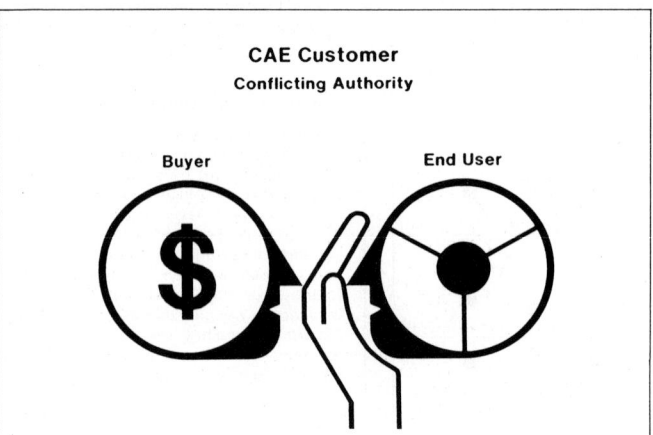

Unfortunately, it is going to be difficult for vendors to address the marketing traditions that encourage separate components over integrated systems and that discourage compatibility between heterogeneous equipment, particularly because it is often unclear just exactly who the customer is for these new products. When that customer is an individual, then of course he alone passes judgment on the cost-benefit aspects. However, if a group is the end user, then usually some computing organization has the buy authority. Ideally, to facilitate marketing, the buyer and the end user should be one and the same, but purchase authority rarely is structured this way in today's engineering organization.

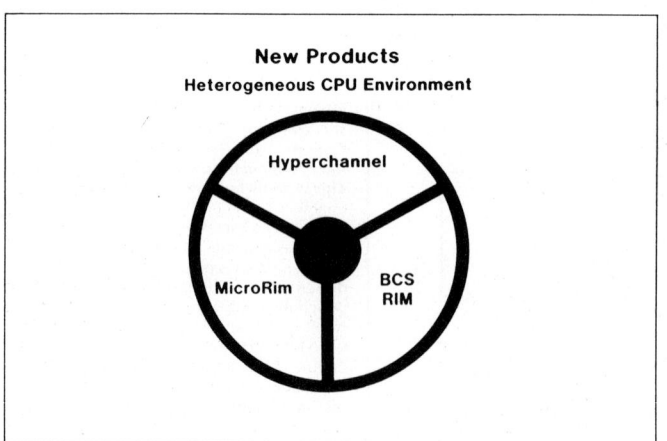

Not only is the sale of integrated systems counter to the traditions of the market, but a heterogeneous environment, composed of products from multiple vendors, is counter to the instincts of the historical computer vendor. Nevertheless, some progress has been made. Some of the new products that function in a heterogeneous CPU environment are Hyperchannel™ by Network Systems, Inc., and RIM, offered on a variety of microprocessors by MicroRim, Inc. Also Boeing Computer Services Company offers BCS RIM for a spectrum of small to large computers.

PRESSURES ON CAE - THE ECONOMIC FORCES

In addition to the problem posed by the tradition of marketing nonintegrated systems, there is the central question, of course, as to whether or not CAE (a large integrated system) is too expensive for the marketplace. The vendors and their customers together will eventually resolve this question.

Vendor and Customer Motivations

Software vendors, especially the large ones, have an established product line which is currently producing a profit even though such products do not constitute a reasonable solution for CAE. The deficiencies observed in the current products tend to constrain the pace of new product development since new development must be integrated with existing products. This need to protect a product base, protect both the vendors' and the customers' capital investment, has generally discouraged the larger software vendors from developing an overall, large-scale CAE solution.

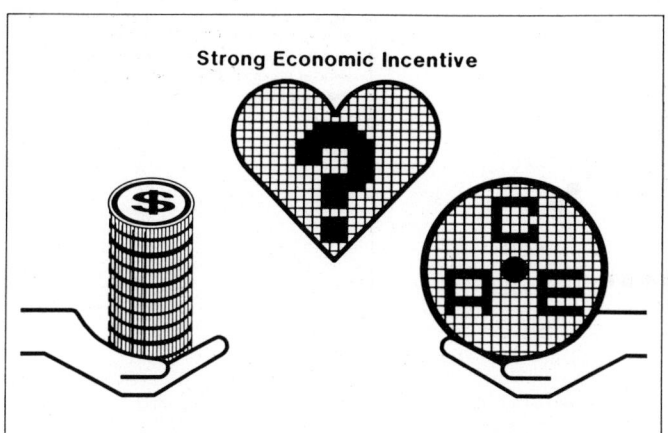

On the other hand, smaller vendors— usually start-up firms—often produce quite innovative products. Unfortunately, with their limited funds they tend to produce products which satisfy a narrow subset of the needed CAE system.

Both vendors and customers have investments in products and equipment to protect; however urgently customers need the capability that an integrated CAE system could give them, they will need strong economic incentives to buy new systems that make their current investments obsolete.

Thus, the marketplace has both customer and vendor interests, which combine on economic grounds to discourage the introduction of large-scale, fully integrated systems.

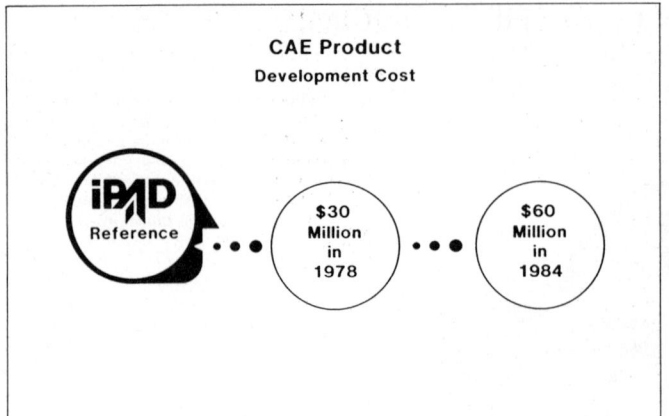

Total System Development Approach

In order to evaluate the total system development approach, let's estimate the development costs so as to provide some feel for the scale of the problem.

Work on the IPAD project in the preliminary design phase provides an experience base upon which to estimate the overall cost and risk of producing a CAE product. In 1978, a full scale IPAD product for CAE purposes was estimated at approximately $30 million. If one accounts for the effects of inflation and development risk on a project of this size, a current 1984 cost would be in the order of $60 million.

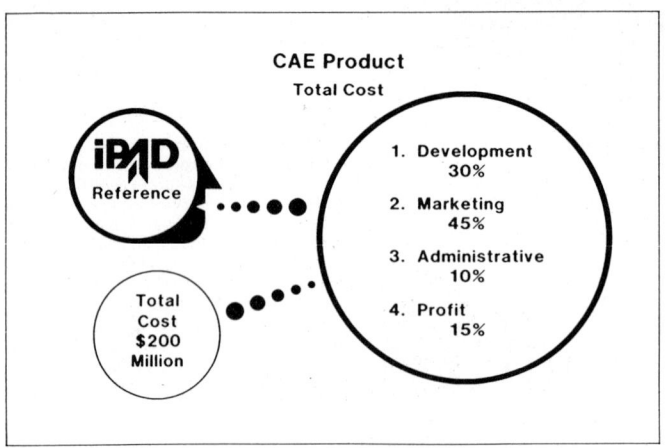

When undertaking a project of this size, a product developer would assume total development investment, over the entire life cycle of a product, to be: 1) development cost 30%, 2) marketing cost 45%, and 3) administrative cost 10%, and would expect a profit of 15%. Based on the estimates from the IPAD development, the total development investment would be approximately $200 million.

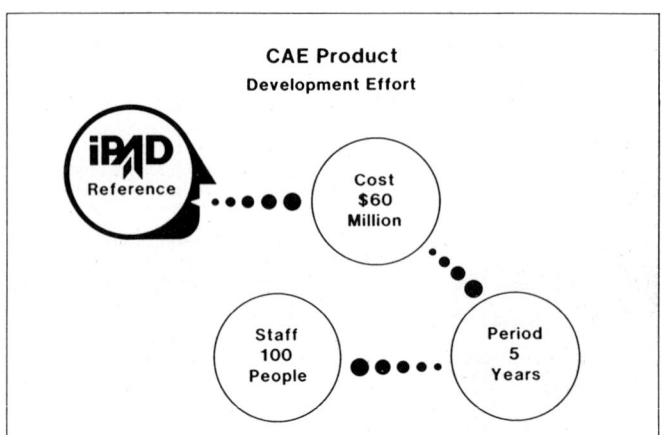

With this development cost of approximately $60 million and a product development period of five years, we would require a development staff of approximately 100 people for the technical development. Such a development time and staff size would be reasonable, recognizing that incremental products could be presented to the marketplace.

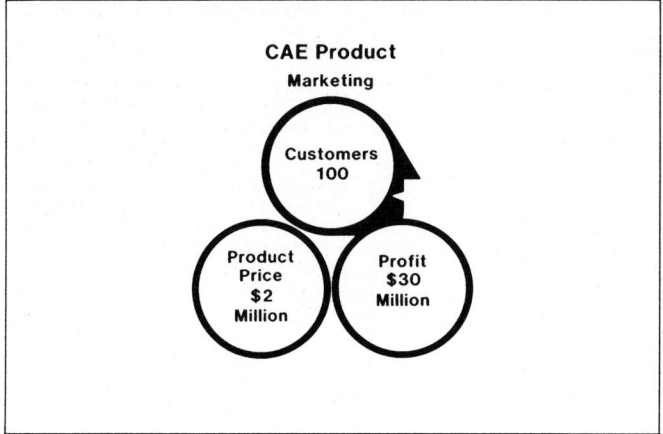

Let us assume that within industry and government organizations there are approximately 100 customers for a system of this size and complexity. This would require a product price of $2 million and would return a profit of $30 million. It is very doubtful, based on current software system prices, that there is a market for a $2 million product.

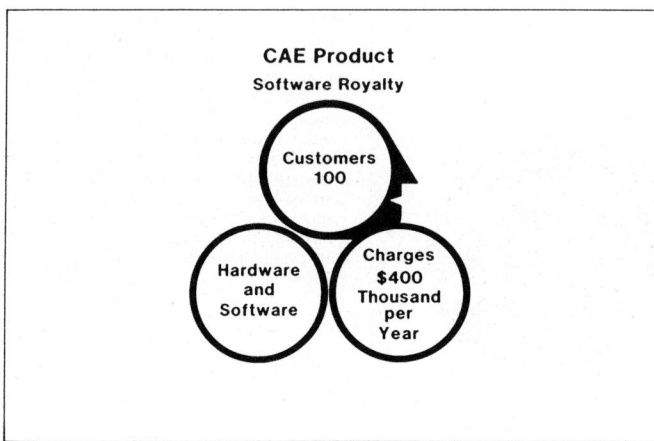

More typically, rather than being sold alone, such software systems are bundled with the hardware offering, and development costs are recovered on the basis of software royalty. If we estimate the same customer base and a product life of five years, then the royalty and maintenance charges for system-level software packages, such as an operating system, data management system, network software, and associated utilities would need to be approximately $400,000 per year. Software royalties of this magnitude are approximately three times current revenues for comparably sophisticated software.

Thus we are led to the conclusion that our new CAE system must have a benefit value to future customers which is three times greater than present systems, which is not unreasonable. Or alternately, we have to triple our customer base to profitably offer such a product.

Modular Development Alternative

An obvious alternative is to produce the CAE system in a modular fashion. Such an approach could take advantage of emerging new technology in hardware and supporting system software as well as reduce the risks associated with marketing a product as complex as a total CAE system.

At the end of 1983, there were about 100,000 major computers in government and industry, not counting minicomputers. Assuming 20 percent of this market are candidates for various CAE modules, then vendors could sell individual modules for approximately $25,000 to $30,000.

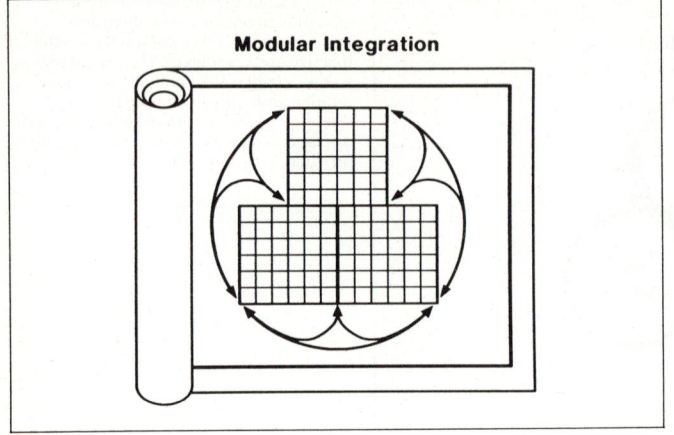

However, a serious obstacle to producing modular CAE lies in the requirement to integrate the emerging modular products into an everchanging, workable whole. Our earlier analysis showed that in the current environment this integration is left to the user. Such an integration, however, would be feasible if a system engineering task had preceded development, thus providing the blueprint for the entire system, the pieces and the interfaces. As discussed earlier, commercial vendors have not produced and are not likely to produce such an entire system blueprint.

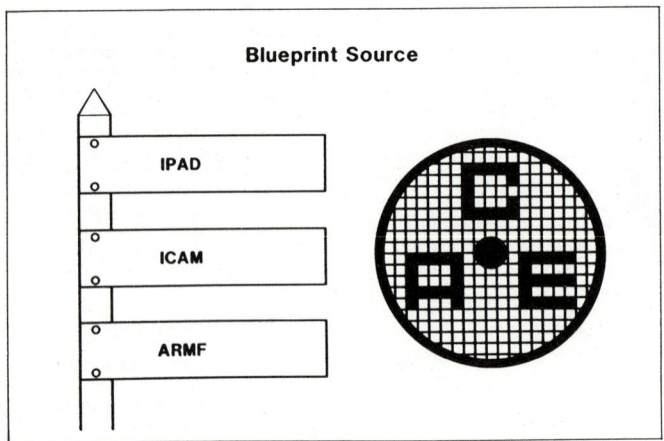

The blueprint is obviously the key to the modular approach. National projects in the United States such as IPAD, ICAM, and ARMF (ref. 15), in the aggregate do point the way towards a total CAE solution. These programs have been unbiased relative to vendor hardware, are oriented toward the required heterogeneous solution and have shown proof of creative thought and useful prototype development. Unfortunately, a national mechanism has not emerged to coordinate these diverse, large-scale projects into a viable whole to provide a CAE blueprint.

CONCLUSIONS

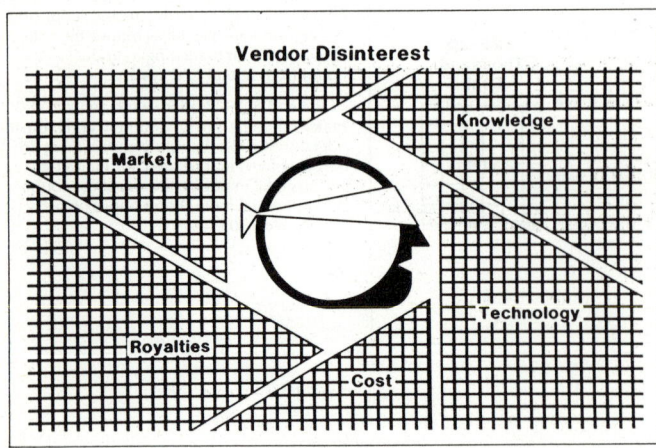

Even though 1) the state of knowledge is adequate to describe the desired CAE system, 2) the IPAD prototype software developments have established a feasible technology for engineering data management and heterogeneous computer communications, 3) the probable software development costs of $60 million ($200 million investment) are reasonable, and 4) an adequate market with reasonable software royalties could be expected, the likelihood that a single vendor would venture such a broad and innovative product is very small, due to the technical complexity of the product and the unresponsiveness of the market.

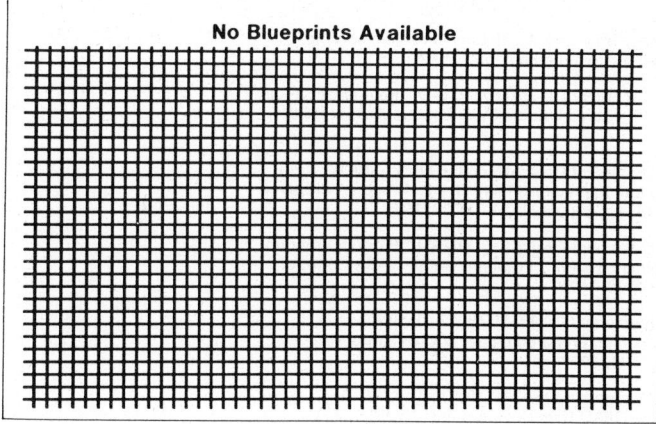

Total system development, through a modular approach by multiple vendors, would require an overall system blueprint to guide the heterogeneous development of modules and to permit their integrated functioning. In spite of the aggressive national programs, IPAD, ICAM, and ARMF, it is unlikely that such a national blueprint can be developed in the near future.

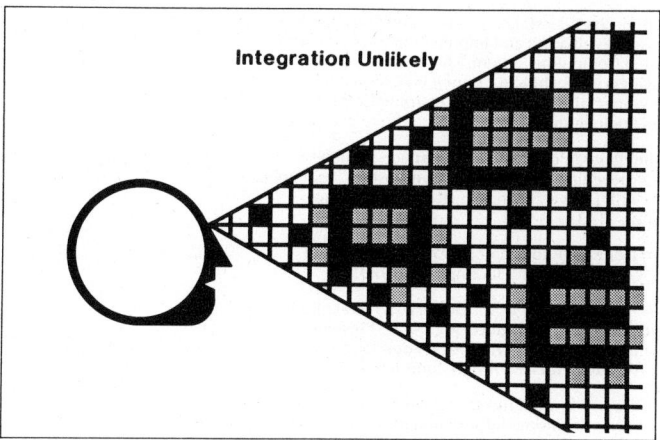

One is led to conclude from the above discussion that, even though very desirable from the end-user point of view, a total CAE integrated system, especially in a heterogeneous environment, is very unlikely in the near term in the United States.

As a consequence, we shall no doubt continue to find in the marketplace a proliferation of modules, all ill-designed to work together as a whole, with the integration of these modules left to the buyer.

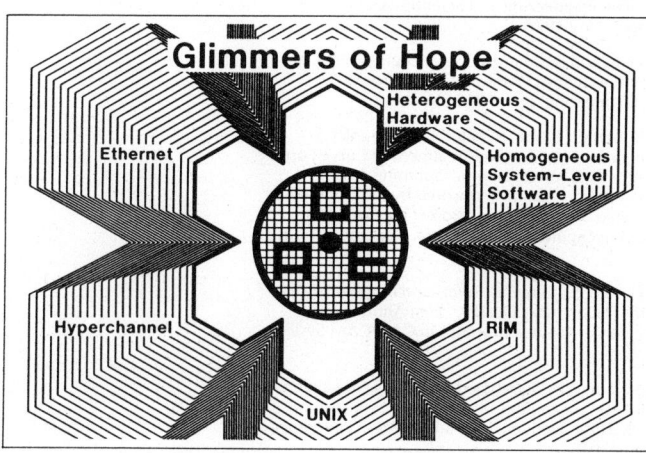

Alternatively, there are some glimmers of hope that a total CAE system could emerge built around heterogeneous hardware but utilizing homogeneous system-level software. RIM, as discussed earlier, provides an engineering data management capability across a large spectrum of heterogeneous hardware from the personal computer level to the super computer level. UNIX™ (AT&T), as an operating system, is also beginning to make inroads on a spectrum of heterogeneous hardware. Aggressive product development in the areas of Hyperchannel, or Ethernet, if extended to recognize the communication between data bases and an intrasystem executive function for task initiation and control, could fill the main remaining gap in such a strategy for a future CAE.

Acknowledgements, The Boeing Company

Graphic Designer: Yukio Tazuma
Technical Writer: Bob Whitehorn
Computer Artist: Susan Henry and Lori Caldwell

This slide show and booklet were prepared by the Seattle Support Services Graphic Organization of Boeing Computer Services. A computer aided graphics system was used to create the art for this presentation.

REFERENCES

1. Newell, A.; Simon, H. A.; "Human Problem Solving." Published by Prentice-Hall, Inc. 1972.

2. Boeing Document D6-IPAD-70010-D, "Reference Design Process." Prepared Under NASA Contract NAS1-14700, 1977.

3. Boeing Document D6-IPAD-70011-D, "Product Manufacture Interactions With Design Process." Prepared Under NASA Contract NAS1-14700, 1977.

4. Boeing Document D6-IPAD-70035-D, "Product Program Management System." Prepared Under NASA Contract NAS1- 14700, 1977.

5. Boeing Document D6-IPAD-70013-D, "IPAD User Requirements." Prepared Under NASA Contract NAS1-14700, 1977.

6. Boeing Documents D6-IPAD-70016-D-1, -2, -3, "First-Level IPAD User Requirements," Vols. 1, 2, 3. Prepared Under NASA Contract NAS1-14700, 1980.

7. Burner, Blair; Ives, Fred; Lixvar, John; and Shovlin, Dan; "The Design Evaluation and Implementation of the IPAD Distributed Computing System." American Society of Civil Engineering Technical Council on Computer Practices, Specialty Conference Electronic Computing and Civil Engineering, June 1978.

8. IPAD—Integrated Programs for Aerospace-Vehicle Design, Proceedings of a National Symposium, September 17- 19, 1980, Denver, Colorado, NASA Conference Publication 2143.

9. Comfort, Dennis; "Requirements for an Engineering DBMS." Computer World, October 29, 1979.

10. Balza, R. M.; Bernhardt, D. L.; Dube, R. P.; "Data Base Technology Applied to Engineering Data." Presented to Second International Conference on Foundations of Computer- Aided Process Design, June 1983.

11. Comfort, Dennis L. and Erickson, Wayne J.; "RIM—A Prototype for a Relational Information Management System." Presented at and in Proceedings of the NASA Conference on Engineering and Scientific Data Management, May 1978.

12. Johnson, H. R. and Bernhardt, D. L.; "Engineering Data Management Activities Within the IPAD Project." Database Engineering, June 1982 issue.

13. Barnhill, Robert C.; Dube, R. Peter; Little, Frank G.; Schweitzer, Jean E.; "A Unified Treatment of Curve Forms for Geometry Data Management." Submitted to CAD/CAM VIII Conference Sponsored by the Computer and Automated Systems Association of the Society of Manufacturing Engineers (CASA/SME), November 17-20, 1980.

14. Crowell, Harold A.; Dube, R. Peter; and Magedson, Robert; "IPAD Geometry Design," First Annual Conference on Computer Graphics in CAD/CAM Systems, April 1979.

15. IPAD II - Advances in Distributed Data Base Management for CAD/CAM. Proceedings of a National Symposium, April 17-19, 1984, Denver, Colorado. NASA Conference Publication 2301.

Very Large Application Systems Development

Development of the Banking System in Japan

Eiichi Ueda
The Mitsui Bank, Ltd.

INTRODUCTION

In Japan, the financial industry is playing a leading role in computerization and in the establishment of networks with the development of the Japanese banking system, which is a forerunner of the future Japanese information society. This banking system consists of two basic parts, a system for money transactions and a system for banking information.

These systems are now rapidly developing both the remote function and networking function, both functions are described in detail with explanations of their development of banking systems.

However, there are still many problems that must be solved in the development of a more efficient, large-scale banking system: security, software productivity, and education. Solutions are discussed, but concrete measures are still to be decided. For example, concerning education, the lack of qualified programmers will prevent enhancement of software productivity and will deter realization of the information society unless something is done immediately.

STRUCTURE AND DEFINITION OF THE BANKING SYSTEM

Since the financial industry has taken a leading role in accelerating computerization in Japanese business and in social life, it seems natural that the future direction of its banking system will also have a significant influence upon the development of the information society in Japan. Therefore, the author's description of the development of the banking system in Japan will help you foresee the general future trend of the financial system, as well as of the information society.

Before going into the details of the banking system. I shall explain its structure and define some of the special terms that are commonly used in the financial community. The so-called electronic Japanese banking system can be divided into two large systems: the financial settlement "delivery" system and the banking information "delivery" system. The financial settlement system is further divided into two parts: a "cashless delivery" system, in which settlements are made by the transferring of funds, and a "cash delivery" system. The banking information system provides accurate information concerning the movement in depositor accounts, outstanding balances, fund transfers, and remittances through a computer.

In addition to this information concerning accounts, it provides information concerning the many interest-earning commodities that have become very popular in recent years, and information concerning foreign exchange rates between the yen and the dollar, or the yen and the pound; because they are indispensable for the banking industry.

Furthermore, in order to insure much broader activities in domestic and international banking, financial institutions require information on domestic financial, economic movement, American economic trend, or the economic status and structure in West Germany to prepare for investment in its industries. All information on these matters are included in the banking information delivery system to be processed. The following figure (Fig. 1) shows you a general idea concerning the structure of the banking system described above.

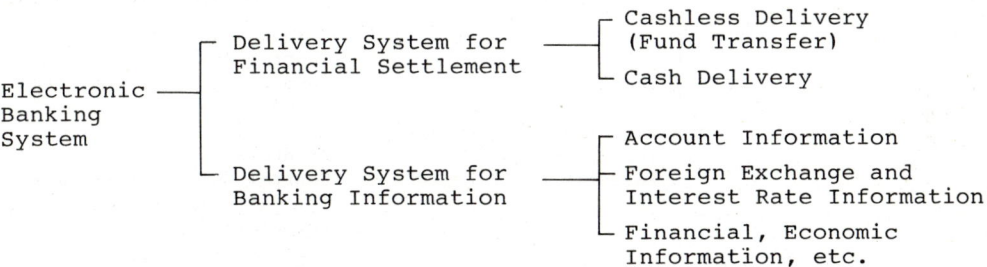

Fig. 1. Structure of the Banking System

Figure 2 is an organization map of the principal financial institutions that will help understanding the structure of the Japanese financial institutions and the relative functions of the system.

REMOTE BANKING AND NETWORKING OF THE JAPANESE BANKING SYSTEM

The Japanese banking system is now rapidly developing both the remote banking function and the networking function. Figure 3 shows the relation between the increase in remote banking and the expansion of networking. The vertical axis represents the development of remote banking (where the term "remote banking" means the execution of a financial transaction with a bank from a location far removed from a teller window or a counter of the bank) and the horizontal axis represents the expansion of networking.

```
Central Bank ──────────────────────────────── The Bank of Japan

Private          ── Ordinary (Commercial) Banks ──── City Banks (12)
Financial
Institutions                                         Regional Banks (64)
                 ── Specialized
                    Financial    ┬── Financial Institution ──── Specialized Foreign
                    Institutions │   for International          Exchange Bank (1)
                                 │   Finance
                                 │
                                 │   Financial Institutions ┬── Long-term Credit Banks (3)
                                 │   for Long-term Credit   └── Trust Banks (7)
                                 │                                  ├─(Banking a/c)
                                 │                                  └─(Trust a/c)
                                 │   Financial Institutions
                                 │   for Small Business     ┬── Sogo (mutual) Banks (69)
                                 │                          ├── Credit Associations (456)
                                 │                          ├── Credit Co-operatives (449)
                                 │                          └── Shoko Chukin Bank
                                 │                              (Central Bank for
                                 │                              Commercial and Industrial
                                 │                              Co-operatives) (1)
                                 └── Financial Institutions
                                     for Agriculture       ──── Norinchukin Bank
                                     Forestry and Fishery       (Central Co-operative Bank
                                                                for Agriculture and
                                                                Forestry) (1)
                                                            ├── Credit Federations of
                                                            │   Agricultural Co-operatives
                                                            ├── Agricultural Co-operatives
                                                            │   (4,345) (47)
                                                            └── Fisheries Co-operatives
                                                                (1,718)
```

Fig. 2. Organization Map of Principal Financial Institutions
        (As of March, 1985)

Fig. 3. Progress in Remote Banking and Networking of the Japanese
        Banking System

The shaded circle in the figure indicates that, although networking had expanded considerably by 1984, the degree of usage of remote banking was still limited; however, the white circle, which represents the status in 1990, shows a steep, rising curve between 1984 and 1990, indicating an acceleration in banking transactions made from remote, bank-branch locations.

Development of Remote Banking

Figure 4 shows the development phases of remote banking in more detail. In the initial stage, banking transactions were processed by a teller. Customers came to the counter (teller window), filled in a form to identify the type of financial transaction - deposit, withdrawal, or so on - , handed it to the teller who processed the transaction behind the counter. This was the starting point, or remote-zero.

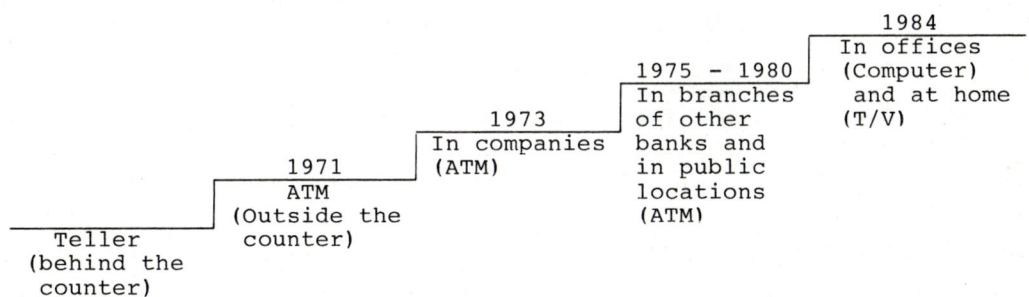

Fig. 4. Development Phase of Remote Banking

In the 2nd stage, ATMs (Automatic Teller Machine) were introduced and placed outside the counter in 1971. They enabled customers to deposit money or withdraw money from their account by themselves, enabling them to make banking transactions from a place slightly removed from the actual banking operation site.

In 1973, revised regulations permitted banking institutions to install ATMs in private companies so that their employees working in factories or facilities that were situated at a great distance from a bank could withdraw money from their accounts during the noon recess.

Between 1975 and 1980, they were permitted to install ATMs in public locations, such as railway stations and department stores. In addition, any depositor of Mitsui Bank could withdraw money from his account not only from a branch of Mitsui Bank, but also from a branch of another bank. From 1984, financial transactions with banking institutions could be made through computers (large and micro computers) and T/Vs in offices and in private homes. This is a brief description of each development phase of remote banking.

Development of Networking

Networking of the banking system started in 1965 as an on-line network within a bank and an on-line system was established among the head office, the main branches, and the smaller branches of Mitsui Bank in 1968. As shown in Fig. 5, in 1973 a small group of special banking institutions established an electronic remittance network on a small scale. This system was called the "Zengin" or the all Japan Banks Telecommunications System.

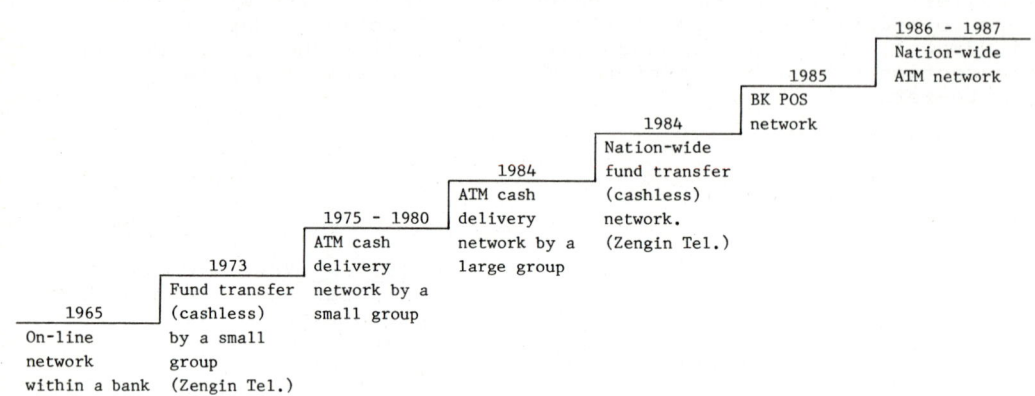

Fig. 5. Development Phases of Networking

Between 1975 and 1980, an ATM network was established by a small group of banking institutions. It made possible cash withdrawals by mutual connection of ATMs among Mitsui Bank, Sumitomo Bank, Sanwa Bank, Fuji Bank, and Mitsubishi Bank only.

By 1984, an expanded ATM network was established through the participation of all the city banks, consisting of 13 banking institutions at that time. The network permitted any cash withdrawal to be made among the branches (head offices) of the 13 different banking institutions.

In 1984, the Zengin cash withdrawal network was expanded to include all banking institutions in Japan. The institutions participating in the network are shown in the organization map of the Japanese banking system.

In 1985, POS (point-of-sales) terminals were installed in department stores and retail stores to form a network that connected these register terminals to the banking terminals. A customer could pay for the commodities purchased at any one of the stores by automatically transferring funds from his/her bank account to the store's account.

By 1986 or 1987, it is expected that the ATM network will enable customers to deposit money in their Mitsui Bank accounts, and to withdraw money from the same accounts through ATMs that will be installed in other banking institutions, including labor banks and agricultural cooperatives or association banks all over Japan. This is a brief description of each development phase of networking of the Japanese banking system.

DELIVERY SYSTEM FOR FINANCIAL SETTLEMENT

The delivery system for financial settlement consists of two systems; a cashless delivery system and a cash delivery system as shown in Fig. 6.

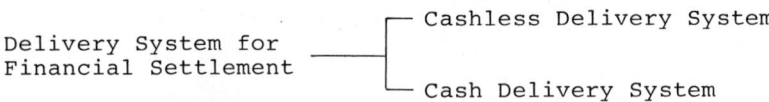

Fig. 6. Delivery System for Financial Settlement

Zengin Telecommunication System

This system is a typical "cashless" delivery system. As shown in Fig. 7, the Zengin computer center is the nucleus of the system, which is consigned to NTT, to which the computer centers of city banks, regional banks, trust banks, and other financial agencies are connected. Local branch banks are connected to their computer centers, because if a customer wants to make a financial transaction from a branch of the Mitsui Bank, for example, that person could have all the necessary information relating to the transaction transferred to a branch or head office of any other bank.

In the case of some foreign banks, because the volume of their transactions is small, the data is received off-line by tape or on paper and then it is reinputted and sent to a foreign bank.

Another example is the network for "Sogo" banks (mutual financing banks), which is shown here for that of the Kyushu area, where all the branches are connected to their joint center. For the network between Mitsui banks and Sogo banks, they have joint regional centers that permit mutual information transfer.

In the case of credit banks, their network is connected to the Zengin computer center through their joint center and two regional centers.

In this way, all the branches of all banking institutions in Japan, large or small, are connected to the large center computer in the Zengin computer center through networking.

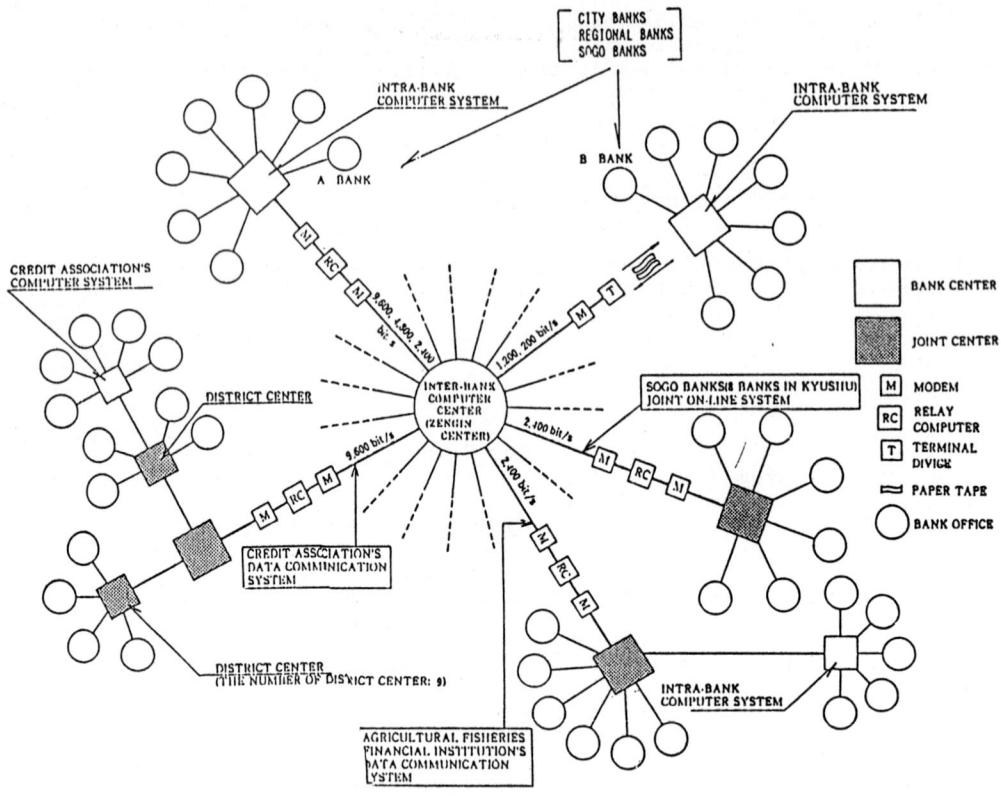

Fig. 7. Zengin Telecommunications System (Cashless Delivery System)

Transaction Volume Increases

There are as many as 5,539 financial institutions all over Japan, including agricultural cooperatives and laborers' credit cooperatives, that are participating in this network and the total number of their offices and branches amounts to about 40,000. This does not include post offices, which are run by the government.

The volume of customer remittances to be processed by the network has been increasing dramatically. In 1983, the volume of remittance transactions that was processed was about 254 million, which amounted to about 366 trillion yen, and it is estimated that 1,312 million transactions will be processed in 1995. Assuming the volume processed in 1973 as the basic unit of comparison, Fig. 8 shows the increase in the volume processed in units of the 1973 volume.

The figures on the vertical axis on the left indicate the volume of remittances in units of the 1973 volume; therefore, the 1983 volume will be slightly less than five times and the sharp curve upwards

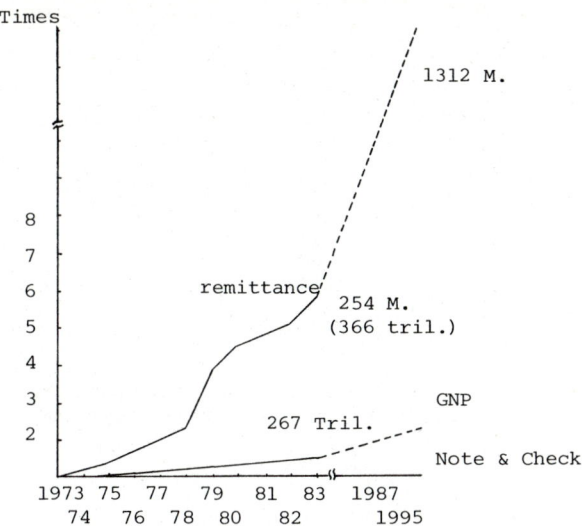

Reference: Volume to be handled.

Average   1,000 Thou./day
Peak   1,600 Thou./day   6.5 working hour/day
Total   366 Tril./year

Fig. 8. Volume of Customer Electronic Remittances to be made Mutually Among Financial Institutions

indicates that it will reach a volume of from 25 to 26 times the 1973 volume in the next ten years. The growth of the GNP of Japan is shown in the same manner for comparison. The trend of transactions being by notes and checks is also shown, but it indicates little or no increase through the years.

The overall conclusion reached from this figure is that Japan is rushing steadily into a cashless society without the use of paper as far as banking transactions are concerned. The volume of transactions to be processed will reach an average of 1 million transactions per day on the average and about 1.6 million per day during peak periods. This is because notes and checks are not commonly used by the general public in Japan.

Enhancement Plan of the Zengin Telecommunications System by 1987

The Zengin System is currently planned to be enhanced to have a much higher capability by 1987, the system structure of which is a remote transferring and editing system using magnetic tapes as shown in Fig. 9.

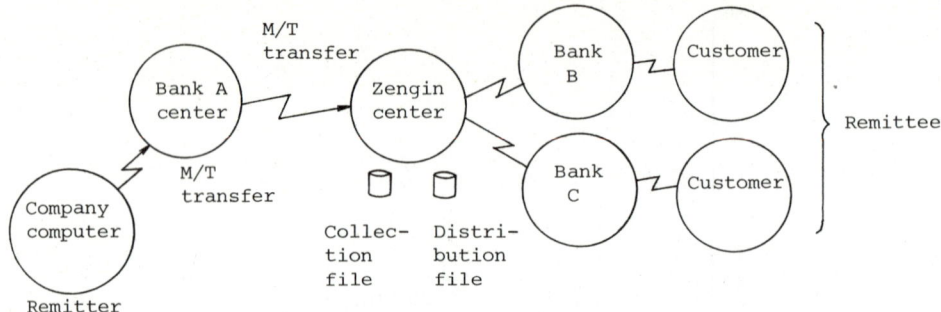

Fig. 9. Enhancement Plan of the Zengin Telecommunications System
- M/T Remote Transferring and Editing System -

According to the enhancement plan, a remitter will be able to transfer the contents of a magnetics tape (M/T) to Bank A's center through telephone lines by means of a computer installed at a company or a personnel computer owned by the remitter.

A DDX line is also available for the transfer of a large volume of transactions.

Bank A will then transfer them directly to the Zengin Center, which will transfer them to the designated Bank B or C through the collection file and the distribution file prepared at the center. Thus, Bank B or C will finally result the money to the account of the customer (remittee).

The system on the left half of the figure shows the functions of the new enhancements and that on the right half shows the current real-time, on-line system. When the new system is completed, remittances will be processed on a real-time basis with no manual operation from the remitter's site to a customer's account.

The customer's requirement following automatic remittance is that he wants to know immediately how much was remitted into his account. It requires full mechanization of a delivery system for banking information. Japan is ahead in this type of mechanization, because no country other than Japan has a system that permits direct remittance into a customer's personal account on a real-time basis. Therefore, this system might well be said to be one of the foremost unique systems.

Cash Delivery System (ATM System)

Regarding the each delivery system, a cash dispenser (money withdrawal only) system within a banking institution was started in 1972. Then it was enhanced, it was converted into ATM (automatic teller machine) by which both deposits and withdrawals became possible.

In 1984, recycling-type ATMs were introduced to permit recycling of paper currency from the deposit function to the withdrawal function. This eliminated much workload required for loading and unloading the paper currency.

At present, these machines can handle paper currency only and in the future, although they may be made to handle coins as well, paper money will continue to be the major currency handled by ATMs.

The media for operating ATMs are magnetic-stripe cash cards. Since the recording format of the magnetic stripe has been standardized by Zengin (Intra-Banking Association), any of their magnetic cards can be used for an ATM located at any branch bank.

ATMs have various functions such as deposit, withdrawal, and recording to the pass book. In 1982, a new function of connecting a customer's deposit account to the Zengin System was added, which permitted intra-remittance or interbank remittance.

In the case of Mitsui Bank, an average of five sets of ATMs per branch were installed. Table 1 shows the installation ratio of cash dispensers (CD) (including ATM) - 99.7% for city banks (branch base), 91% for regional banks, 9.9% for agricultural cooperative associations, and 0% for fishery cooperative associations. The figures indicate that ATMs are installed at almost all branches of the city banks, and that the larger the banking institutions becomes, the more ATMs are installed.

Table 1. Installation of CD (ATM included) at Banking Institutions

	No. of Banking Institutions	No. of Branch (A)	No. of Branch Installing CD (B)	% B/A	Total No. of Installed CD	ATM
Nation-wide Bank	86	9,659	9,072	93.9	22,466	10,360
City Bank	13	2,952	2,943	99.7	11,264	6,251
Regional Bank	63	6,298	5,779	91.8	10,713	3,763
Trust Bank	7	350	350	100.0	489	346
Long-term Bank	3	59	0	0.0	0	0
Sogo Bank	71	4,227	3,663	86.7	5,155	1,317
Credit Bank	456	6,507	5,276	81.1	6,486	3,149
Credit Union	468	2,750	298	10.8	329	77
Agricultural Bank	4,345	15,658	1,552	9.9	1,588	548
Fishery Bank	1,786	2,244	1	0.0	1	0
Labor Bank	47	555	143	25.8	143	42
Total	7,259	41,602	20,005	48.1	36,168	15,493
Post Office	1	23,469	1,724	7.3	1,724	approx. 1,350

Concerning the volume that can be processed, one ATM can process 140 transactions, equivalent to the workload of one teller. As a result, personnel expense can be reduced and, since an ATM can be used for 6 to 7 years, a considerable amount of man-power can be saved by a banking institution. At present, 87% of payment transactions is processed through ATMs and 73% of the customers use these machines for deposits. This description indicates how remarkably the cash service system of Japanese financial institutions has advanced so far.

As remote banking was accelerated, cash dispensers were first introduced and in 1984, the Ministry of Finance permitted depositing by ATMs so that customers did not need to come to the bank for performing financial transactions.

The development and functions of cash delivery systems are summarized in the following Table 2.

Table 2. Development and Functions of the Cash Delivery System

1. Development

    - Cash dispenser (withdrawal only) in 1972
    - ATM (deposit and withdrawal) in 1978
    - Recycle ATM (deposit and withdrawal) in 1984

2. ATM Functions

    - Deposit and withdrawal
    - Recording on a passbook
    - Remittance possible (intra interbanking) in 1982

3. Installation of ATM

    - 5 per branch in average (Mitsui Bank)

4. Transaction Volume

    - Average 140 transactions per day
      (equivalent to workload of one teller)
    - 87% of payment transactions
    - 73% of receipt transactions

5. Magnetic Cash Cards (debit card)

    - 88 M pieces

Accelerated Networking

NCS (Japan Cash Service Company) was established as a typical banking network in 1975 by a group of banking institutions. Common cash dispensers were installed in public places, such as railway stations and department stores. Although banking institutions without branches in the metropolitan areas did not join in NCS, 54 banks consisting of city banks, regional banks, and Sogo banks with main offices in the metropolitan areas jointly installed their common CDs.

The following Table 3 shows the summary of NCS.

Table 3. Status of the CDs Network

1. NCS

    - 54 banking institutions established NCS in 1975
    - Common CDs were installed in public places.
    - Charge: 100 yen/one transaction
    - Transaction volume: 11 M/year

2. NCS Network

    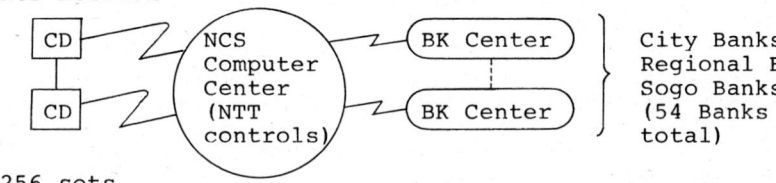

    256 sets

3. Network by Banking Institution

    - City Bank: 13
      BANCS in 1984 ← TOCS 1980 –
                      SICS 1980 –
    - Regional Bank: 63
      ACS in 1980
    - Sogo Bank: 71
      SCS in 1980
    - Credit Bank: 345
      SKS in 1983
    - Transaction volume of BANCS:
      2,600 K/month (1,910 K/month before BANCS) – Large effect of networking
    - Charge: 100 yen/one transaction

4. Structure of BANCS Network

    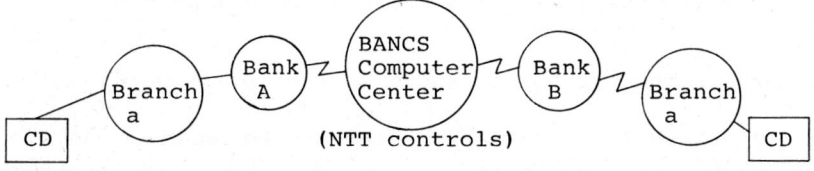

5. Development to all Banks Consolidated Network System 1986 – 87

The system currently consists of 256 sets of CDs. NCS entrusts the control of the network to NTT, and the centers of the participating banks are connected to the NCS computing center through 2,400 lines. (Refer to Table 3, Structure of Network)

The transaction volume is 11 M per year. Customers do not need to go to a bank to withdraw cash, but can withdraw it from a CD on the way to a department store.

Another network to provide cash service was established by the banking institutions. Thirteen city banks established BANCS to insure mutual connection of all ATMs owned by them. In 1980, 63 regional banks established ACS to enable all their CDs to be connected to one another.

The 11 Sogo banks (mutual financing banks) and the 345 Credit banks established SCS and SKS respectively in 1983 for the same reason as described above.

In 1980, there were two different systems for two groups of the city banks - TOCS and SICS. In 1984, these two systems was consolidated into one system, BANCS. Before consolidation of the systems, they processed only 1.9 M transactions per month, but after consolidation the consolidated system immediately processed 2.6 M transactions per month. Thus, customers of Mitsubishi Bank can withdraw cash at any branch of Mitsui Bank and vice versa. The result proves that the more widely the network is expanded, the more convenient it gets.

By 1986 or 1987, it is expected that all the systems now being blocked by the bank institution will be consolidated into one network. As the cashless systems were merged into a nation-wide telecommunications system at an earlier stage, the systems for cash service will also be consolidated into one network system within the next two or three years.

Connection with Consumers Through the Bank POS System and the Videotex System

The day is now approaching when consumers will be able to make a line connection with the computer centers of the banking institutions, wherever they may be, at a department store, a supermarket, a retail store, or even from their homes. Until today, remote banking and networking were promoted by means of CDs and ATMs installed in the bank branches, company offices, factories, or public places. In near the future, however, remote banking and networking are approaching much broader aspects of social life.

There are two approaches to realize the above idea - one is POS and the other is a videotex system.

The summary of the systems is shown in the following Fig. 10.

In Fig. 10, there is a bank A's center in the square, and is their network around the center, and it is connected to their branches, the other banks and customers. At the lower right below corner of the figure, CAPTAIN (Character & Pattern Telephone Access Information system), that is, videotex system is shown, which NTT planned to start in Tokyo and Osaka in November, 1984.

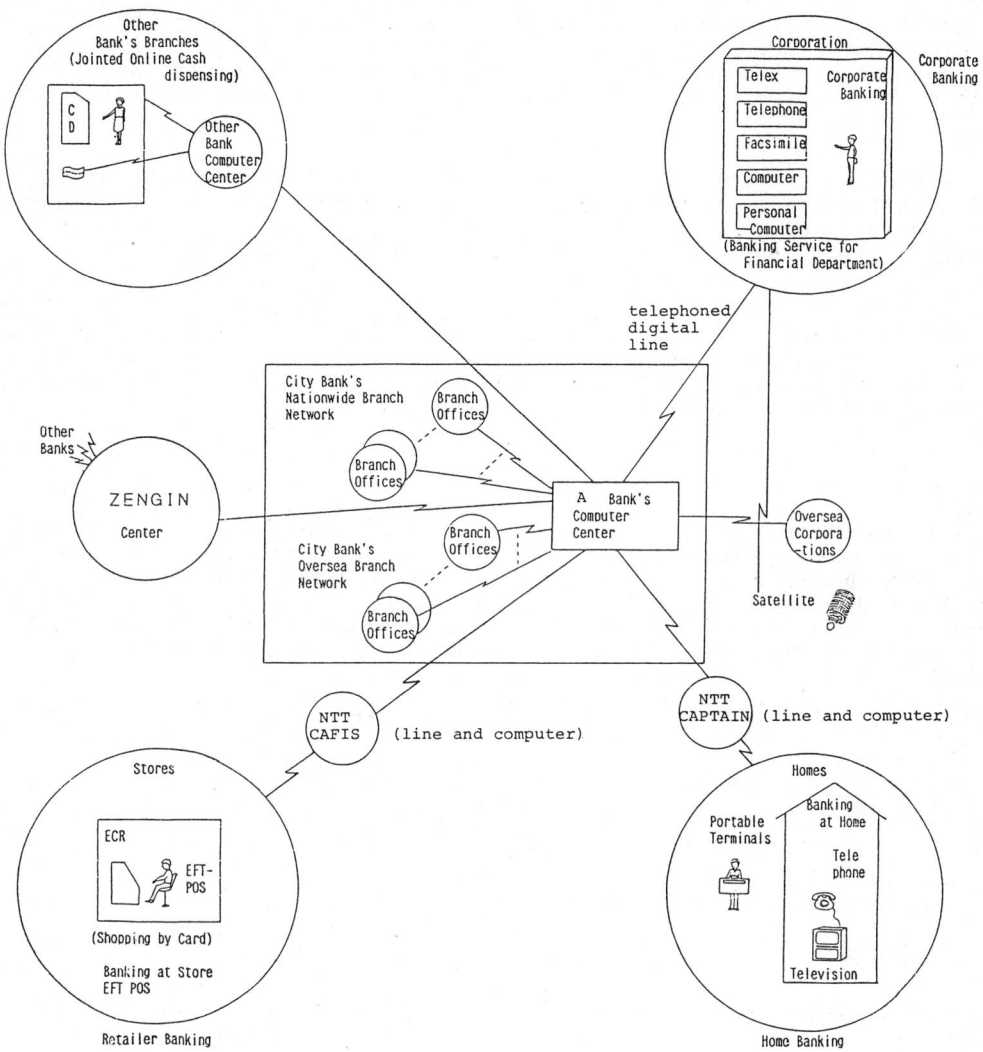

Fig. 10. Connection with Consumers

The system is connected to home T/Vs. Thus, home banking and home shopping are realized consumers could purchase commodities at home looking at the information given by providers of information over T/V. Payments will be made through the computer centers of their banking institutions and Zengin Telecommunication System.

At the lower left below corner of the figure, CAFIS (Credit and Finance Information Switching System) is shown, which is connected to POS terminals of supermarkets, department stores, and retail stores through the network constructed by NTT. If a customer purchases some commodities with his/her cash card or credit card at the stores and requests them to make a cashless transfer from his/her deposit account, CAFIS transfers the money to the stores' account through the whole network connected to the center of the bank and the Zengin Telecommunications System. If a customer makes a purchase with his/her credit card, the information is transferred to credit center A to establish credit for the customer.

As system connection with consumers is enlarged, the banking system will be increasingly required to enhance the capabilities of remote banking and networking. The bank POS for purchase at stores is called store banking and the home shopping system is called home banking.

Cooperative banking is shown in the right upper corner of the figure. It is connected directly to the computer, personal computer, facsimile, telex, and telephone at an enterprise via a NTT line. In this way, the network makes possible the settlement of payments and the delivery of information as described later in Chapter 4.

Paradox of the Electronic Banking System

As described before in this paper, the banking system has accelerated its remote banking and networking function in both aspects of cashless delivery system and cash delivery service system. However, this seems to be a great contradiction. On the one hand, a system to facilitate cashless delivery was developed, and on the other hand, a system to promote cash delivery was also developed. Figure 11 describes the contradictory situation in the banking system.

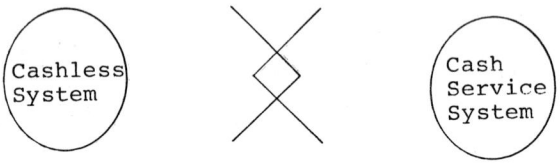

Fig. 11. Tug-of-War between Banking Systems

This situation can be illustrated as a tug-of-war and it is very difficult to foresee which side will win. Japanese, particularly as individuals, have a special attachment to cash. As indicated in the forecast (Fig. 8) that the remittance volume will increase by 26 times within the next 10 years, most of the Japanese companies use cashless fund transfer even at this moment. Salaries for employees are paid into their bank accounts.

However, once the salary is received by an employee, he spends the money in the form of cash as a consumer. Consumers in Japan spend 80% of their money in the form of cash. On the other hand, companies do business by means of cashless (transactions). Therefore, the Japanese banking institutions are forced to develop and implement two systems - cashless and cash delivery - to satisfy the requirements of both companies and consumers. About 10 years will be required to decide the definite winner of the tug-of-war.

As the rope is pulled to the left side (cashless), consumers will eventually quit using cash for payments; however, this will not come for some time. The banking institutions may have made consumers prefer cash by implementing a system convenient for cash payments.

The financial practices and uniqueness in the Japanese society are shown in Table 4 for the reference.

Table 4. Financial Practices and Uniqueness of Japanese Society

1. Settlement Method (transaction basis)

    . Cash .... Extremely frequent for individual
                         (77% of expenditures)

    . Fund transfer ..... Extremely frequent for company
                         (80% of expenditures)

    . Check .... 100 M sheets/year, infrequent

    . Credit card .... 17% of individual expenditures,
                         increasing

2. Payment of Salary

    . 67% of Japanese companies use payment into employee's account.

    - Classification

        Big companies over 5,000 employees:
                        97.1% - mostly M/T

        Small companies less than 30 employees:
                        32% - mostly paper

3. Settlement of Public Fees and Credit Card.
   Pre-Authorized Direct Debit System as below is adopted.

    . Agreement in advance between payer (User) and payee (Company).

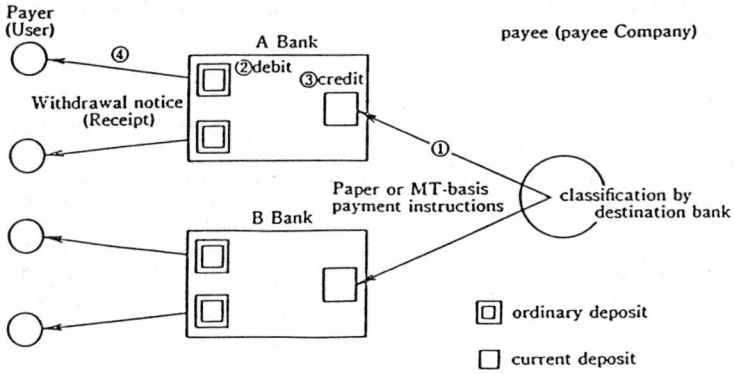

- Pre-Authorized direct debit of major public utility rates - as of March, 1981.

    (Unit = 1,000)

Company	TOT Users(A)	Users of Pre-Authorized Direct Debits(B)	B/A(%)	Year of Introduction of M/T Presentation
NTT Tel. Rate	40,010	30,525	76.3	1969
NHK TV Fees	28,553	11,961	41.9	1972
Tokyo Elec. Power Rate	15,764	10,390	65.9	1971
Tokyo Gas	5,442	3,717	68.3	1970

- Note Book

    - Note book is used for ordinary deposits. (Contents of the computer file are recorded.)
    - Outstanding balance (Credit, debit)
    - Remitter of automatic payments (Credit)
    - Automatic withdrawal accounts (Debit)
    - AMT can record.

DELIVERY SYSTEM FOR BANKING INFORMATION

In this section, the delivery system for banking information is described briefly. The system contains accounting information service (accounting information and information service) and economic/financial information. The accounting information includes information relating to the deposit balance, money transfers, movement of money receipts, loan status, foreign exchange transactions, etc. The more remote banking and networking are enhanced, the more information the customers desire to have.

Such information can be obtained through an on-line data transfer system. If a customer wants to have the information on his personal computer, the required information can be sent to it. If a customer wants to have it through his facsimile, it can be sent to the facsimile. If a customer dials the telephone number to get the information, the computer of Mitsui Bank will respond audibly "such and such amount of money was payed into your bank account" through the automatic telephone. Table 5 shows the kind of information services that is provided by the Mitsui Link Service. The circled items are available at present.

Table 5. List of Information Services
(Mitsui Link Service)

		Services	Classification of Services	Media to Provide Services						
				On-line Data Transfer	P.C.	Fax	Telex	Tel.	Floppy/M/T	Slip
Mitsui Bank Own Services	Accounting Information Services	Deposit Balance	Deposit balance		O	O		O		
			Foreign current deposit balance		O	O				
		Receipt of Money	Information on receipt			O	O	O	O	O
			Inquiry on receipt		O	O		O		
			Inquiry on receipt in advance		O					
			Information on receipt in advance	O						
		Deposit & Withdrawal	Inquiry on unrecorded ordinary deposit		O	O		O		
			Inquiry on todays deposit & withdrawal of current account		O	O		O		
			Information on deposit & withdrawal	O						
			Inquiry on deposit & withdrawal per month		O	O				
		Account Report Service	Intergrated transfer of deposit & withdrawal statement	O	O			O		
		Loan Transaction	Information on loan transaction	O						
			Inquiry on loan transaction per month		O					
		Foreign Exchange Transaction	Information on foreign exchange transaction	O		O				
			Inquiry on foreign exchange transaction		O	O				
			Inquiry on foreign exchange transaction per month		O	O				

		Services	Classification of Services	Media to Provide Services						
				On-line Data Transfer	P.C.	Fax	Telex	Tel.	Floppy/ M/T	Slip
Mitsui Bank Own Services	Fund Information Services	Due Date	Inquiry on due date		O					
		Control for Collection Bill	Control of collection bill due date							O
		Receipt of Money	Integrated money receipt	O	O				O	O
			Receipt of salary	O	O				O	O
		Pay by Phone	Money receipt, transfer					O (Purchase)		
		Local Tax Payment Proxy Service	Local tax payment						O	
		Computer Format for Bills Payable	To provide format for bills payable							O
		Auto Centralization of Fund at H.O.	Balance adjust of fund	-	-	-	-	-	-	-
		Quick Money Receipt	Money collection	-	-	-	-	-	-	-
		Acc't Transfer, Auto Collection, Mitsui Finance Service	Money collection proxy	O					O	O
									O	O
	Economic and Financial Services	Market Information Service	Information on foreign exchange rate		O					
			Inquiry on foregin exchange rate (future)		O	O				
			Inquiry on found rate		O	O				
			Information on economic, financial bond issue		(O)					
		Economic, Financial Information (Bond Link Service)	Inquiry on economic, financial, foreign exchange, bond information		O					

	Services	Classification of Services	Media to Provide Services						
			On-line Data Transfer	P.C.	Fax	Telex	Tel.	Floppy/ M/T	Slip
Others	Failure of Acc't Transfer	Inquiry on auto acc't transfer failure		(O)					
	Data Process by P.C.	Data processing, programing		(O)					
	Mini-Fax	Information service by Fax			(O)				

The system structure shown in Fig. 12 is the Integrated On-line System of Mitsui Bank.

It consists of 6 sets of large-size IBM computers and a UNIVAC computer for communication, which runs the entire banking system of Mitsui Bank.

For the information system and settlement system, UNIVAC NCU at the right side and IBM NCU at the left side are mutually connected. Each center of BANCS, NCS, and Zengin (shown in the lower right corner) is connected to the UNIVAC computer through NCU, and further connected to the IBM computers. The UNIVAC computer is used for the processing of teller-window business and data communications, and the IBM computers for the data base. In the dotted areas, customer service, personal computer, facsimile, telephone, audio-response, T/V, and customer's CPU are connected to the computers of Mitsui Bank through public lines, CAFAIS, CAPTAIN, DDX, or telephone lines.

Three lines of leased circuits are used exclusively for BANCS, NCS, and the Zengin system. Public lines are used for CAFAIS, CAPTAIN, DDX and TEL. Audio-response in the TEL is used mostly for notification of money receipts. This system responds audibly to the customer's inquiry concerning receipt of his/her money. The number of calls to Mitsui Bank is approximately 1 M transactions per month. Replies to inquiries are made partially through facsimiles, but most of them are done by telephone. In order to enlarge the information for foreign countries, such a system is now under construction to provide the information to clients by connecting Reuter's market information system to the Mitsui Banking system.

The above is a brief description concerning the development of the systems for settlement information and banking information in Japan. In the future, the banking system will continue to expand the capabilities to respond to customer's needs. At the same time, all banking institutions are now encountering a variety of services problems that must be solved. These problems will be discussed in the next section.

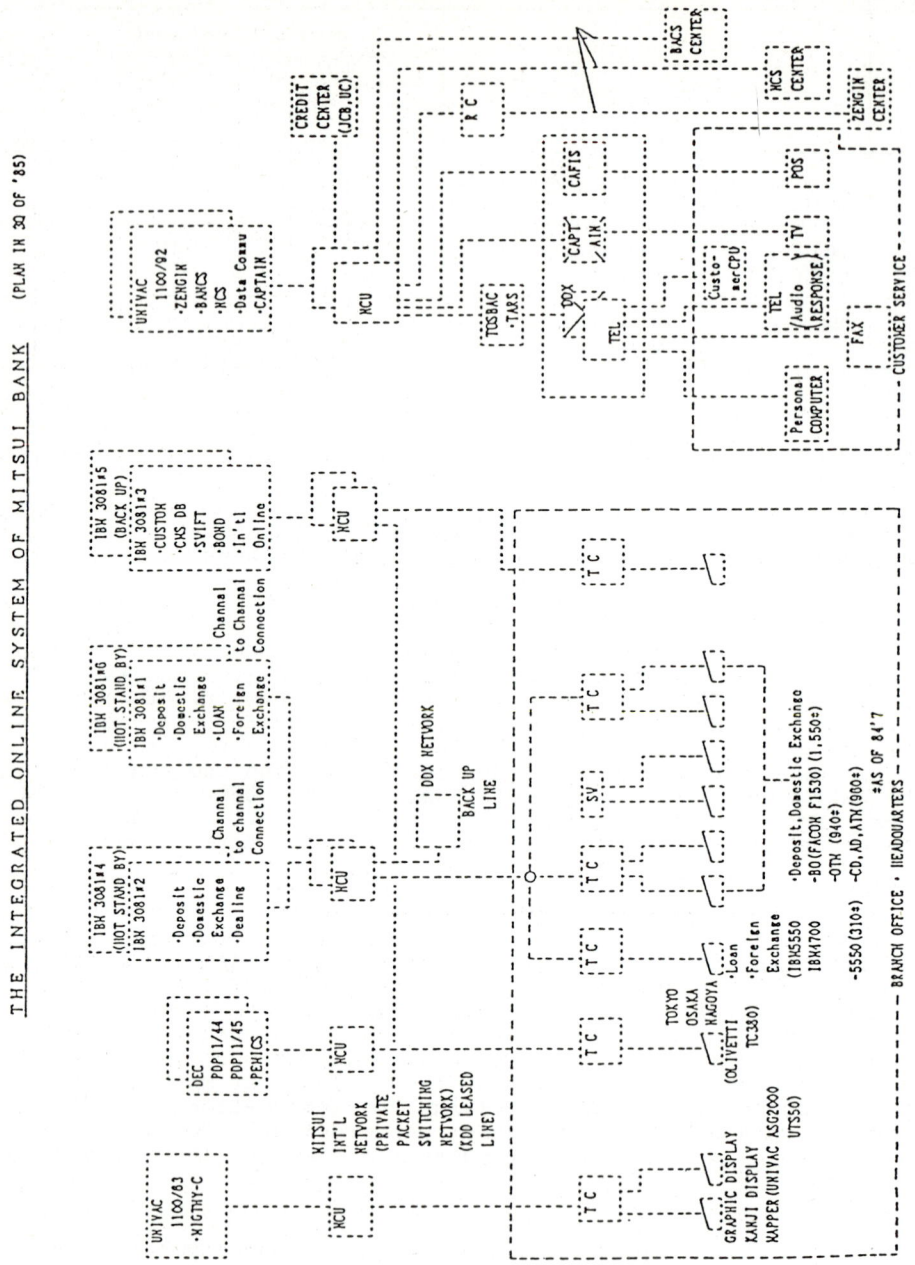

Fig. 12. The Integrated On-Line System of Mitsui Bank

PROBLEMS OF THE BANKING SYSTEM TO BE SOLVED.

There are three major problems to be solved for the development of a more efficient large-scale banking system: security, software productivity, and education. In an advanced information society, it is vital that the three elements of information, money and materials are computerized in a well balanced manner. In other words, only when each system to deliver information, to transfer money, and to transport materials has been synchronized, can it be said that the information society has been realized in the truest sense.

Information and money can be effectively handled through computer and communication. However, materials must be carried by men or vehicles. Now various kinds of quick home delivery service systems are being constructed in Japanese society. Even if the information that a customer has purchased some commodities through his home T/V may instantly reach the department store, the effect of CAPTAIN will be lost if he physically receives the commodities a few days later.

This is why the carries in Japan are largely trying to shorten the physical delivery lead time of the commodities, as seen in the fields of information and money. When the material can be delivered to the client within 24 hours, a real high information society with zero inventory will be realized. In this information society the banking institutions will play a more vital role as the backbone to support it.

Security of the Banking System

The larger the system gets in scale, the more serious the impact will be to the society if it fails to run, or if it runs out of order. The security system for is now being developed as shown in Fig. 13.

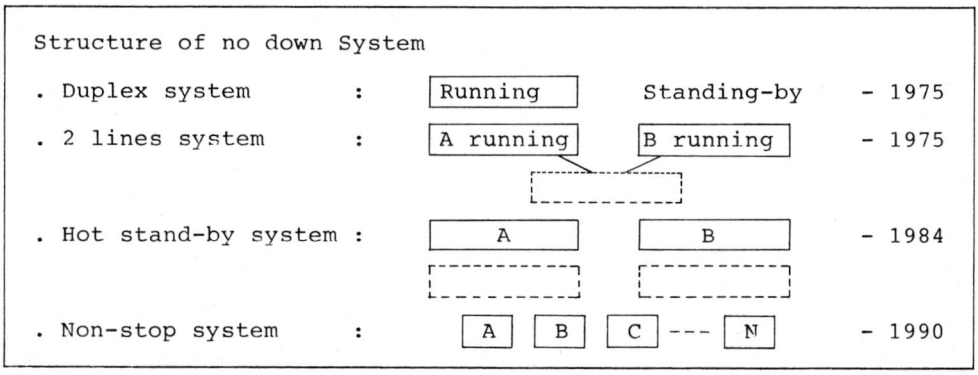

Fig. 13. Security System

In the early stage of 1965, there was the duplex system consisting of 2 systems of "running" and "stand-by". In 1975 two lines of systems were completed. They were two independent systems - terminals (A group) connected to computer A and terminals (B group) connected to computer B. When the A group of terminals fail to operate, the B group of terminals can backup. In 1984, the "hot" stand-by system was installed in addition to the two lines of systems. When computer A is down, there are computer B and the "hot" stand-by for backup. They are designed to start operations within a few minutes after failure of computer A. In 1990, a non-stop system will be introduced, as shown in Fig. 13. Then, even if one of the computers fails to run or is damaged, the operation will be switched to computer N for immediate recovery. The early introduction of a non-stop system is required for public organizations, such as banking institutions, to minimize the impact on the society at the time of a computer malfunction.

As for application software, protection measures have progressed as shown in Fig. 14.

Year	Application Software	Impact
1965	PGM	One program - one task (Trouble of one task affects, whole system.)
1975	PGM → Task 1 ⋮ Task N	One program - multiple task (Trouble of one task affects the same task only, but trouble of basic parts affects whole system.)
1985	PGM 1 → Task 1 ⋮ Task i ⋮ PGM m → Task 1 ⋮ Task n	Multiple program - multiple tasks (Trouble of one task affects the same task only, and trouble of basic parts affects the partial system only.)

Fig. 14. Development of Security by Software

As early as 1965 when one program - one task was common, trouble in one task affected the operation of the whole system. In 1975 when one program - multipile tasks was adopted, any trouble in one task impaired that task only; however, trouble in a basic portion affected the whole system. When multi programs - multi tasks was adopted in 1985, any trouble in one task or one basic portion only impaired the corresponding task or portion. These are the software problems caused by accident.

Another problem is an intentional one, such as stealing of data or information. Some devices must be protected to prevent data or information from being stolen. One of the devices is to encrypt all data against theft. It was not necessary until today to encrypt the data, because the data was transferred within facilities of a banking institution or through exclusive leased circuits for intra-bank transactions. When public circuits are to used in the near future, it is mandatory to adopt encrypted data and information. Every banking institution is now looking forward to an economic and errorless method of encryption.

The next problem is security for the computer system during natural catastrophes. Since Japan is notorious for frequent earthquakes, every banking institution has a backup center or a subcenter in the Osaka area, assuming that its computing center located in the Tokyo area is completely destroyed. Therefore, the same data (for example, the same contents of M/T) are maintained both in its Tokyo and its Osaka center. This is the case of city banks. Smaller scale banking institutions are not yet prepared well enough to cope with such a situation.

Cost and effect in security is another important aspect of the banking system, although it is very difficult to solve completely. According to the guidance of the Ministry of Finance, the frequency that a system is malfunctioning is a yardstick for the measurement of security. This frequency of down-time for a large-scale Japanese computer is, in general, once every 460 hours; that of a major system is once per 520 hours, and that of a banking system once per 620 hours. This shows that banking institutions maintain longer continuous running hours with more money invested.

In this way, the banking system is the most secure system in Japan. The problem is the time a system is inoperable. The latest statistics indicate 70 minutes of system down-time for a general system and 25 minutes for a banking system. From experience, customers can endure a down-time of about 30 minutes; however, if it continues over 30 minutes, they begin to get restless and are ready to storm into the bank.

At present, Mitsui Bank has managed to keep their down-time to an average of 25 minutes. A specific security structure, such as a duplex system, two lines of systems, or a "hot" stand-by system, as described above, has been introduced by each banking institution depending upon its special situation. In 1984 Mitsui Bank was at the stage of two lines of systems, but in the spring of 1985 the hot stand-by system was installed to reduce the down-time to three or four minutes.

At present, there are some cases where 25 to 30 minutes are required for complete system recovery. In the case of smaller banking institutions, the down-time is much longer, sometimes 45 minutes to one hour; therefore, it is a very serious decision for banking institutions to decide how much money they should invest to keep their down-time within what time limits.

A questionnaire revealed that the amount of money invested for security maintenance is less than 3% of the total of computer-related investments. The banking institutions invest an average of 4 to 6% of the total computer-related investments for security. If recovery is to be maintained within one minute or a no-trouble installation is to be created, it would force banking institutions to invest more money than they could afford. The Ministry of Finance is now investigating the reasonable and acceptable down-time.

In conclusion, banking institutions desire to be provided by the computer suppliers with "non-stop" OS and an OS for trouble recovery, because users can do nothing about OS problems. Furthermore, the following are the capabilities that they require additionally.

- Complete unmanned operation
- Systems designed to prevent wrong insertion of M/T or any other human errors
- Reasonable enciphering
- Strengthened access control to data bases

As a common, public-circuit network will be used by the banking system, in the near future, the above problems must be solved for quick enhancement of the system.

Software Productivity

The next significant problem is software productivity. Mitsui Bank has about 30,000 programs with approximately 9 million steps that run the entire system shown in Fig. 12. Productivity increased 2.5 times in five years - CGR of productivity is 18% per year. Five years ago the capability to write programs was 450 steps per month using COBOL; the latest standard is 900 steps per month. Using the assembler, it was 160 step per month, but now, it is 400 steps per month.

Productivity has increased for the following reasons:

- One TSO terminal for program development was installed for two programmers.
- 21 kinds of software tools were introduced.
- Intensive training of computer related employees.

For assurance of program quality, the following actions have been taken:

- Establishment of an inspection system
- Modular programs
- Perfect test data base

As for subjects for future study, programless programming is what is most earnestly desired to be developed - direct input of a diagram or a chart automatically generates the required programs. Other required subject for development are automatic tools for program testing and portability of programs, as they are now available in C language or Ada language.

Education

In the above sections, various problems have been cited for immediate solution. The most important matter in accelerating computerization to a high level, however, is the absolute shortage of software manpower. Table 6 indicates this serious trend.

Table 6. Shortage of Software Manpower

Description	1977	1982
No. of SE + Programmer per No. of Computer	8.7	4.1

The number of SEs and programmers per computer decreased sharply to less than a half in five years. It is anticipated that the trend will decline still further in the future. In 1985, only 100,000 programmers were available although 500,000 were required.

Unless this problem is solved, Japanese computerization or a high-level information society will be defective in the aspects of security and quality, no matter how sophisticated the language system may become conceptionally. In the Japanese society, computer education in schools is not well established.

It was reported in an information magazine that a major field for the study of computers would not be established in colleges of the United States. There, students who are majoring in economics or sociology must successfully complete some course concerning computers in order to graduate. Such a direction of education must be emphasized in universities or colleges of Japan as well. Particularly in Japan, computer education in schools is so poor that there are hardly any primary schools or junior high schools that provide their pupils with computer lessons.

CONCLUSION

In the next ten years, into the 1990s, the Japanese society will undergo rapid development in rationalization and efficiency in the services and support areas and, together with the high-level mechanization and rationalization in the production and manufacturing areas, will become one of the most advanced and efficient society in the world.

The banking system will play a leading role in this process.

However, no matter how much improvement and progress is made in computer network technology, a high-level society cannot be attained unless the underlying foundation, the infrastructure, such as security, crime prevention, eliminations of inconsistencies in the current laws and regulations, and establishment of an education system to improve software-development capability, is firmly established. This is just the same as you cannot have an automobile society if the roads and the traffic signal system are poor, no matter how advanced automobile technology has become.

In attaining this information society, the most important task for Japan is to establish a firm infrastructure.

Large-Scale Computer Systems in the Steel Industry

Yoshisuke Inoue
Nippon Steel Corporation

I. STEEL INDUSTRY AND PRODUCTION PROCESSES

1. Introduction

Nippon Steel Corporation (NSC) maintains large-scale computer systems typical of those used by the steel industry. These include the head-office-based sales and production control system and steelworks-based integrated production control systems. This report describes the relationship between the two types of control systems and, mainly with regards to steelworks-based control systems, their necessity — why they should have a large scale — role, structure, development process, and extension in the future.

2. NSC in the World Steel Industry

Annual world crude-steel production has stagnated around 700 million tons since the oil crises (see Fig. 1). The data in Figure 1 also indicates the recent sharp crude-steel production increases in China and slight increase in USSR that has supplied steel to communist nations and a marked decrease of the production in the USA.
 The Japanese steel industry, depending on imports for almost all of its fuel and raw materials, was adversely affected to a large degree by the rise in energy and raw material prices.
However, it has avoided a large reduction of output through extensive improvements on plant operation, including energy-saving activities and a variety of measures to reduce production costs.
 NSC ranks above all steel manufacturers in the world in the production of steel (see Figs. 2, 3, and 4) and has large and small nine steelworks throughout Japan.

3. Outline of Steel Production Process

This section explains the outline of steel production processes (see Fig. 5) that are managed by the large-scale sales and production and integrated production systems being introduced later in this report.
 The first production process is to produce two raw materials; the sintered iron ore made through the crushing, mixing, and calcination of iron ore and the coke made through coal carbonization. The sintered iron ore and coke are alternately thrown into the blast furnace (BF) from its top and hot blast is blown from the furnace bottom to produce pig iron through combustion and reduction.
 Produced molten iron is transported to the BOF (Basic Oxygen Furnace) by the Torpedo car. In the BOF, impurities, including carbon and silicon, are oxidized and removed from the molten iron, Cr, Ni, etc. are added to adjust chemical compositions, and molten steel is produced.
 Subsequent processes involve the making of slabs or blooms which can be done through one of the two methods: the ingot making and primary rolling process, and the continuous casting process.

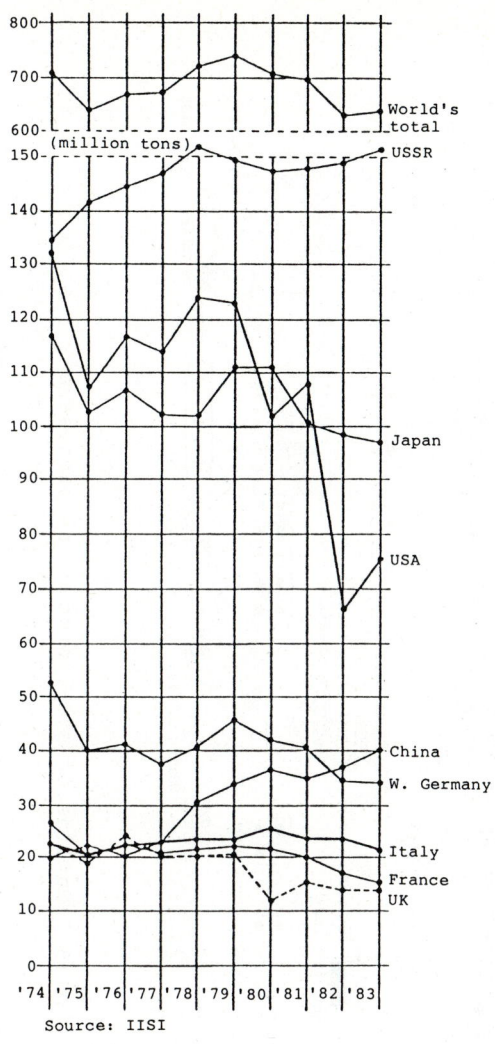

Fig. 1 Crude steel production by major steel-producing nations

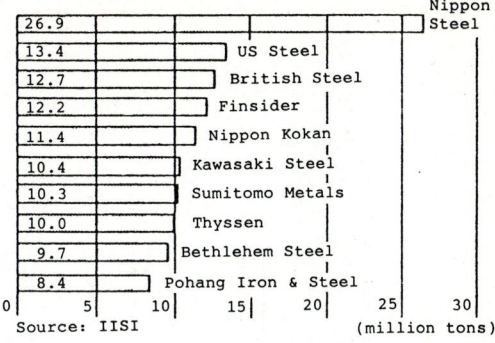

Fig. 2 Top ten steel-producing companies (crude steel in 1983)

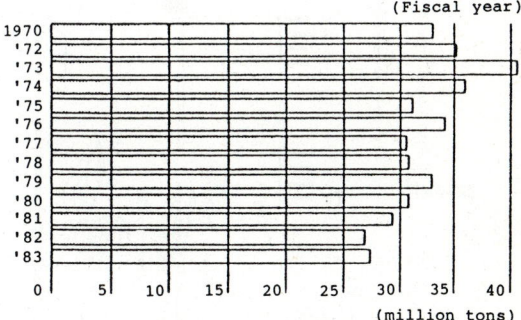

Fig. 3 Crude steel production of Nippon Steel

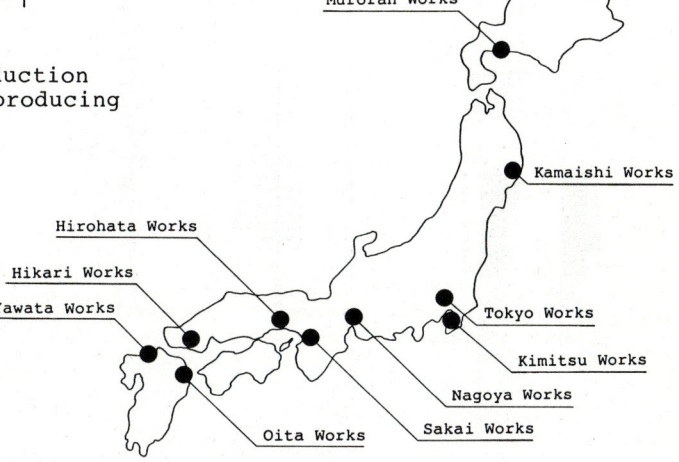

Fig. 4 Location of each works of Nippon Steel

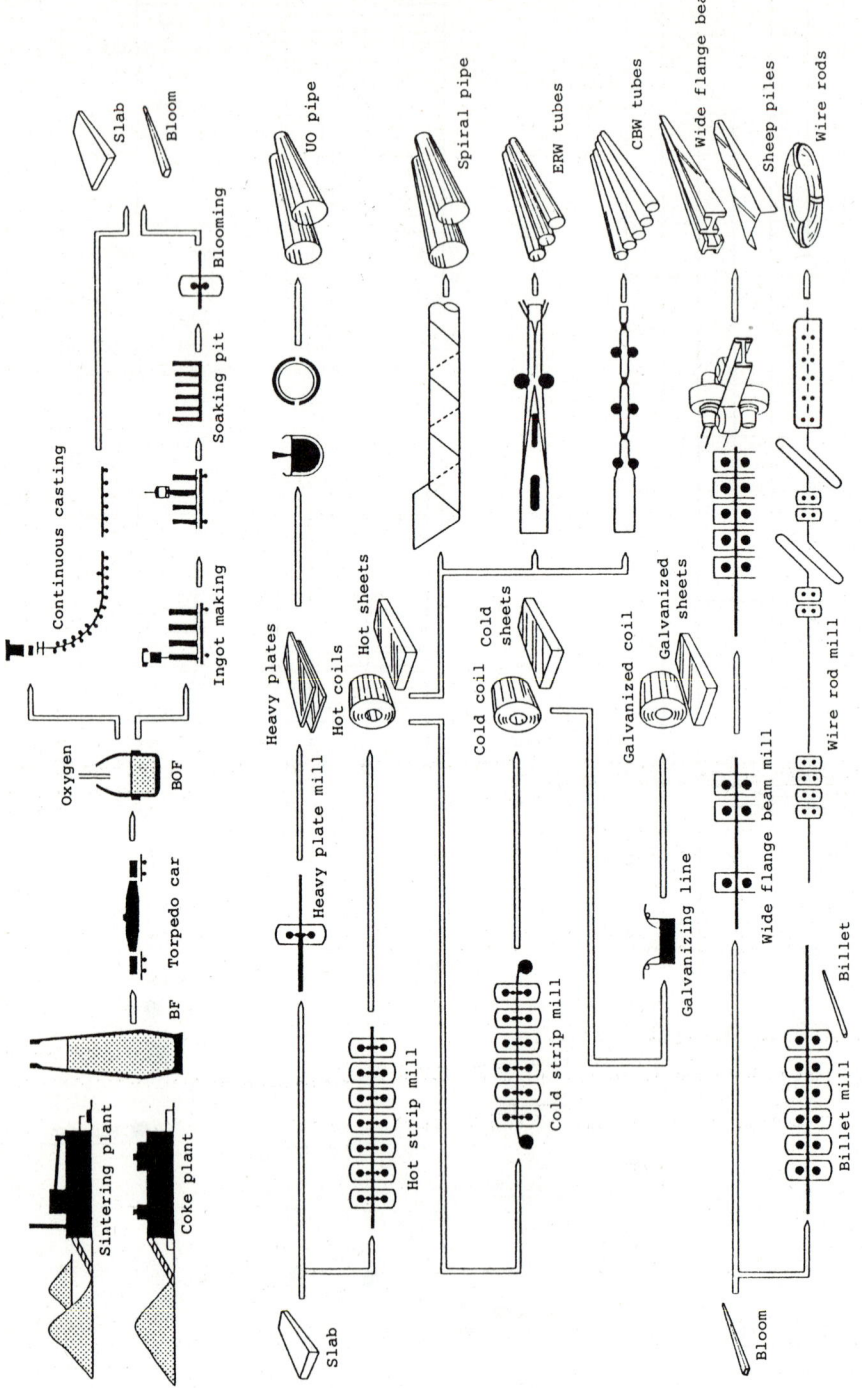

Fig. 5 Outline of steel production processes

The ingotmaking and primary rolling process pours the molten steel in large molds and rolls the molded steel by the primary rolling mill to form slabs or blooms. The continuous casting process pours the molten steel directly into the mold that has a cross-sectional shape of slab or bloom and produces slabs or blooms in one step. The continuous casting process adopted in NCS is a new continuous casting method that has been developed through various technical improvements.

The slabs and blooms are inspected for surface defects and scurfed, then heated at around 1,200°C in a heating furnace, and rolled and processed to become products of varied size and shape, including heavy plates, hot coils, not sheets, cold coils, cold sheets, pipes, sheet piles, and wire rods.

During the steel-production processes, a large amount of combustible gas is produced as a by-product by coke furnaces, blast furnaces, and BOF. The gas is used for the heating furnace in rolling processes, with the remainder being used to supply electrical power to the works. In this sense, the entire steelworks can be regarded as an industrial complex.

As so far described, steel production is an integrated process, including blast furnace, steel-making, and rolling. Consequently, the system to control the processes must be also integrated. NSC, accordingly, uses a management system, including production control system, having a program steps of 10,000K--15,000K for a large steelworks.

4. Recent Changes in Steel Production Processes and Their Influence on Production Control Systems

The fundamental structures of the ingotmaking and primary rolling process, and the production control system based on this process were established in the latter 60's and early 70's. However, steel production processes and their control systems in Japan must have been changed radically because of the two oil crises. The changes include the omission, continuation, and connection of production steps to realize low energy and resource consumption. The innovation of the production process is represented by the shifting to continuous casting and the connection of casting and rolling steps.

Figure 6 shows the comparison of ingotmaking and primary rolling processes, the conventional continuous casting process, and the directly connected process. As shown by the figure, the continuous casting process realizes the omission of soaking and ingot rolling processes and fine slab cutting according to customer's orders. These advantages lead to high yield rates and greatly reduced manufacturing time. Therefore, the ratio of continuous casting process has been steadily increased in major steel-producing nations (Fig. 7). Many steelworks of NCS presently show 95% or more ratios of continuous casting.

The directly connected process was developed, on the basis of continuous casting, to conserve energy more and radically improve the production process, including the reduction of running inventories. This new production process requires the following continuous-casting techniques: (1) making of no-defects casted slabs to eliminate conditioning, (2) a high temperature at the exit of continuous casting to enable continuous rolling, and (3) supply of casted slabs of the size (thickness and width) required by rolling process.

Since the directly connected process executes rolling just after continuous casting, there is the least material buffer in the course of the process in comparison with conventional ingotmaking and primary rolling process. The directly connected process, therefore, requires an accurate control of material and product flow to manage the quality and timing of production according to schedules.

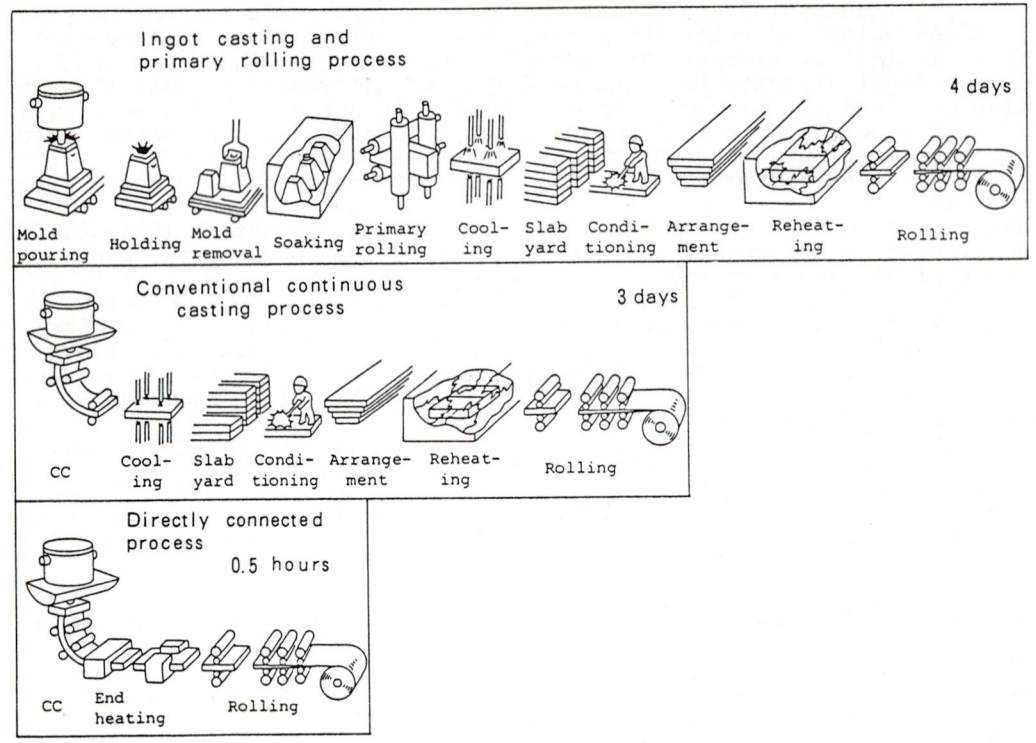

Fig. 6 Comparison of ingotmaking and primary rolling process, conventional continuous casting process and directly connected process

Consequently, the new production process requires the development of an integrated production control system combining control of the process from steel-making to hot rolling and hot-rolled product quality control techniques. This necessitates a control system which is large in scale and diverse in content, filling an important role in the steel manufacturing industry.

5. Characteristics of Steel Industry and Its Production Control Systems

The computer systems to control steel production processes introduced above are characterized by the factors as follows.
(1) Characteristics of steel industry
　　(i)　Heavily equipped process industry for mass production with complicated production steps
　　(ii)　Absolute on-demand production
　　(iii)　Large production lots and small order lots
　　(iv)　Keen competition in keeping delivery date and product quality
　　(v)　Discontinuous manufacturing processes and breakdown structure

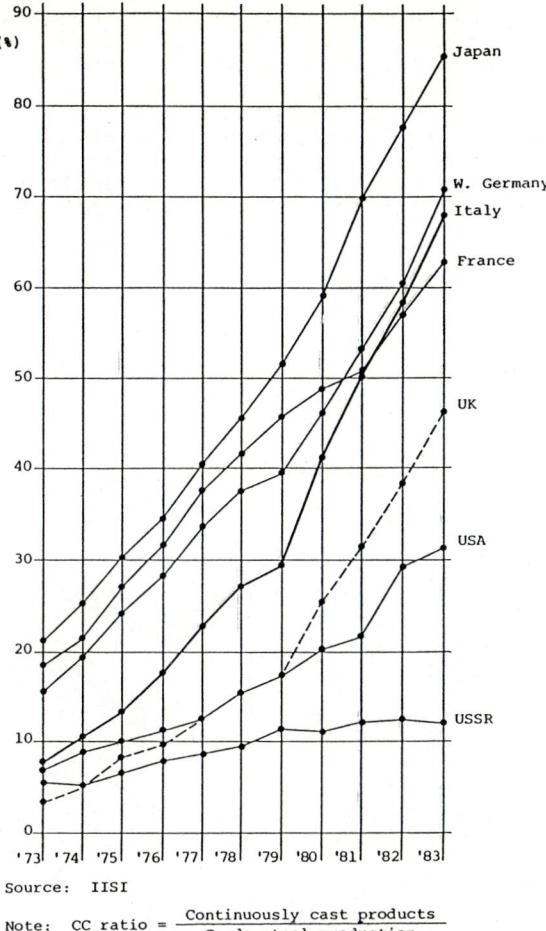

Fig. 7 Continuous casing ratios of major steel-producing nations

(2) Features of computer system
 (i) Total production control from order reception to shipping high-level use of computers to improve office-work efficiency
 (ii) Typical hierarchical structure using batch system, on-line realtime system, and process computers
 (iii) Batch production scheduling system requiring appropriate processing speeds for bulk data processing
 An on-line realtime system (operating 24 hours/day) requiring large-scale traffic, high reliability as well as security, and quick response

Particularly, the steel industry characteristics "(1) (v) Discontinuous manufacturing processes and breakdown structure" determine the features of the production control system. Steel production processes are arranged in a breakdown structure, and the steel products made therein must pass through numerous noncontinuous lines while undergoing changes in their properties and shapes. If, therefore, trouble occurs in any intermediate process, corrective action must be taken all the way back to the first process.

This characteristics results in an extremely complicated data processing for management work, including computer systems. This also marks a fundamental difference between the steel production process control and the process control for the automobile industry, electric-apparatus manufacturing that have "build-up type" assembly system (See Fig. 8).

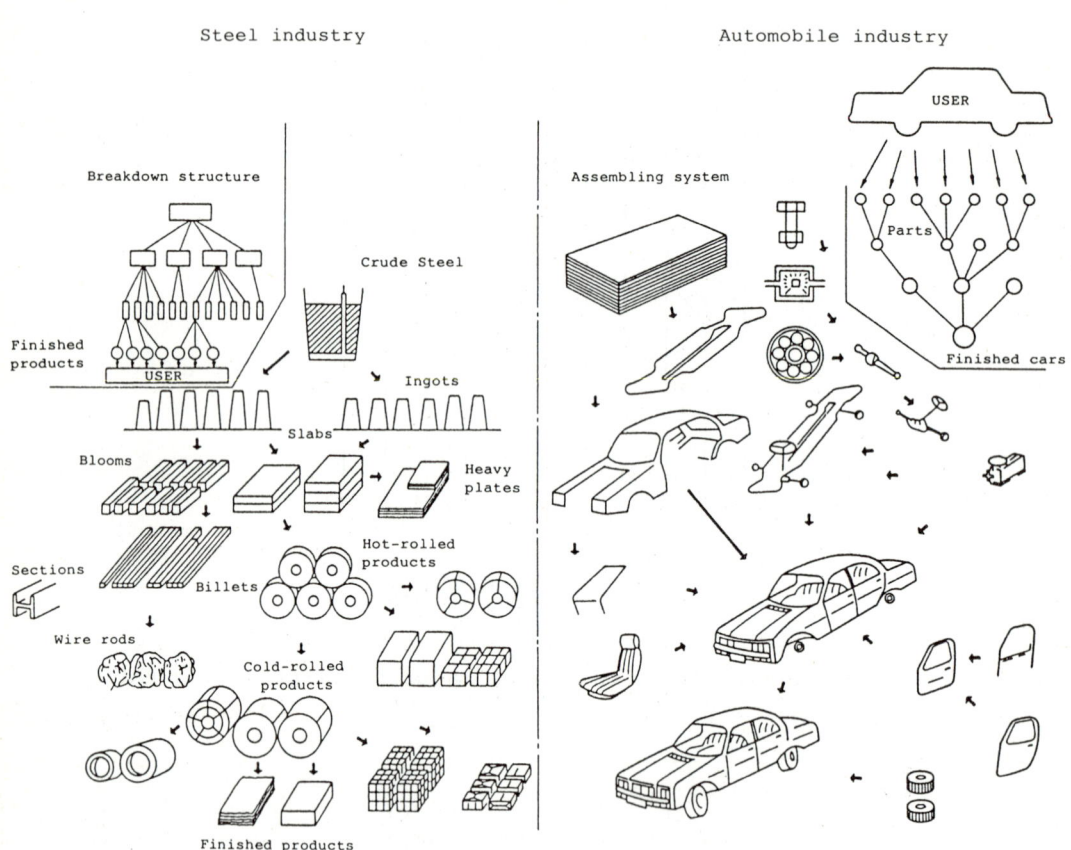

Fig. 8 Features of production control in the steel industry

II. LARGE-SCALE INFORMATION SYSTEMS IN THE STEEL INDUSTRY

1. Conceptual Diagram of Information Systems in the Steel Industry

An information system used in the steel industry consists of multiple systems covering all functional fields of strategy, administration and planning, and operations level. Figure 9 shows the conceptual diagram of the integrated information systems.

This chapter explains a large-scale integrated information system by introducing the outline of NCS's sales and production control system and the details of an integrated production control system used in NCS's steelworks.

2. Sales and Production Control System and Integrated Production Control System

The two systems are connected each other to form a hierarchical structure as shown in Fig. 10. Individual hierarchical ranks provide the functions as follows.

(1) Order entry (head office ⟶ steelworks)
(Order acceptance)
Order contents of every customer are received through a trading firm.
(Quality design)
Manufacturing and quality conditions are determined on the basis of NCS's guaranteed quality standard.
(Order entry control)
Product types and capacity of every steelworks and delivery date and amount of order are checked, and production commands are given to the steelworks.

(2) Process routing (steelworks/business computer)
Operation conditions for production lines and steps are divided according to the compositions, materials, and shapes of ordered products. Thereby, required amount of processing in each step is decided.

(3) Scheduling of all processes (steelworks/business computer)
Similar orders (orders of same chemical compositions for BOF process and orders of same sizes and shapes for rolling process) are gathered together to improve production efficiency, and, at the same time, schedules are made for the shipment on delivery date.

(4) Work instructions (steelwork/business computer -- process computer)
Detailed work instructions are issued to every process computer and the operator at the work site.

(5) Process control (steelworks/process computer)

(6) Operation and quality data collection (steelworks/process computer ⟶ business computer)

(7) Quality check (steelworks/business computer)
Quality checks are performed in each step on the basis of order specifications, NCS's guaranteed quality standard, and manufacturing and inspection data.

(8) Shipment information and billing (steelworks/business computer ⟶ head office)
Shipment on appointed date of delivery is informed, and payment is requested.

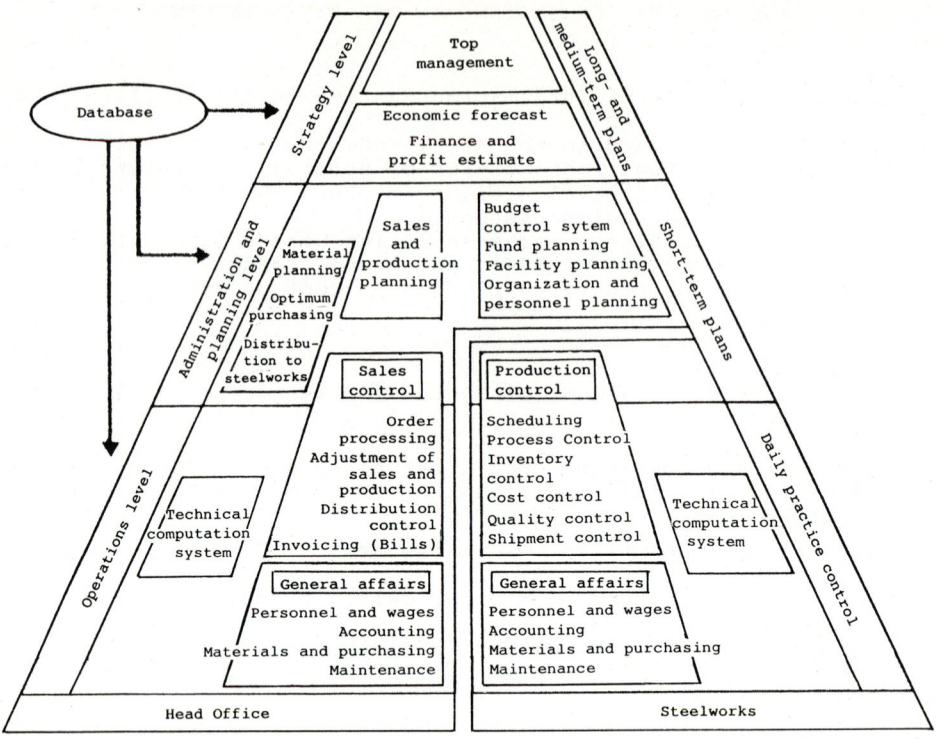

Fig. 9   Information systems in the steel industry

3.   Outline of Sales and Production Control System

NCS presently uses a batch-processing type sales and production control system that started operation in 1972 and has nearly one million program steps. Figure 11 shows the major functions of the system. The current system, which is planned to be reformed to a large-scale on-line system, will start functioning one and a half years from now and the whole system will be completed one year after that. Figure 12 shows a future system that adds other development to the new system.

4.   Integrated Production Control System

(1)   Total structure of the integrated production system

Figure 13 shows the total structure of the integrated production system in steelworks. According to the sales plans provided by the head office and basic principles of production activities, planned values (tons/hour, yields, unit consumption, etc.), various standards, and control indexes are combined to prepare annual, quarterly, and monthly production schedules. These schedules include not only the production plans according to mills, product types, and product sizes, but also the plans for the operation and maintenance of production facilities and manning. Order details based on the sales schedule are sent from the sales department of the head office to the steelworks everyday.

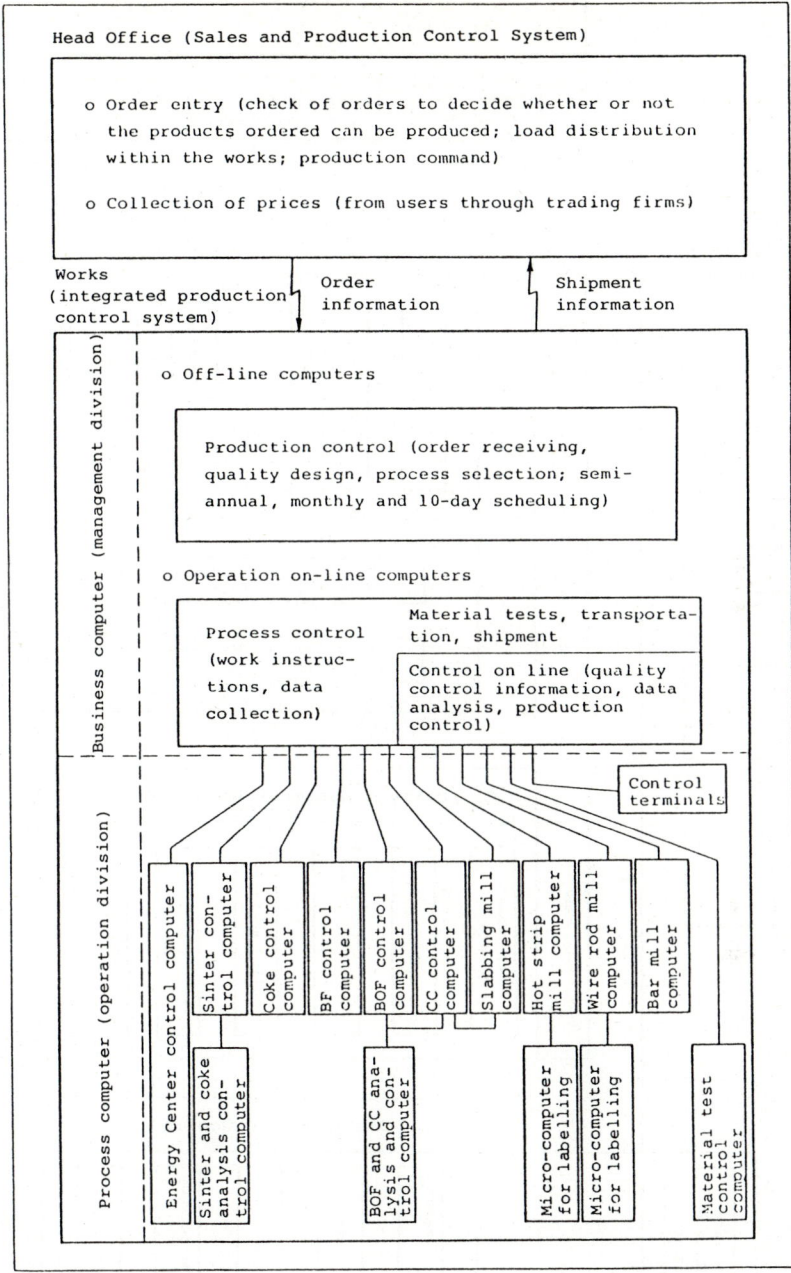

Fig. 10 Hierarchical structure of sales production control system and integrated production control system

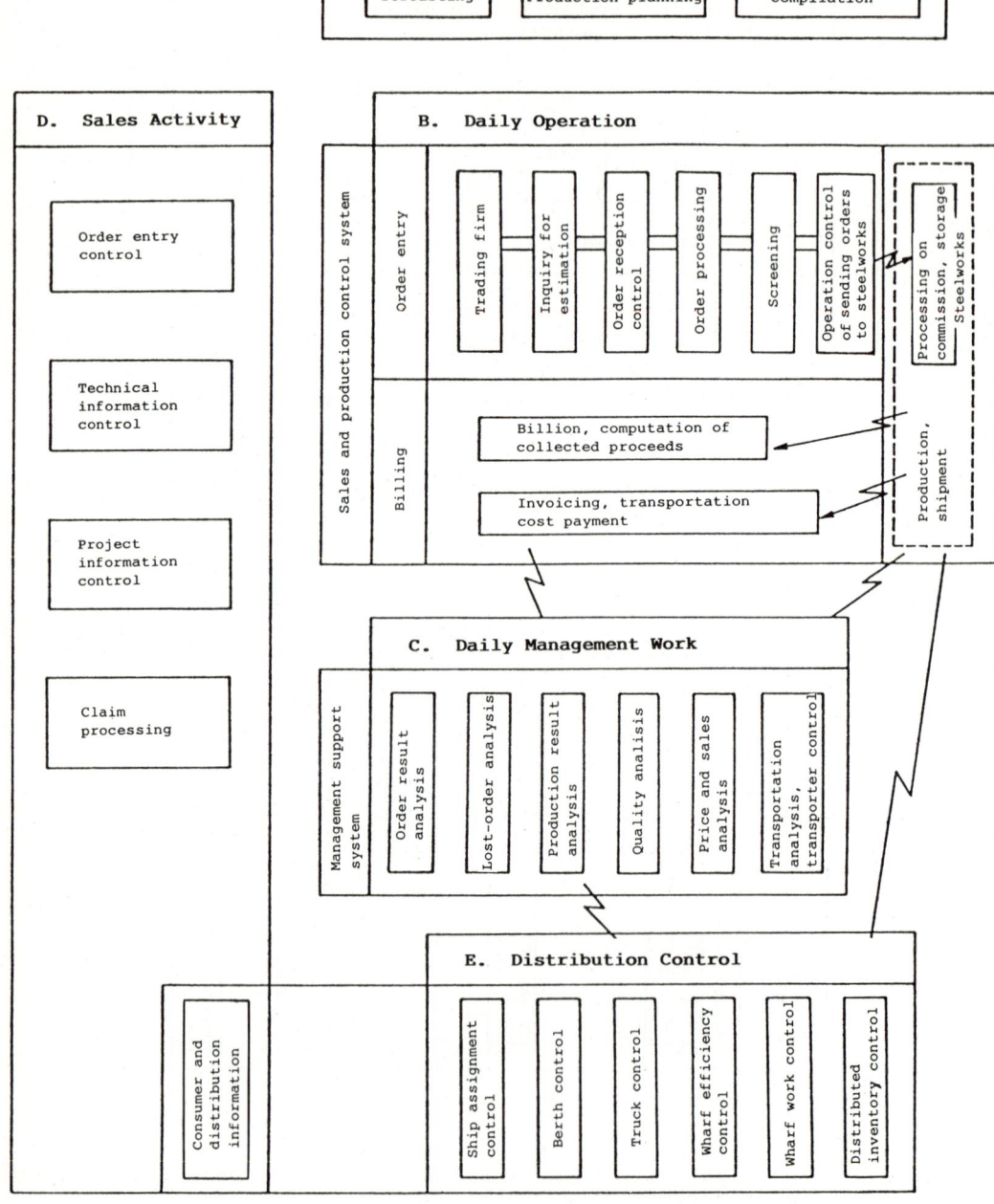

Fig. 11 Conceptual diagram of sales and production control system

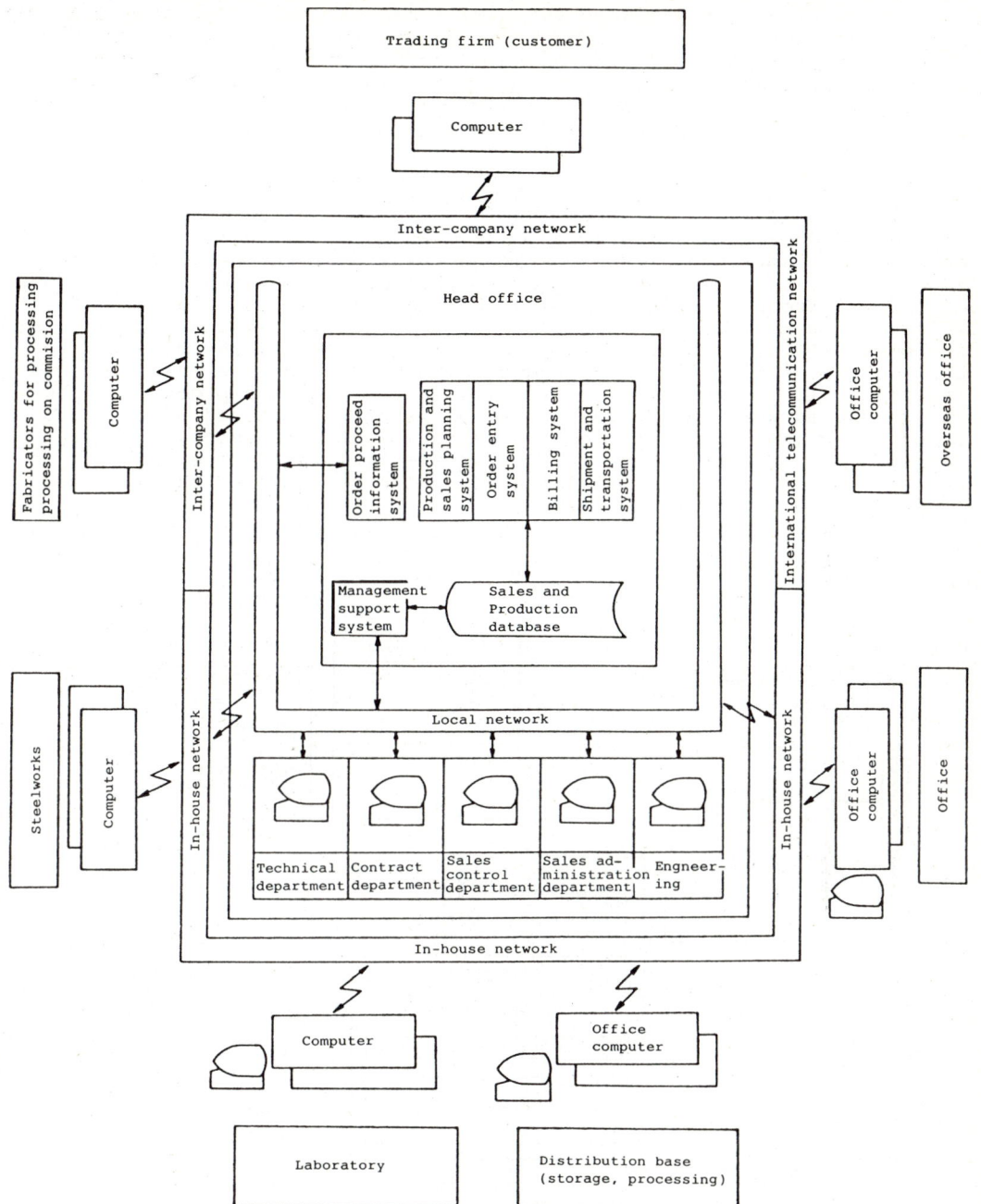

Fig. 12 Future concepts of sales and production control system

Steelworks receive the orders and make a ten-day day-by-day schedule, which covers steel making process through rolling process, in the frame of predetermined monthly plan once in ten days. The ten-day schedule is rescheduled once in two or three days. The day-by-day schedule offers the basis to schedules for each work shift that include work sequences and instructions, and are displayed on pulpit video terminal in each mill. Production activities are performed (often with process computers) according to the work instructions. As a by-product of the work instructions, operation results are collected; the instructions will be automatically collected as the results when the work is done as instructed, but actual operation data must be input when the work is not done as instructed. The operation results are compared with the schedules by the computer, and, if they do not match each other, causes of the difference are analyzed to make countermeasures. If the material for an order is insufficient, the material at the entrance of the mill will be used as a substitute of the lack and the material to be used as substitute will be requested otherwise or the insufficient material will be requested to the mill in the previous process. When the scheduled work and actual operation have been matched with each other, a "plan-do-check-action" cycle is executed to encourage the "challenge to production activities at higher levels" (see Fig. 13) in the subsequent quarterly or half a year plan as shown by the outside "feed-forward loop" in the figure.

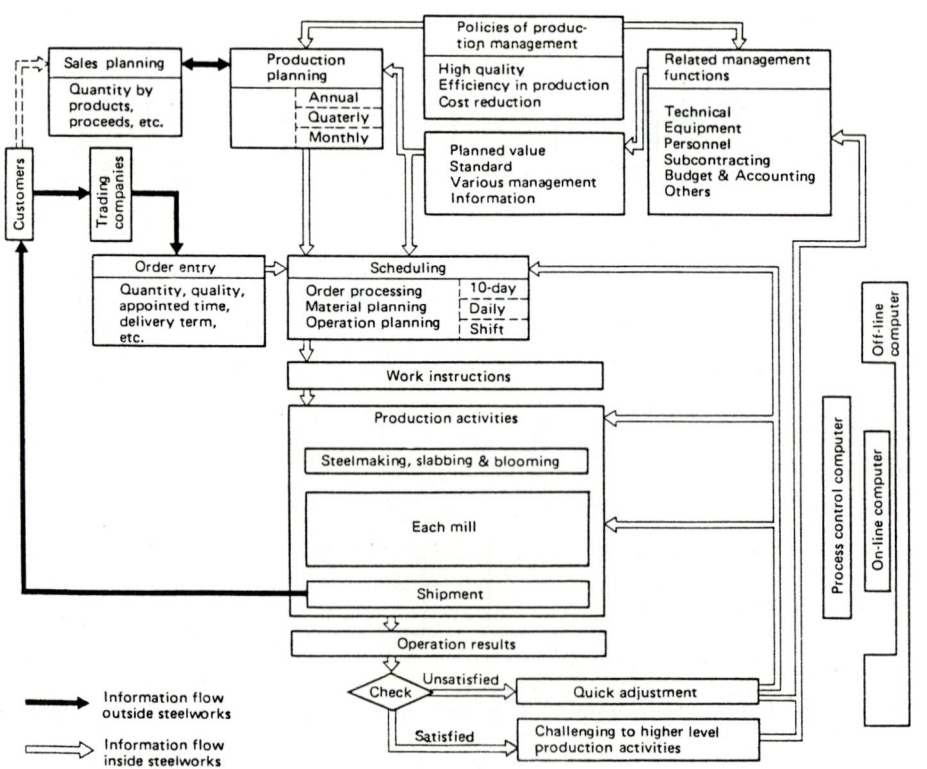

Fig. 13 Concept of production management

Figure 14 shows an example of computer system hierarchy for hot rolling mills. In the administration center, production control clerks use the business computers in the off-line batch and office on-line systems interactively to create ten-day, day-by-day schedules on the basis of predetermined quarterly and monthly schedules. Day-by-day schedules are divided into shift schedules by the operation on-line system, and the shift schedules, as work instructions, are sent to

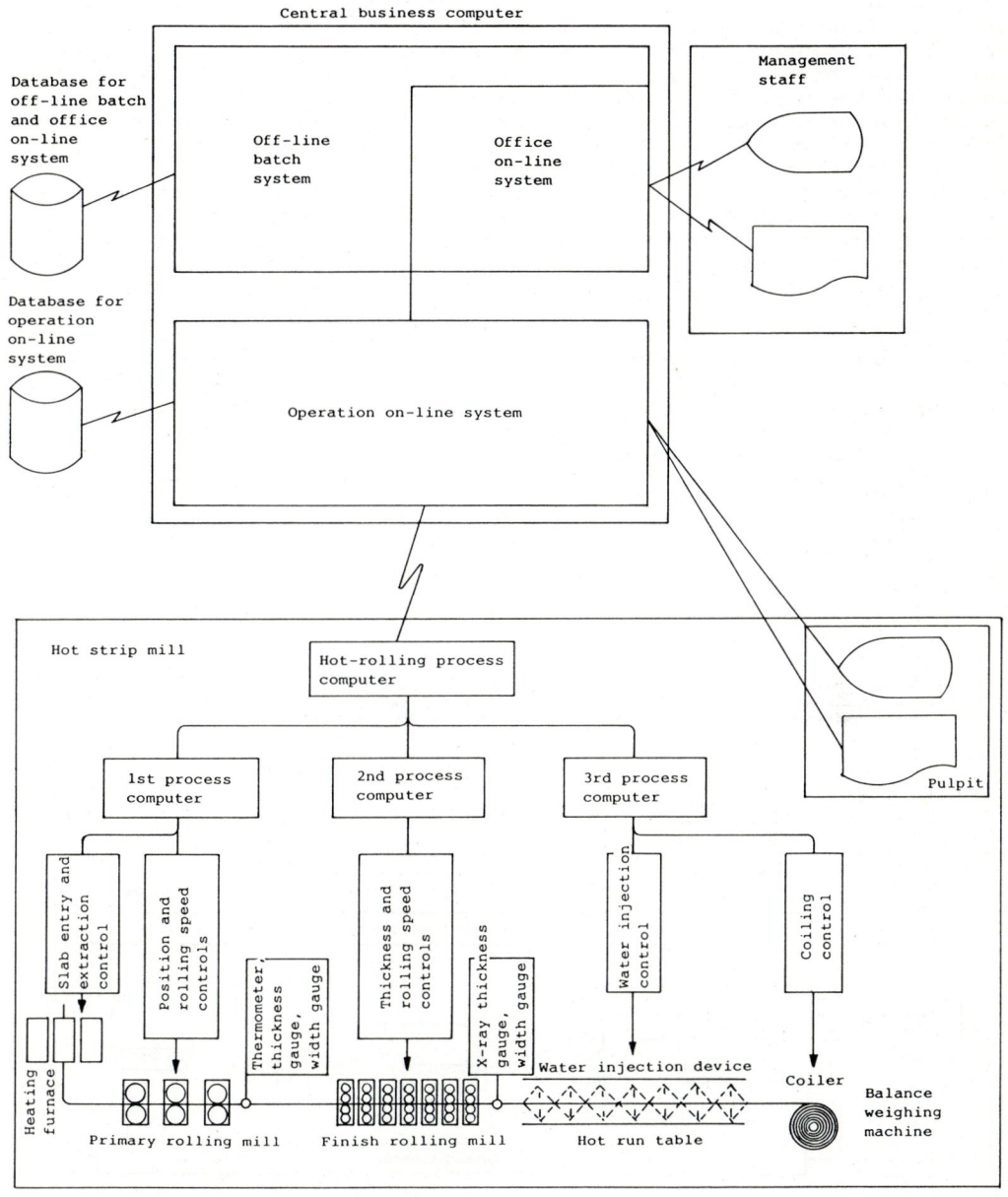

Fig. 14 Hierarchical structure of computer system

Pulpit display directly or via the process computers. Also, operation results are collected in realtime. The process computers are installed in the mills. In this example, the hot-rolling process computer controls the operation of whole works, the 1st process computer covers from the inlet of heating furnace to roughing stands, the 2nd process computer controls the finishing stands, and the 3rd process computer controls coil cooling water supply and coiling.

(2) Ten-day day-by-day scheduling concept

As shown by the upper row of Fig. 15, a ten-day, day-by-day schedule is made for each order as follows. After the material request and route selection, production quantities in each mill and process are calculated from corresponding yield rates. With predetermined standard leadtime, the production processes are simulated by computer in the reverse order of actual process flow by calculating the operation terms of every mill and step. Thereby, the types and quantities of steel products to be manufactured day-by-day in coming ten days are calculated. And, as shown in Figure 16, day-by-day schedules for the coming ten days for all processes are prepared, by production control clerks operating computer inter actively from the schedule of steel making plant, operation results of every process, and leadtime of every process. The created schedules are stored in the data file for each process.

The schedules created by the procedure as shown in Figure 15 are called "plan-oriented" schedules. On the basis of these day-by-day schedules, work instructions for every shift are created, as shown by the lower rows in Figure 15 and by Fig. 17. The work instructions sent in realtime via operations on-line computers to each Pulpit or process

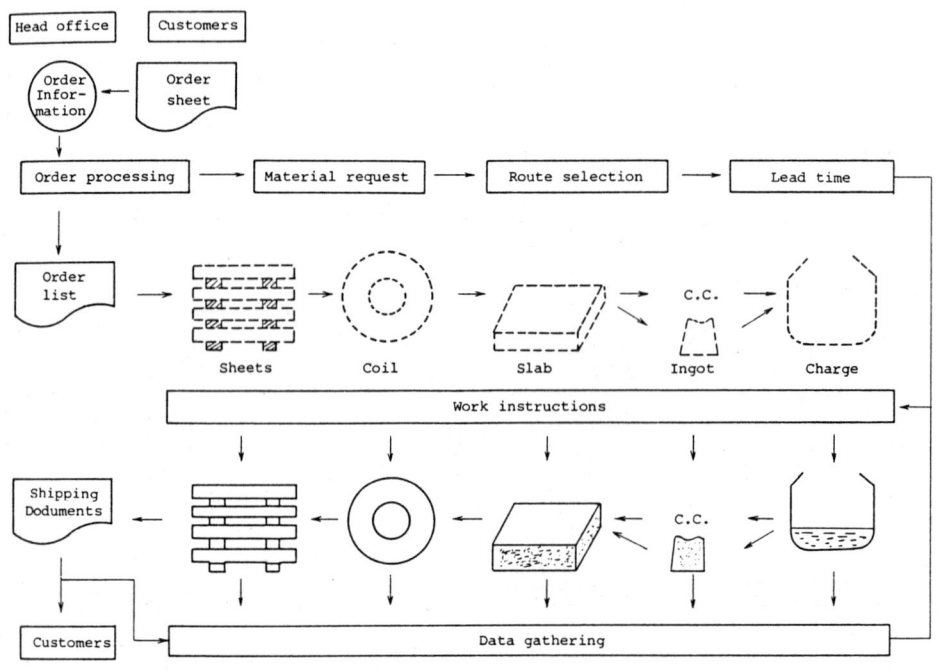

Fig. 15 Concept of Plan Oriented system

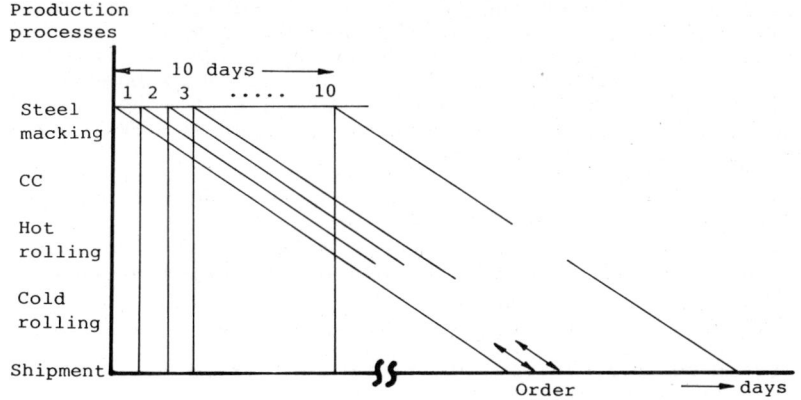

Fig. 16. Conceptual diagram of ten-day day-by-day schedule

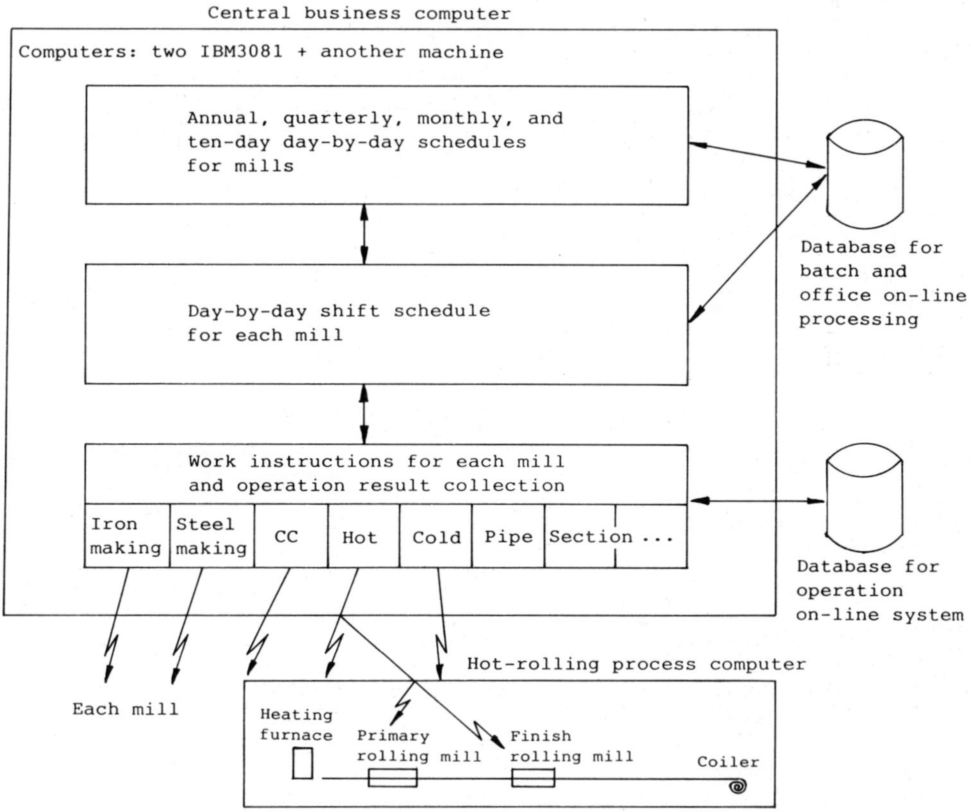

Fig. 17. Ten-day day-by-day schedule, work instructions, operation result collection

computer. The system uses the control by exception: that is, the instruction contents are automatically stored as operation results in operation on-line files when the work has been done as instructed, but actual operation results must be input manually when the work is not done as instructed. The control-by-exception system not only eliminates conventional manual data collection by 1,000 workers, but also ensures accurate data collection.

In addition, the step-by-step process flow diagrams (Fig. 18) are prepared as a train timetable with the computer so that the operators of the mills can see the relationship of operation schedules for all mills to maintain the schedule of their mills. The timetable is displayed on the video terminals in the administration center and major mills.

(3) Concepts of the integrated production control system design

The integrated production control system was designed in the early 70's with the concepts as follows, many of which are maintained even now.

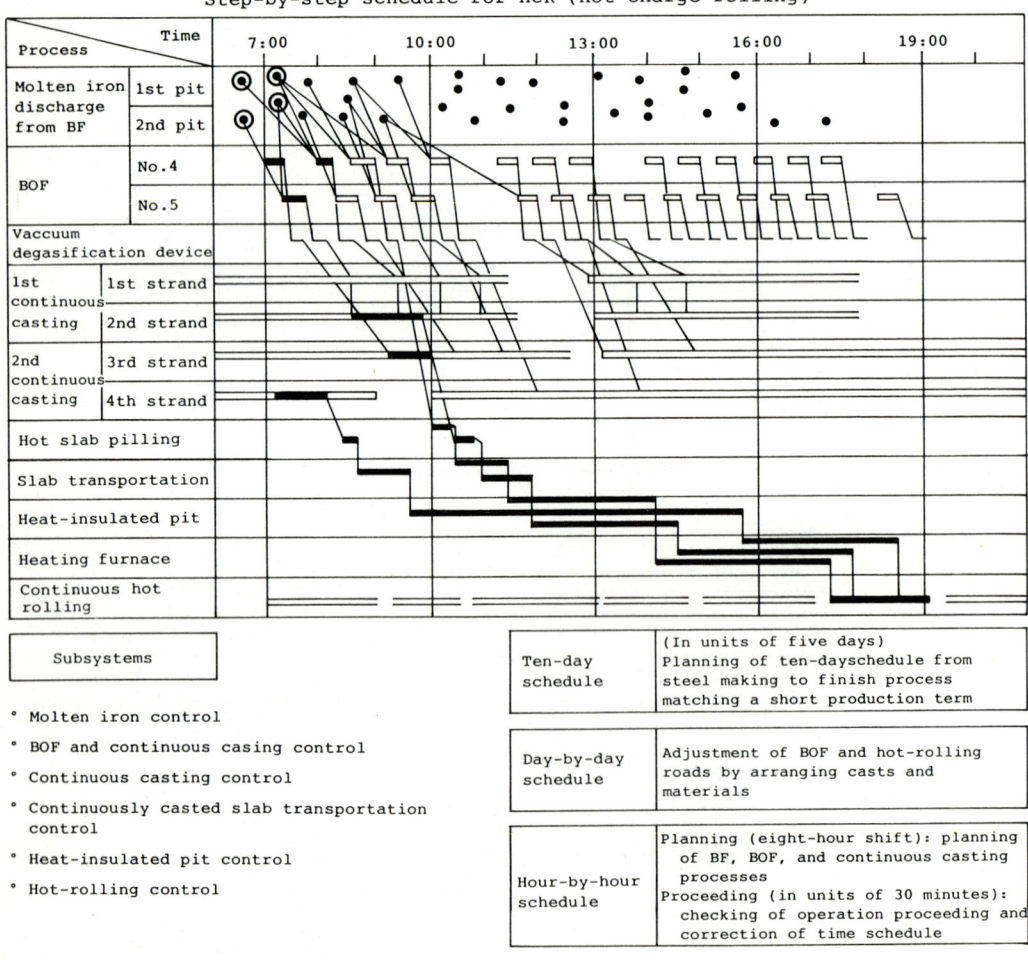

Fig. 18. Schedule diagram for Hot Charge Rolling

i) Simultaneous start-up of computer system and production equipment
Use of computers was planned at the same time with the planning of mill layout, facility automation, organization, and personnel assignment; the computer system was designed to be an indispensable element of the overall works operation.

ii) Emphasis on system life concept
For a prolonged life — two times or longer than the term of its development — the computer system was designed by considering expected changes of facilities, mill operation, and market conditions.

iii) Plan-oriented system
With a stress on a control cycle of "plan-do-check-action," the system was designed on the prerequisite that the operations were executed as they were planned. This designing concept aimed to offer the advantages as follows.
- Reduced amount of information processing due to the use of control-by-exception
- Reduced result data input and input error prevention because the work instructions themselves are referred to as operation results
- Quick actions and radical reduction of number of operators through the use of on-line terminals

These advantages can be feasible only on the basis of various management systems, the facilities, and operational techniques that support stable operations.

iv) Hierarchical structure based on the concept of "man-machine" system
Even with a highly computerized system, men make important exceptional decisions. The computer served only as a tool for such decisions. The system should be such that its structure and every part could be fully understood by operators and the flow of its work could be easily followed. Therefore, the whole system should be considered and not in detail, and its parts should be considered narrowly in details. Also, the total optimization had the priority; the partial optimization of individual components should be studied within the frame-work formed by total optimization. The system should be designed so that a local problem could be solved locally as much as possible without affecting the whole.

v) Linkage to the head office system
As shown in Figure 9, the order entry, production, and billing systems were to be integrated.

vi) Appropriate use of business and process computers
To make the most of the characteristics of business and process computers and achieve data transfer between the two types of computers, the business computers at the administration center were used for production control which requires a large amount of information processing, and the process computers installed separately at each mill were used for automatic control requiring quick response, operator guidance, and technical data collection.

(4) Importance of technical levels of mill operation and production facilities maintenances as the fundamental conditions for plan-oriented integrated production control system

As described before, NCS's integrated production control system has the base on the plan-oriented concept and has so far achieving excellent results. It should be remarked that the production control system has operated on condition that there have been the skills to execute operation plan accurately and the maintenance techniques to avoid unexpected troubles of production facilities. In addition, this production control system aims to give a motivation to workers so that they intend to innovate such techniques as described in subsection 4, (1).

(5) Scales of large systems and related personnel

Head office presently uses and is improving sales and production control systems as well as material purchase control system, personnel affairs system, accounting system and so forth. Each steelworks also operates and is developing various systems related to integrated production control. In a large steelworks, the system uses programs of 10--15 million steps and about 3,000 terminals. NSC, as a whole, uses programs of as many as 50 million steps. These programs have been developed in the last 20 years and written with assembler, COBOL, and PL/1. The personnel required to maintain the existing programs and develop new systems for coming several years amount to about 1,200 (except process computer staff). There are also about 1,000 outside software house personnel working on NCS's software.

(6) Past and future of the production control system development in steelworks--in case of the Kimitsu Works of NSC

The Kimitsu Works of NSC developed and operated the world's first on-line production control system in 1967 when the operation started in a new location of the works. During the approximately 20 years since then, the systems in each mill has been restructured according to the changes in sales situations, operation conditions, and computer technology. At the same time, the systems have been integrated to match the introduced continuous casting process, and the operation on-line systems have extended step-by-step to office on-line systems.
Figure 19 shows the historical changes. It should be noted that the processing capacity (MIPS) and external memory of computers in 1965--1969 have become 67.5 times and 426.5 times higher, respectively, in 1983. And after 1975, when these computer performance factors showed sharp growth, on-line systems also became widely used. In the future, the on-line systems must be applied for further extended purpose in the field of office work.

(7) Future outlook for the production control systems

The concepts of future system development are as follows.
(i) Automated and unmanned production and improved management levels
    - Advanced use of process control system
    - Matching to directly connected process
    - Matching to diversified and high-quality products
    - Positive use of AI
(ii) Positive construction of office systems
    - Development of decision support system for every hierarchical rank
    - Improvement of office work efficiency
(iii) Computer application extended to engineering projects
(iv) Connection to network

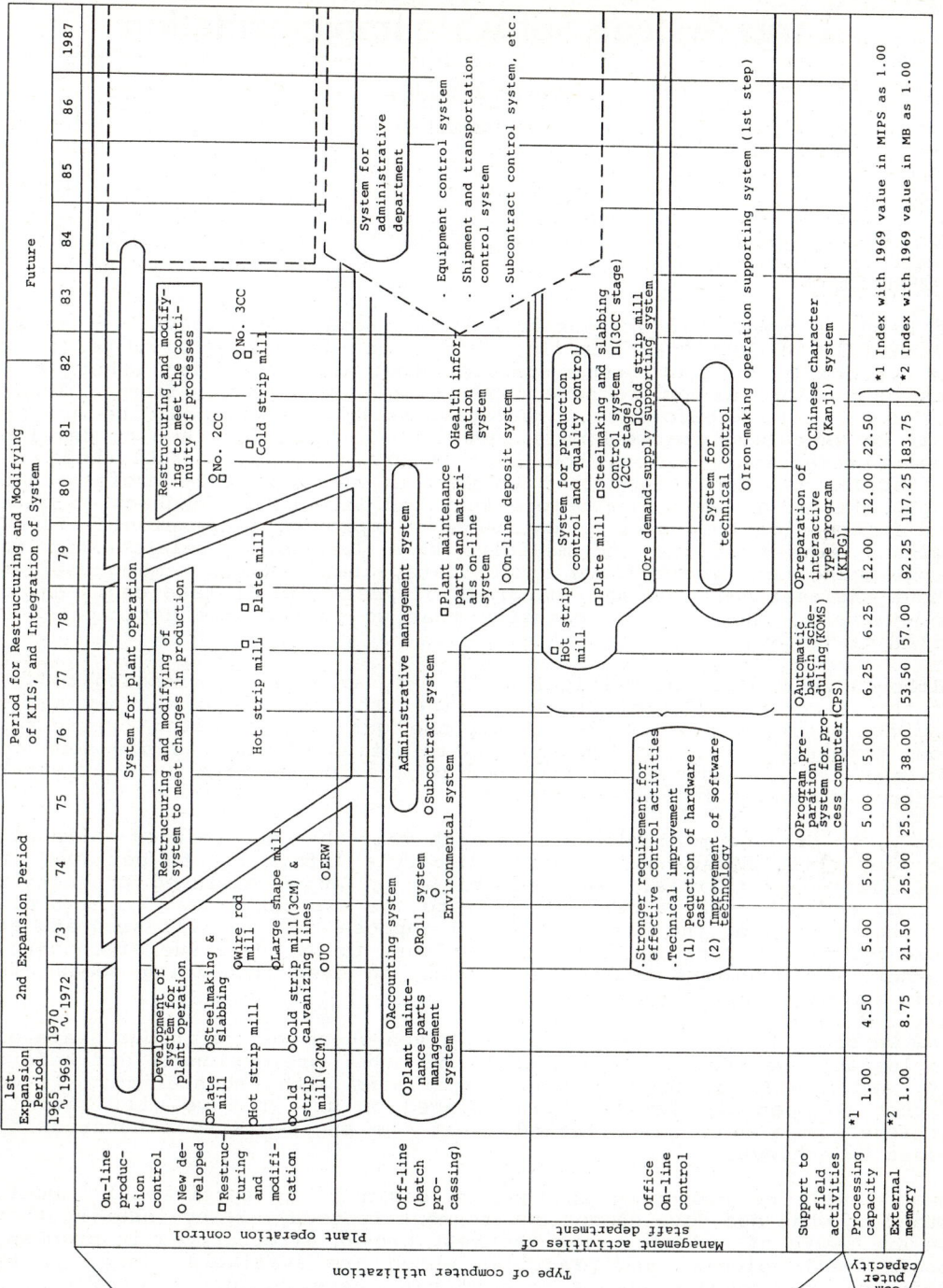

Fig. 19 Transition of KIIS

Large Systems Software Implementation

Ronald A. Radice
IBM Corporation

INTRODUCTION

We in the software professions have a long road yet to travel before we reach full professional maturity. Yet, in the last thirty years we have probably evolved faster towards that maturity than other professions before us and we have done it in a briefer timeframe. Think how long it took before the practitioners of law and medicine had to wait to be called professionals, how long it took for them to develop their professional ethos, how long before they had established the methodologies that helped them to perform to their best professional capabilities, and how long before they had integrated feedback of data and information to foster a continuously improving profession. If you perhaps find that law and medicine have less of a relationship to programming than you believe appropriate, then ask the same questions about engineering. How many hundreds of years did it take engineering to evolve to the level of maturity that we in software anticipate we will soon enjoy.

Although we can all be critical about many areas in programming today, we should not be so critical as to forget that we are both a new and a rapidly maturing discipline. I suppose our self directed criticisms come primarily from the hope and anticipation of what we are soon to become; i.e., a mature, disciplined profession wherein we can better control the products that we create.

Today, we produce more software in each year than we produced certainly in the entire first half of our industry's life, if not, perhaps, in the first two third's. We seem to be on an accelerated production curve, and yet even at our present rate, we are challenged to meet all of the demands of the marketplace. I suppose that we will saturate at some point, but for now that does not seem to be the issue of concern. Rather, our problems today are meeting an ever increasing demand in both quantity and quality.

How we begin to solve these problems given where we are in our state of evolution? We begin by doing what the other professionals did before us, and I will especially draw comparisons with engineering. If we expect to meet the demands being placed on us, then we will have to achieve, in a very short period, what the engineers took hundreds of years to achieve.

What I will be proposing is that we begin 1) by specifically understanding what our development environment is today, 2) by insuring that we are aware of and are using the best proven alternatives in process, tools, methodologies and practices which are available today, 3) by defining our development process within these bounds to establish a solid baseline position from which we can evolve and 4) by evolving to a target development process which will fully exploit the coming technologies and methodologies and which will place us firmly in a defect prevention position for the creation of our products.

Let us begin then by looking at what we in IBM Large Systems Programming are calling our Process Architecture. Specifically, let us focus on what it is, why it is necessary, and how it relates to the Development Process.

We will explore the development process from a series of stages, and then discuss a series of attributes that we can assess and analyze against those stages. We will also explore the programming development practices. I'd like to then shift the discussion to configuration management, or as I've learned to call it in my experiences, Change Control. I will define what I mean by Change Control, why it's necessary, when does it begin, and a series of classifications that will define change across the development process. Then I'm going to shift once again and go into multi-site development coordination for large systems. I would be remiss if I did not discuss metrics in the software development process, so we will spend some time on this subject. Here I will discuss a series of topics that I have dealt with over my years in the industry. These include how we might better measure, and certainly what we must measure in terms of quality and productivity. Finally, I will share with you some of my opinions on the challenges in the future of large system development.

Although I will be speaking about a Large Systems Programming perspective, I believe all of what I will discuss applies to any large programming effort. Clearly, however, Large Systems Programming will have to face all the issues first, as the problems here are larger.

I. PROCESS ARCHITECTURE

The Programming Process Architecture is the framework that provides guidelines for the definition and management of a disciplined process for software development. Later we will explore the development process itself. It should be noted that the term "architecture" is used explicitly because this perspective is not meant to provide a clear "how to" list; rather, it is a set of guidelines for "what" one must do during the software development process. Therefore, the essence of the architecture is to provide a structured method of process control using a checklist for all entry criteria, tasks, validation procedures and exit criteria for each activity or stage in the process. By stage, I mean one of a series of discrete phases within the development process. In each of these essential phases we will always address entry, task, validation, and exit, and we will always maintain a focus on process management.

Figure 1, illustrates the architecture essence in which we map the entry criteria, form there moving to the task level where there is an interactive flow of subtasks into and through the validation level, and then, upon completion of the validation criteria on through and to the exit criteria. This recurring subprocess of entry criteria, task, validation, and exit criteria, is called an activity. Each activity within a stage has associated with it a checklist of entry criteria, a task description that describes what is to be accomplished, a validation procedure to verify the quality of the work items produced within the task, and a checklist of exit criteria that must be satisfied before the activity is completed. This architectural essence applies at any level within the development process. One can view it at the activity level, at the task level, at the stage level, and at the entire development process level in aggregate. In all cases, however, the checklist for the entry criteria, and at the exit criteria are understood well in advance by the programmers and managers who will be using the architecture.

As mentioned earlier, the architecture is not intended to give a "how to" list; rather a "what to" list. A particular project, or a particular center or site for software production must define its own detailed process. It must delineate its own environmental differences, constraints, and aspects in order to successfully map its detailed process against the process architecture as it has defined it. In cases where the development process for a particular product is believed to need to deviate from the defined architecture, then the site or product must be able to substantiate the reasons for its deviation. Deviations should not be <u>ad hoc</u> or random. They should be based on some substance, fact, or fundamental reason. The data that has been gathered and analyzed should show that, in fact, the product in deviation warrants the deviation. Where the product can not justify its deviation, then it should not be permitted to vary against the architecture.

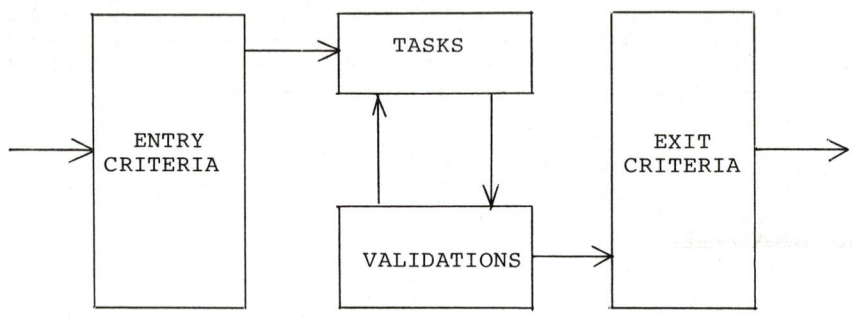

Figure 1

The architecture is not meant to be a static, once defined situation. Rather, it is meant to be a living definition that incorporates the latest state-of-the-art technologies and methodologies available to the software community. The architecture is built firmly on the roots of data gathering and data analysis which reflect for each process stage specific entry criteria, tasks, validation procedures, and exit criteria.

An organization setting out to define for itself the process architecture by which it believes it should be producing its software must come to terms with the pragmatics of its environment. It must decide whether to define an architecture that is rich in the aspects of all those things that we would like to see done; aspects like code generation wherein we can eliminate the coding or implementation stage; or aspects like continuous flow of languages from requirements through detail design where we have only one language for programming. I could continue to add a long list of aspects, in the making or being refined, which we all believe will be part of our future programming environments. Alternately, the organization could be very pragmatic and choose to define the architecture only from the aspects of what is available <u>today</u>, and not wait for the invention of technology or methodology.

What I suggest to you is that you begin from that very pragmatic viewpoint, and first define a current architecture which is a description of a process that can be achieved using the best of today's available tools, methodologies, procedures and practices. When this is done, then you will have established a framework whereby each of the products in your environment can define their own very explicit process definition which is a "how to do" each of the tasks and each of the validation procedures, you can do this without having to wait for new technology and new methodologies.

Given then that you have defined a current architecture, and as I mentioned earlier, the architecture is not meant to be static, we then must look towards some target architecture that we wish to implement over some strategic period. I would anticipate that this target architecture will emphasize to its fullest capabilities defect prevention, and it will be based on all of the presently available state-of--the-art thinking. The target architecture is also not to be viewed as some static once defined process, rather, it too will and must continue to evolve. In order to bridge from the current architecture that we define, and a target architecture we will need a series of steps that we might call intermediate architectures. These architectures will, in fact, allow us to create a technology road map from what is available today to those methodologies or technologies which we believe desirable, although we may not yet have proven them. Therefore, we have to establish this bridging network or technology roadmap. We will then have a more controlled way of evolving the industry from where we are today, to one of defect prevention.

We have done this in IBM and I want you to understand that getting the architectural definitions, getting the specific product or site process definition, and getting everyone to agree to them is not an easy task. I will take time, effort and dedication, but the rewards will amply demonstrate the effectiveness that can be achieved with a good foundation in process management.

As I mentioned earlier, the architecture has a direct relationship to the definition of the development process which defines the "what is required" for each of a series of stages. In the architecture that I am familiar with, we have defined twelve stages, and each stage is viewed by our people as a state of evolution of the product over a particular time period. These stages life cycle stages are defined based on major activities that occur during the stage of program evolution. For example, in our first stage we would be dealing with requirements and planning. Then we would see typically three generic levels of design decomposition from product level design to component level design and finally on to module level design. Then we would go into two stages that are concerned directly with implementation. These are the coding, and then the first level of testing which is the unit testing of that code. Then we would go into a cluster of recompositions during the testing phase and we call these stages the function verification test stage, the product verification test stage, and the system verification test stage. From there, and upon completion of the system verification test, we would go into a package and release stage and then on into an early support program. Finally, we have general availability, which is the actual use by the user of the product that we have created.

It should be understood that while I have spoken about these stages as if they occurred serially, they can, and indeed must, overlap to insure high productivity. This is the challenge, for if we overlap too much, we become vulnerable to losing the very controls we had planned to achieve.

II. THE DEVELOPMENT PROCESS

Time does not permit exploring with you each one of these twelve stages in detail. However, I would like to show you one stage in one more level of detail. Let me choose module level design.

During this stage there are six viewpoints which will be proceeding concurrently:

- Program Development
- Build and Integration
- Testing
- Publications
- Marketing and Service
- Process Management

There are actually a small number of additional viewpoints which apply across the development process.

PROGRAM DEVELOPMENT

Within this program development viewpoint, the programmers create and complete the detailed design, or lowest level of design, for each module, macro and data definition needed in the product.

BUILD AND INTEGRATION

Within this viewpoint the Build Plan, which is the complete project roadmap is refined to reflect any changes.

TESTING

Within this viewpoint a Unit Test plan and Product Verification Test plan are created and completed. The Functional Test plan is also completed. An installability walkthrough is performed, and the Performance Test plan is completed.

PUBLICATIONS

Within this viewpoint, the Publication Plan is completed. The publications measurement plan is completed, the first drafts are created, and a multi-national language support plan is completed.

MARKETING AND SERVICE

Within this viewpoint, the Early Support Program plan is completed, a Service Plan is completed, a Service Training Plan is completed, service education requirements are updated, and serviceability walk-through is performed.

We could continue on through more levels of detail where each task is broken into subtasks and relationships are defined, but I think the point has been illustrated. The process definition for any specific product requires enough detail to represent all the tasks and relationships necessary to complete a programming project. I want to accept the word "necessary", and I know that you understand that this will vary among products for two reasons:

1. The project environments are different.
2. There are generic product types which do differ in tasks needed to complete them.

I am not going to discuss this second reason, as it is fairly obvious.

III. PROCESS ATTRIBUTES

There are eleven process attributes that I wish to discuss. Each of these attributes applies to all of the twelve process stages.

The first is the attribute of "Process". By Process I mean the systematic flow and relationship of tasks and information to produce a product. You can see in Figure 2 an example for the code stage. This representation is not meant to imply that I think that this code process is exemplary. For example, it lacks a validation step for the rework. It is nonetheless a definition that was used by some programmers within their environment. We can see three main elements in this process: there are inputs, there are tasks, and there are outputs. Some of these tasks are a set of validations for other tasks.

The second attribute is Methodologies, which is defined as the set of systematic procedures and techniques used to accomplish a task. An example might be a code inspection or code review during the code stage.

The third attribute is Adherence to Practice. This is in two parts. First is that a practice be properly defined and commonly understood as a "proven" ethic by the members of the team. Secondly, we are concerned with how consistently that ethic is followed. Examples might be code change flagging standards, how well or how consistently the code reviews or inspections are performed, and whether or not all test plans for a particular unit test are verified.

Another attribute is Tools, which we define as the automated support for tasks methodologies and practices. Examples for tools would be compilers or static analyzers during the code stage.

The fifth attribute is Change Control which is defined as those methods by which change to the product is controlled. Examples of this would include all changes that might occur and which could impact the code activity stage. These might be design changes, they might be fixes to prior releases, and they might be program trouble reports that are written during a later test phase.

Data Gathering is the collection of appropriate data. We are not concerned with all data necessarily, rather we are interested in the collection of appropriate and meaningful data and information which can effect the process of data gathering that we might be concerned with during the coding phase would be formal recording to the inspection review findings.

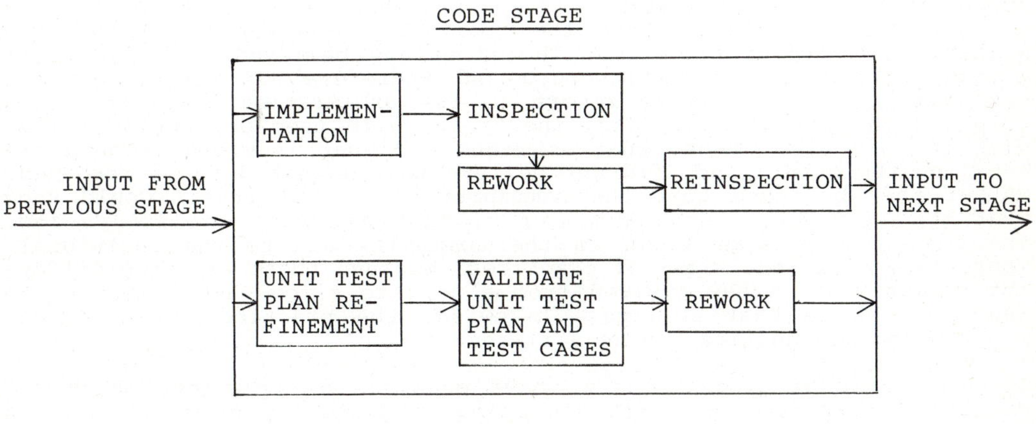

CODE STAGE

Figure 2

Communication and the Use of Data is the effective analysis of data or information and communication of this data or information to control and improve the process. An example would be using the inspection data, which was recorded, to modify unit test plans, system test plans, etc.

The eight attribute is Goal Setting, or the establishment and use of targets for the purpose of improving the process. We could consider the number of defects that we would target to remove during a code review or inspection as an example. The percentage of lines of code that we might reinspect as a result of some attributes or characteristics that were apparent at a proper inspection would also be an example.

Ninth is Quality Focus; i.e., the pursuit of product excellence at every process step. Here we mean, specifically, that we are striving not only for zero defects, but for other aspects of quality. This is true not only for the end users, but for the next programmer in line. For example, from a coding stage we would be concerned with effect on the people who would be performing the first test.

Tenth is Customer Focus; i.e., the customer needs and requirements in the product at every process step. Examples of this attribute could be usability reviews and how these are incorporated during the code stage.

Finally, the attribute of Technical Awareness, which is the people part of the equation. Here we are concerned with the technical knowledge of the state-of-the-art products and processes in use within the profession. Examples are the knowledge of competitive product offerings and the knowledge of software engineering alternatives for a particular programming community.

When these attributes are combined with the twelve stages, it becomes necessary to decompose and partition the software cycle into cells upon which we can focus. Then we can insure that each cell is receiving the proper business attention to move the process and its product from stage to stage in a more controlled manner.

IV. PRACTICES

The attribute of practices is particularly important because it is these definitions that embody the ethos by which the professional community performs their work. If you are familiar with the history of ancient Greece you know that the Spartans and Athenians viewed the mechanism of law differently. The Athenians believed that for their purposes the law must be recorded and therefore all members of the community had something to resort to for arbitration. Specifically, this was the set of laws, and, therefore, with them one could decide, with the aid of a judge, what is right and what is wrong. The Spartans, on the other hand, did not believe that laws had to be recorded. Rather they believed that the community in and of itself would know what the laws were. When a member of that community violated a law, the consequences were known to the community and to the individual. These were different views of practices, because these were two different communities. They evolved from a need to allow each community to function to its best ability. Likewise, the practices in different programming communities can be different.

In many areas of programming we have examples of practices which are defined, or written, but are not being consistently executed. So in these instances we have a part of the Athenians, but we lack the follow through. We lack the actual performance of the practice. We might as well not have them. What we wish to achieve is a clear understanding within these communities of why practices are important. Why is structured programming preferred over unstructurred programming is an example. Why do we wish to create a plan of execution that allows us to be in better control while we build a product? Why do we wish to have cost estimates? Why do we want to have goals that we review against targets? Until this ethos and understanding is instilled in the community, until the community truly believes that performing inspections or code reviews is better than not performing them, we are going to stumble and we are going to continue to make the same mistakes that we have made in the past. Until the community believes that doing cause analysis, taking a defect back to its root cause to understand why it was not found earlier, why it was caused, and what we must do to prevent this occurrence in the future, we are going to continue to make the same mistakes and continue in a defect detection posture, rather than evolving to one of defect prevention.

So I think you would agree with me that not only is it important to define the practices in a written form because we can do that, but we must also adhere to them. Those of us who have been working in software engineering know what the practices should be because we have some evidence to suggest the value of one way of doing business versus doing it another way. More importantly, however, is that the community see, know, and understand what those practices mean for them and why they are the preferred why of doing business. If we achieve this, we will have achieved significant strides.

V. CONFIGURATION MANAGEMENT

Let me propose a hypothetical programming shop with a well defined development process, where the state of the art tools and methodologies are used, where the defined practices are indeed practiced by the programming community, where process data is gathered, where goals are set, where feedback loops to management and the personnel are dynamically maintained, where there is excellent quality and customer focus, and where the professionals are technically aware. Let me now propose that within this excellent environment that there are no process control provisions for changes brought about by new requirements or by defects in the work product. What do you think will happen to that excellent product under development once changes begin to flow into the system? Well the answer is obvious isn't it. What had started out under control results in a product out of control.

What I am arguing then, is that we need to have a change control or configuration management process that is at least as good as the mainline process. To do otherwise is futile, if not foolish.

What should we be considering within this Change Control Process? Let's Start by looking at a spectrum of programming environments from the simplest to the most complex. Figure 3 shows that representation. This list is not ordered in any priority of level of significance, Let's go through the list of aspects in what I am calling a simple programming environment. By simple I mean that the chance for error is at its lowest. By complex, I mean that the chance for error is at its highest.

In the environment where only one part or module is being created, there is a clear advantage over those where multiple parts are needed. This environment I believe becomes pronouncedly complex when more than 1,000 parts must be managed, but if you believe a lower number of parts is equally as complex to handle, I won't disagree.

A one person project environment is less likely to cause errors that an environment in which a hundred or more programmers with a normal distribution of capabilities have to work together and communicate daily to insure they are all moving the product in the same direction. This has been demonstrated by Books in his <u>Mythical Man Month</u>.

PROGRAMMING ENVIRONMENTS

SIMPLE	COMPLEX
ONE MODULE	>1,000 MODULES
ONE PERSON	>100 ACTIVE PROGRAMMERS
OUTSTANDING TECHNICAL CAPABILITIES	NORMAL SPECTRUM OF TALENT
ONE SITE DEVELOPMENT	MULTI-SITE DEVELOPMENT
EXPLICIT REQUIREMENTS	FUZZY REQUIREMENTS
FIXED REQUIREMENTS	DYNAMICALLY CHANGING REQUIREMENTS
ONE VERSION	>2 CONCURRENT VERSIONS
NO DEFECTS	>1 DEFECT/KLOC IN TEST
NO MACHINE CONSTRAINTS	ARBITRARY MACHINE CONSTRAINTS
ONE MANAGEMENT TEAM	>1 MANAGEMENT TEAM
COMPLETE TOOL SUPPORT	MINIMAL TOOL SUPPORT
COMPARATIVE PROCESS DATA	NO HISTORY
WELL DEFINED MAINLINE PROCESS	AD HOC, VAGUE PROCESS DEFINITION
WELL DEFINED GOALS	VAGUE EXPECTATIONS
INTEGRAL METHODOLOGIES	RANDOM USE OF METHODOLOGIES
WELL DEFINED CHANGE CONTROL PROCESS	AD HOC, VAGUE PROCESS DEFINITION
DEFINE PROGRAMMING PRACTICES	RANDOMLY ADHERED TO PRACTICES

Figure 3

How many of you have had to try writing a program where fuzzy requirements were given and then were changed randomly and dynamically? Of all the areas in our development process where we still need critical focus, it is here. How simple, on the other hand, it is to write a program where the requirements are explicitly defined and NOT changed after work begins. Although it may be naive to assume that requirements should not change, it is poor management to expect and plan as if they will not change. We must expect change, and we must have a process which accommodates change.

Most programming environments today still work on only one version of a program in any given time cycle. However, this is rapidly changing, such that we see more and more environments where multiple version are under development at the same time. When one version is in delivery mode, another is in test, while another is in implementation, another is in design, and yet another taking form in the requirements stage. This is one of the most challenging problems that a programming shop must face, if they have not already. Unless properly anticipated, it is almost destined to cause failure the first time encountered.
We would all like to live in a programming environment where no defects exist, but we have not quite gotten there. However, as the volume of defects to be driven out in test increases, so does the complexity. In turn, the quality of the delivered product decreases. This is certainly true because of the volume, but it is additionally true and perhaps more critically so because fixes to defects are known to cause yet other defects. As these bad fixes increase, the complexity of the environment increases and the controls to ship a quality product on time decrease.

Concerning machine constraints: I view this as any computing facilities and terminal capabilities needed to develop the product. As these constraints increase, are arbitrarily brought about, or are missing because of insufficient planning, they will then lead to a more complex environment.

A project with more than one team brings with it the same problems we see in multi-site development and in a large programmer community; i.e., the need for communication and synchronization. I will discuss multi-site development in more detail shortly.

The remaining aspects on my list: tool support, process data, defined process, defined goals, methodologies, change control and practices are all ingrained in the basic attributes I talked about earlier. In all these instances as we move from a fully mature and integrated use environment to an ad hoc, random use environment, so moves the issue from simple to complex.

Let's discuss now some factors which can <u>negatively</u> affect this already complex environment.

A. <u>LARGE MODULES</u>

It is one thing to have 1,000 modules which are 100 LOC (lines of code) each and another in which the modules are in excess of 1,000, 2,000 or even 5,000 LOCs each. The volume and capacity of what the library system can handle is clearly impacted more in the latter. This will have also a pronounced affect on how fast systems or test drivers can be built, and this in turn impacts schedules. Where quality is kept constant in delivered products, then as the number of modules and their size increase so do the machine requirements to store and process these modules increase, and this in turn can only elongate the product cycle.

B. <u>MODULE OWNERSHIP</u>

If more than one programmer can make changes to a module during the same timeframe, then these programmers certainly must communicate to insure that they are not tripping over each other and causing defects in areas outside of their domain of responsibility. I am a definite proponent of one person module ownership, as I have seen and lived with both, and in all cases one person module ownership leads to reduced complexity and better quality.

C. CHANGE CUT OFF DATES

Changes seem to come in two basic ways: change to requirements and change as a result of defects in the product development. The latter can be controlled to a lesser degree, but can and indeed must be controlled as the product progresses towards delivery. Change in requirements, if not controlled, can cause erosion in schedule and quality. Therefore, a cut off date beyond which no changes to requirements will be accepted needs to be established. The date may, in fact, vary based on impact of change criteria, so there may be a number of cut-off dates, but there must be at least one.

D. DEFECTS FROM USERS

If a version of the product is already being used by users or customers, then any problems found must be fixed not only in this shipped version, but in the version under development also. If they are not, then the possibility exists of having the customer find the problem again and clearly this is not desirable.

The impact to the version under development varies with the magnitude of the fix from the user version. It also varies with the time the defect is found prior to the delivery of the new version.

E. TOOL SUPPORT

Minimal tool support means different things to different people, and I do not want to try to answer at this time what minimal means. Rather let us say that if the processing of modules and changes to them has a high degree of manual effort in your environment, then you should be looking to change this as fast as possible. Tools are already available, others will be available imminently, while some others are taking shape as we talk. The only issue is how to acquire, use and control them.

F. MULTI-VERSIONS UNDER DEVELOPMENT

If multiple versions are under development at the same time, then without sufficient tools control of the integrity of each level will become a people driven nightmare.

G. NO DATA HISTORY

Without data history, establishing goals at a detailed process level will be frustrating, complex, and probably not done. However, until goals are established you can only guess at where and how you might improve your change control process. In some cases you will be right, in others wrong. The only answer to improve your batting average is to begin to gather data. Today more data is probably gathered about major league baseball players, than is gathered in most programming development shops.

H. ASPECTS TO CONTROL COMPLEXITY

You may think I have described a bleak picture for complex environments, and indeed I have in the sense that these environments do require their just attention. We all know that. How then do we get control of this complexity?

Foremost is that the attributes defined earlier all be pursued to their best level of availability today. There is no need to wait for invention or for a more elaborate solution to the problem of change control. You can begin quite handsomely with alternatives which have already been proven. We must not wait for magic, for we will be disappointed. Rather focus on the proven available processes, tools, methods and practices, and integrate them into an architecture.

KEYNOTES

I have taken you through a number of areas under the topic of change control or configuration management. In summary, I would like to leave you with the following basic ideas:

- ° PLAN FOR CHANGE
- ° DO EVERYTHING TO APPROACH A SIMPLE ENVIRONMENT
- ° SET CHANGE CUT OFF DATES
- ° ANALYZE THE PROCESS DATA TO REDUCE THE COMPLEXITY
- ° EXPLOIT ALL ELEVEN ATTRIBUTES IN EACH PROCESS STAGE

VI. MULTI-SITE DEVELOPMENT COORDINATION

What is it that is different between a one site development effort and a multi-site development effort? I have tried to dissect this problem into a set of aspects will be address and may shed some light towards an answer.

- ° METHODOLOGICAL SYNCHRONIZATION OF WORKFLOW
- ° COMMUNICATION BETWEEN PERSONNEL
- ° COMMUNICATION BETWEEN MANAGEMENT
- ° COORDINATION OF DEPENDENCIES
- ° TECHNOLOGY FOR CONTROL AND INTEGRATION OF PRODUCT PARTS

I know there are other aspects which would complete the set for this problem, but if we focused on these five I believe we would be well on our way towards solving the general concerns. I anticipate that the lessons we have learned should be helpful to others who are about to begin multi-site development efforts. Let us take a brief look at each of the aspects I have mentioned.

A. METHODOLOGICAL SYNCHRONIZATION OF WORKFLOW:

We saw earlier that if we have a defined development process which capitalizes on available and best proven alternatives in processes, methodologies and practices that we are well on our way towards controlling both the quality and quantity of what we produce. This requirement for a defined process becomes accentuated when a product is developed across more than one site. In fact, it has been my experience that as the number of sites and the distance between them increases the problem expands dramatically. In order to diminish this additional vulnerability to problems we must very clearly insist upon and require a well defined mainline and change control process.

B. COMMUNICATION BETWEEN PERSONNEL:

I suppose people communication is an obvious aspect of the problem, but the solution is not so obvious, and yet it is in fact trivial to incorporate. A key to solving this problem is to ensure that responsibilities at each site are defined and assigned to specific individuals, and to ensure that others know who the responsible individuals are. With a defined process and pre-established goals, targets and triggers, the personnel have a framework within which they can evaluate their own progress on a detailed level specific to their contributions.

As these goals, targets and triggers are visible, the personnel will take advantage of all opportunities to insure that information they need to successfully complete their tasks will be made available to them. If the information they need exists at another site, they will find a way to get it.

C. COMMUNICATION BETWEEN MANAGEMENT:

Everything that I said about communication between personnel applies here too, for as I was once told "managers are people, too". However, the problem is different in that management may be more optimistic than is sometimes warranted, and, therefore, they seem at times to lapse into false senses of security too readily. Often this seems to occur just before the problem begins to surface and when it is most easily avoided. Again, the key is to have visible goals and triggers, but here it would probably be at a product or function level. Goals and triggers if discussed and reviewed at frequent meetings between the management teams will go a long way to either solving the problem or at a minimum, keeping the issues within the problem of multi-site coordination visible and, therefore, more manageable.

D. COORDINATION OF DEPENDENCIES:

Ideally we would divide the work across the sites such that there were no dependencies. In the real world this never happens. Therefore, we know the problem will occur and only vary by degree. We can do some things to isolate the dependencies. First we must make it known and visible where the dependencies exist. Next we must insure that at the periodic management meetings the dependencies are always given their own part of the meeting agenda and are tracked individually against project milestones. This tracking should include more than the process data tracked for all the other parts of the product. It should include periodically scheduled tutorial presentations which discuss the technical and personnel aspects of the dependencies.

E. TECHNOLOGY FOR CONTROL AND INTEGRATION OF PRODUCT PARTS:

As the parts are developed in each of the sites, they will have to be brought together to work as a whole. In the worst case, the different sites will have different product management systems to control their own design, code modules, and publications. If the differences are pronounced, the integration of the parts will produce many problems with compatibility. Where the sites are using the same product management technology there still remains the problem of bringing the parts together into the final integration level. It is from this level that systems for test and delivery to the user will be created. What will be needed then is a project management system that allows shipping of parts from many sites to one central site, a system that allows for multiple and evolving levels to exist during the driver testing stages, and one that allows for quick and controlled turnaround for fixes to problems found during the test stages. The less automated this system, the more the problem of maintaining control at the central site and the more the potential of delivering an inferior level of product quality.

VII. METRICS CONSIDERED IN SOFTWARE DEVELOPMENT

I want to spend a little time now on the subject of metrics, and in particular I'd like to explore the following aspects with you:

- ° What is meaningful to measure.
- ° At what level should we measure.

- ° What kind of meaning must be included in the measure.
- ° How do we measure from a common attribute perspective and from a normalization perspective.

A. <u>LINES OF CEDE AS A METRIC</u>

In software we produce networks, systems, programs, functions, modules, and lines of code. Any yet, when we measure what we produce we go to the most detailed aspect, the lines of code metric. Furthermore, we use this metric in both our productivity and quality measures.

There are a number of problems with this metric of lines of code. One of them being the language level constraints, in that, high level languages are penalized. For example: If you are writing in an assembler language and it takes 1000 lines of code to complete a function and then you decide to move to a higher level language, you can write the same function in say 200 hundred lines of code. You have obviously produced less lines of code in approximately the same amount of time for the entire project time cycle. This is true because we may save some time in coding, maybe some time in test, but we will not necessarily save during requirements or design. We will certainly not decrease the project cycle by a factor of 5 as we did in the lines of code reduction. Why then move to a higher level languages? Capers Jones has discussed this in detail in his productivity papers.

Another problem with the language level issue is the macro facility. Do you use macros? How well do you use macros? We can include in this aspect the control block mapping facilities, the extent that they are used in the language, and the extent to which they are counted. High macro use will lead to fewer source lines of code, so why write macros if you are penalized by the counting rules?

Additionally there are inherent language differences which we must face. This has been discussed and demonstrated by Halstead and by Weinberg and shows that some languages are inherently more difficult and, therefore, possible more error prone than other languages.

Assuming that we could overcome these language level constraints we still have the problem of defining what specifically is a line of code. Are we counting executable compiler comments? There are many different views on this subject. We clearly know that these different views are not equivalent and yet the industry seems to focus on that one universal solution called the line of code metric.

I want to now focus on the denominator portion of the equation. For example, in productivity tracking the programmer months or dollars expended to complete a project could be a measure on which we would want to keep data. What's constable to a product? What dollars should we be including in the programmer months we count? How do we know that these are consistent between organizations? For example I can include direct and indirect costs. Direct means those costs that are specifically attributable to a product and include activities like product development and maintenance. Indirect costs could be overhead items. Yet, you would be surprised at what some organizations might try to include in overhead. You can see from Figure 4 that one can seemingly create the same output with different costs. This example shows that apparent productivity is distorted by allocating costs into indirect pockets, which would not be counted against the product directly. Thus, once again, unless we specifically understand the definition of productivity used in different organizations, we may be misdirected if we compare them.

Finally within the focus of this line of code emphasis, we are creating a situation where we reward a programmer or management team for producing more lines of code per programmer month. This seems appropriate. Then it should be no surprise that in fact programmers may produce more lines of code for a given function which should indeed require less lines of code.

	PRODUCT A	PRODUCT B
OUTPUT UNITS PRODUCED	1000	1000
ACTUAL DIRECT COST	$100	$ 90
APPARENT PRODUCTIVITY	10 units/$	11 units/$
ACTUAL INDIRECT COSTS	$ 10	$ 20
ADJUSTED PRODUCTIVITY	9.1 units/$	9.1 units/$
INDIRECT TO DIRECT RATIO	10%	22.2%

Figure 4

B. MICRO VS GLOBAL VIEW

Let's take a look at another problem which I call the micro versus global view. In Figure 5 you see that the productivity costs are distributed across the design, code, test, and maintenance phases. The question I face is, if design in this example improves from a measure of 15 lines of code/programmer units to 20, is this good. The answer depends on how and why design was able to improve. If the design was hastily done and passed on to the developers, then global productivity may in fact be impacted. However, the perspective from the designers is that they were more productive. What we want to do is to measure productivity from a complete global perspective, while we measure it in each project development activity at a micro level.

MICRO VERSUS GLOBAL VIEW

PREVIOUS PRODUCT PRODUCTIVITY (LOC/PM)

DESIGN	CODE	TEST	MAINTENANCE
15	20	15	50

SCENARIO:

1. DESIGN IMPROVES TO 20 LOC/PM

2. IS THIS GOOD OR DESIRABLE?

Figure 5

C. QUALITY LEADS TO PRODUCTIVITY

Let us now turn to the relatioships between productivity and quality, schedules, costs, and estimates. I want you to understand that I firmly believe that we need to focus on quality first and not productivity. Focusing on quality we will certainly get the productivity.

For example, in Figure 6. You can see that for the cost distributed in the first hypothetical instance that we are producing 100,000 lines of a product for a cost of almost 16 million dollars. By increasing our quality focus in the earlier stages of defect removal through in-

QUALITY

ACTIVITY	ERROR DETECTION RATE (E/KLOC)	$/ERROR	K$ FOR ERROR WORK		K$ FOR PRODUCTION WORK		K$ TOTAL
HLD	8	50	40	+	300	=	340
LLD	12	100	120	+	300	=	420
CODE	18	200	360	+	400	=	760
TEST	22	1,000	2,200	+	2,200	=	4,400
MAINTENANCE	5	20,000	10,000	+	0	=	10,000
							15,920

100,000 LOC PRODUCT

Figure 6A

QUALITY

ACTIVITY	ERROR DETECTION RATE (E/KLOC)	$/ERROR	K$ FOR ERROR WORK		K$ FOR PRODUCTION WORK		K$ TOTAL
HLD	10	50	50	+	600	=	650
LLD	15	100	150	+	600	=	750
CODE	20	200	400	+	800	=	1,200
TEST	16	1,000	1,600	+	2,000	=	3,600
MAINTENANCE	4	20,000	7,000	+	0	=	7,000
							13,200

QUALITY IMPROVEMENT BY INCREASING ERROR DETECTION IN EARLIER ACTIVITIES

Figure 6B

spections, for instance, we could change the relationship in the costs to remove the defects. In fact we can reduce the cost to 13 million dollars, and, therefore, the productivity has increased. What we see is that we spend 1 million dollars earlier in this example and we saved 2.72 million dollars later because we focused on increasing quality. Note that the productivity took care of itself by addressing quality.

D. NORMALIZATION IS NECESSARY

My basic issue with lines of code is that they can be misleading and certainly require some kind of normalization if they are going to be used in comparison with products in different environments. Unless we fully understand the environmental aspects which include factors such as the experience level of the personnel, the skills capability of the personnel, the complexity of the programs that were being compared, etc., we will not have a fair or just comparison. Other factors include: Do we have a rewrite situation using an existing design versus a first release situation with a new design? Are we dealing with new code or modified code? What are the average module sizes on the products that we are comparing? What kind of group interaction do we have in the products under cosideration? What tools and methodologies are we using and are we bounded by. What dependencies and requirements exist between products that we interface to? What are our maintenance strategies?

What I would prefer to see and what I expect we would all prefer to see is more of a functional unit measure as a solution to the problems that the measurement of lines of code creates. however, what we have seen where alternatives to the measures of lines of code have been proposed is, a strong correlation between the alternatives and lines of code. This does not mean we should stop searching for a better measure, but it does say we may be forced to live with lines of code as our basic production measure for a long time.

Within one product, within one environment where we compare similar products, all of these alternatives seem to work equally successfully. However, as we start to cross the boundaries and try to compare a product in one environment versus a product in another environment and unless we are sure that we have defined lines of code and programmer months the same way, then we may be comparing two different perspectives and, in fact, comparing them invalidly. So we must be very critical of cross environment comparisons on productivity.

The same concern applies to quality. What do we count as a defect? When do we count the defects? Unless the definitions are consistent and equivalent then we're going to be belaboring needlessly the issue of one product having a higher quality rating than another.

VIII. CHALLENGES IN LARGE SYSTEM DEVELOPMENT

In this section I'd like to share my opinions in the area that need better answers as we proceed through the evolution of our industry.

The first of these is paramount because it's the bottom end of that iceberg that keeps growing beneath us. It is the old technology of the old code problem as Les Belady called it. We have it around us, beneath us, and it is upon this old code base which we are building much of our future programming and operating systems. Specifically what I am talking about here are all those millions of lines of code that exist today and that were created using lower level languages and unstructured disciplines. They were created without capabilities like stepwise refinement, decomposition, higher level languages or data abstraction.

All of these old programs we know are very difficult to maintain, hard to change, yet they represent clear business assets. How then do we apply the newer technologies that have been created in the last 5 or even 10 years? In the 1970's aspects like structured programming, became visible across the industry. Yet we see millions of lines of code that are unstructured today because it is costly to transform an unstructured module into a structured module or an unstructured program into a structured program. Business may be unwilling to accept the cost of making something structured which already works and which only needs a small functional change made to it. The cost to redo some of our large programs would be astronomical. The difficulty is that we keep making small changes from year to year to year. We never remove the old technology. In fact, there is often a tendency to create old code out of new code.

A similar difficulty exists with the newer technologies and methodologies that we see coming out of the late 70's and the 80's. How can we begin to apply data abstractions in an environment of millions of code that was written ten or fifteen years ago? Well, I suppose the answer is one of two choices. You begin by chipping at it, that is, doing it piece by piece. I suppose this is the most pragmatic and the most businesslike way of approaching it. The second is more revolutionary and costly and that is to redo in total. I remember speaking to some people about the concern of old code and I drew an analogy to a situation that amazed me when I first read about it. Lower Manhattan in New York City, was one of the first, if not the first, areas to be wired and serviced with substations that Thomas Alva Edison had designed and manufactured to produce electricity for homes and business. Edison, as you know, was a proponent of Direct Current. In fact, he thought Alternating Current an evil proposition. For years he and George Westinghouse had technological battles over which was going to be the dominant and prevalent form of electricity in home and business. As we know, Westinghouse clearly won on that count. Alternating current was much more efficient. Edison's concern was that it was also much less safe. nonetheless, while that debate was going on, Lower Manhattan was wired to produce and still produces Direct Current for a large number of its subscribers. Clearly someone with a refrigerator or a washing machine in that environment is in the technological minority. There is cost to the user because the environment has not adapted to a newer technology, i.e., alternating current. My concern is how do we in software face these similar technology problems. Will we be carrying around our own population of Direct Current users for years at an obvious cost to both us and the user. My point is that we are facing the same problem and dilemma in software. There are no easy answers at this time, yet we have to face these questions daily as managers.

The second problem that I want to discuss with you is our apparent inability to deliver more and more to meet the demands of the marketplace. As you all know, the marketplace is growing at a significant compound growth rate. Yet we are certainly not producing the computer science specialists at that same rate and we have not made any significant breakthroughs in productivity to meet the demands with a limited population of programmers.

This raises the concern that I want to call the "limits of software productivity". In Figure 7, you see a hypothetical example of cost breakdown. You see costs distributed from high level design all the way through to maintenance. A thousand unit distributed as they might be costed out through those activities.

Now, the question is how can we improve these costs. What I suggest is that by reducing the cost of quality, by reducing the cost of maintenance and inspections; i.e., the defect removal activities prior to testing, and by reducing the costs of testing, that we could reduce these 1000 units by 780, or 78%, as this example demonstrates. Now I'm also a firm believer in the fact that if you create a new environment, there is a cost for creating it. Let me suggest that the cost for being able to remove inspection, testing, and maintenance; i.e., doing it right the first time will cost 150 units. If you want to make it bigger, please do. If you want to make it smaller, fine. But I think you will all agree that there is a cost for doing it right the first time. It doesn't come by magic.

What we see is that we have reduced the cost of producing an equivalent amount of function from 1000 units to 370 units. What's left? Maybe we could evolve to some code-generation tools. So we could further diminish the cost of coding. However, as Robert Heinlin says in The Moon is a Harsh Mistress - TANTAAFL. This is not Dutch, it means, "there ain't no such thing as a free lunch". There's a cost. There's a cost for creating an environment with code generation capability. Let me argue that this cost could add another 20 units. With 50 units reduced for reuse and/or code generation, we now have a base of 340 units. There's only one other area where we could possibly find improvements and that would be in user publications. Let's say we have removed 10 units for that, taking it down to 330 units. What this represents is a 3:1 improvement in productivity and possibly a limit to the software productivity. If we are looking for some larger increases of 10:1, 50:1, 100:1, we may be overly optimistic. Maybe we could get it to 4:1. The point is, there appears to be some saturation level given where we are today. We have got to do the work of thinking about what it is we create. We have got to do some verification of what it is we create. We have got to provide the user with documentation. We have got to do some packaging of those pieces, even if they are reusable pieces. These are costs that will be there simply because we are creating a product, not manufacturing a product.

I want to leave with you the thought that we may be limited in the productivity of software production - somewhere in the neighborhood of 3 to 4:1, and this assumes that we could implement all of those improvements that I mentioned. Doing it right the first time, making no defects, defect prevention, code generation capabilities, and improvements in how we create our publications for our users. These are not easy process solutions to adopt, so we are further limited perhaps to something less.

LIMITS OF SOFTWARE PRODUCTIVITY

HYPOTHERICAL WORK DISTRIBUTION

HIGH LEVEL DESIGN	60
LOW LEVEL DESIGN	60
CODE IMPLEMENTATION	50
TESTING	280
PUBLICATIONS	50
MAINTENANCE	500
	1000

Figure 7A

I.   IF IT COULD BE DONE RIGHT THE FIRST TIME

```
            1000
       -     280      INSPECTIONS AND TESTING
            ────
             720
       -     500      MAINTENANCE
            ────
             220
       +     150      TO MAKE IT HAPPEN
            ────
             370
```

II.   CODE GENERATION

```
             370
       -      50      CODING
            ────
             320
       +      20      TANSTAAFL
            ────
             340
```

III.   PUBLICATION IMPROVEMENTS

```
             340
       -      10
            ────
             330
```

A 3 TO 1 IMPROVEMENT

Figure 7B

Let's take a look at it another way. If we wanted to achieve that 4:1 improvement level, and we improved at a continuous compound growth rate per year of 7% in our productivity, it would take us 21 years to reach a 4:1 improvement level. 7% per year is a good honest productivity improvement rate, and is consistent with other industrial rates.

I hope that I am wrong, that I am not being overly pessimistic in this limit to software productivity. I believe that a good part of our answer is in reusability of design and code. However, we have not defined the term reuse adequately in our industry. I can take ten different programmers and ask them what they mean by reusability or reuse and get at least 5 different answers. So (1) if we can arrive at a consistent definition of reusability and reuse, I think we will have made a significant stride. (2) If we can provide the technology, the methodology, the practices that will allow us to do reusability and reuse to a higher degree than PROCESS MANAGEMENT. We talked about process management earlier. The reason why I mention it as a challenge is because we in the industry have not yet applied a complete focus on process management. We've applied it to product management. What is Process Management? What are the principles of process management that I believe we need to focus upon:

1. The process must be actively, continually, consistently managed to achieve a consistently and improving quality and increasing productivity.
2. Consistent management of a process requires that the process be defined and decomposed into its parts; that it have criteria for entry, validation, and exit. Process data should be regularly reviewed and analyzed and used for process improvement.

3. Each work item must be validated before being included in the product or its associated user publications.

4. The problems with the product or process must be recorded and analyzed for cause effective improvement.

5. Changes to the product or process must be controlled. They should be recorded, tracked and evaluated for effectiveness.

6. Data capture, analysis, feedback and goal setting are essential to the improvement of both the product and the process.

7. Management and the programming community must know what the process is before beginning to develop a product.

The ways to implement these principles are: 1) to incorporate a firm process management foundation by having an operational process definition with central support organizations, process and project management systems for tracking and analysis of the data; 2) to have a visible and measure product quality plan; 3) to have process post-mortems at the end of each major stage; 4) to have a controlled rework process; and 5) to insist upon process management dynamics during the development cycle, not after.

Names and Addresses of Contributors

Rudolf Bayer
Institut für Informatik
Technische Universität München
Arcisstr 21
8 München 2
F.R. Germany

David A. Field
Mathematics Department
General Motors Research
Laboratories
Warren, MI 48090-9055
USA

Yoshisuke Inoue
Computer System Center
Nippon Steel Corporation
6-3, Otemachi 2-chome
Chiyoda-ku, Tokyo
100 Japan

Hidenobu Ishida
Software Division
Mitsui Engineering &
Shipbuilding Co., Ltd.
6-4, Tsukiji 5-chome
Chuo-ku, Tokyo
104 Japan

Yahiko Kambayashi
Department of Computer Science
and Communication Engineering
Kyushu University
10-1, Hakozaki 6-chome
Higashi-ku, Fukuoka
812 Japan

Tosiyasu L. Kunii
Kunii Laboratory of
Computer Science
Department of Information
Science
Faculty of Science
The Univeristy of Tokyo
3-1, Hongo 7-chome
Bunkyo-ku, Tokyo
113 Japan

Hideo Matsuka
Robotics Systems
Japan Science Institute
IBM Japan Ltd.
5-19, Sanban-cho
Chiyoda-ku, Tokyo
102 Japan

Ralph E. Miller, Jr.
Dr. Ralph E. Miller, Jr. and
Associates
17220 Eighth Avenue Southwest
Seattle, WA 98166
USA

Yasunori Mochida
Nippon Gakki Co., Ltd.
10-1, Nakazawa-cho
Hamamatsu
430 Japan

Jurg Nievergelt
Department of Computer Science
University of North Carolina
Chapel Hill, NC 27514
USA

Terumoto Nonaka
R&D Laboratory
Nippon Gakki Co., Ltd.
10-1, Nakazawa-cho
Hamamatsu
430 Japan

Michio Ohnishi
Software Division
Mitsui Engineering &
Shipbuilding Co., Ltd.
6-4, Tsukiji 5-chome
Chuo-ku, Tokyo
104 Japan

Ronald A. Radice
IBM Corporation
Dept. D14/901 Building
P.O. Box 390
Poughkeepsie, NY 12602
USA

Phyllis Reisner
IBM Almaden Research Center
650 Harry Road
San Jose, CA 95120-6099
USA

Yukari Shirota
Kunii Laboratory of
Computer Science
Department of Information
Science
Faculty of Science
The University of Tokyo
3-1, Hongo 7-chome
Bunkyo-ku, Tokyo
113 Japan

Nan C. Shu
IBM Los Angeles Scientific
Center
11601 Wilshire Blvd.
Los Angeles, CA 90025
USA

Ib Holm Sørensen
Programming Research Group
Oxford University Computing
Laboratory
8-11 Keble Road
Oxford OX1 3QD
England

Rob Strom
IBM T.J. Watson Research Center
P.O.Box 218
Yorktown Heights, NY 10598
USA

Shunji Tsuchiya
Higashifuji Technical Center
Toyota Motor Corporation
1200 Mishuku
Susono, Sizuoka
410-11 Japan

Eiichi Ueda
The Mitsui Bank, Ltd.
1-2, Yurakucho 1-chome
Chiyoda-ku, Tokyo
100 Japan

Shaula Yemini
IBM T.J. Watson Research Center
P.O. Box 218
Yorktown Heights, NY 10598
USA

Taketoshi Yoshida
Robotics Systems
Japan Science Institute
IBM Japan Ltd.
5-19, Sanban-cho
Chiyoda-ku, Tokyo
102 Japan

Subject Index

abstract computation model 43
abstraction 127
action 100
All Japanese Banks Telecommunication System 312
analog information 81
analog data file 81
Apple Lisa 123
application database 27
application development system 2
application system 27
application user 38
architecture 355, 356
- phase 41
ATM 311
- networking 312
- system 316
attack 189
audio visual 78
authorization mechanisms 65, 74
automatic performance function 192
automatic teller machine 311, 316
AV devices 78

BAL 208
BANCS 320
banking system 308
bitmap 114
blood-alcohol level 208
Boeing 291
- Commercial Airplane 283
- Computer Services 283
Boolean operation 225
boundary representations 227
broadband 79
Build Plan 358

CADAM 298
CADANCE 221
CAE 251, 270, 274, 283
- environment 287
CAFIS 321
cancellation operation 136
capability 41
CAPTAIN 329
cash delivery system 308, 316
cashless delivery system 308
CAVIAR 126
CDC CYBER 298
change control 359
CIM 272
classroom-type tests 168
CMOS 256

code generation 356
command environment 118
component level design 357
computational algorithms 225
computer aided design 222
computer aided engineering 251, 274, 283
computer aided visitor information and retrieval system 127
computer-driven screen 115
computer integrated manufacturing 272
conditions 11
configuration management 355, 361
conference room 128
- booking subsystem 147
constructive solid geometry 225
continuous casting 337
- process 337
Control Data Corporation 298
control scheduling 79
CONVERT 4
credit and finance information switching system 321

CrossoverNet 80
CSG 225
- trees 227
customer focus 360
customized chips 250
customized VLSI 251
CYBER 294
CYPRESS 27

data areas 55
database 27, 80, 166, 267
- management system 83
- query language 167
- schema 27
data compression 67
data environment 118
data flow 52
- language 56
data gathering 359
data history 364
data management 290
data model 268
data processing 54
DB2 298
DBMS 83
decentralized data processing 61
deductive database system 30
deletion operations 136
depositor account 308
development environment 354
device-independent design 79, 82
device-dependent 82
DeviceLevel 84
devices 78

DF language 52
DF-COBOL 60
DF-PLI 60
dialog history 116
diary system 145
difference information 69
digital control information 81
digital data file 81
dinig room 128
- book subsystem 148
diode transistor logic 259
direct field of view 200
distance judgement 202
documents 3
document editor 69
document history 65
document management system 65
driver's response 208
driver/vehicle model 209
driving and seating postures 197
DTL 259
dynamics 277

easy to use 177
editor 65
- instruction sequences 68
electronic musical instruments 188
electrooculogram 203
engineering data management 283
engineering workstation 251

entity classes 27
entity types 27
entity-relationship data model 27
EOG 203
ergonomics 107
EWS 251
exception handler 45

field of view 200
finite element analysis 220
finite element decomposition 237
finite element method 228
FM method 190
FORMAL 2, 4
- compiler 19
formal system 126
forms-oriented approach 3
frequency-modulated 190
functions 58
fund transfer 308

General Motors Research Laboratories 220
genetic specification 130
global conditions 12
GMSOLID 222
goal setting 360
go-to-less 45
graph-type history 66
greatest lower bound 46
guarded commands 45

habitability 197
headlamp 204
hexahedron 229
hierarchical menu 267
hierarchical structure 351
historical database 66
history graph for documents 66
Horn-clauses 30
hotel reservation 128
- subsystem 140
human-computer interface design 114
human engineering 192, 196
- factor 207
human factor 188, 196, 224, 289

IBM 52, 298, 327, 354
- Almanden Research Center 165
- 5080 graphics system 267
- Human Factor Center in San Jose 179
- Japan 266
- Los Angeles Scientific Center 2, 24
- Research Center in Yorktown Heights 174
- Research laboratory in San Jose 24, 166
- T.J. Watson Research Center 40
- Yorktown 40
ICEM 298

icons 88
identical subsequences 68
improved programming technologies 52
IMS 298
indirect field of view 200
Information Network System 122
information types 81
input/output flow 55
INS 122
instantiate 130
integrated computer-aided engineering and manufacturing 298
integrated production control system 341, 342
integrated programs for aerospace-vehicle design 291
integrity constrains 30
intelligent buildings 109
intelligent cities 109
intelligent database querying 38
intelligent support 27
Interactive Proto 87, 96
interface 116
- type 47
interpretation 238
inversion 23
IPAD 283, 291
- Information Processor 292
IPEX 292
IPIP 292
IPT 52
IQ-system 38

Japan Cash Service Company 318
Japanese banking system 309
JVC 109

Kimitsu Works of NSC 352
kinematics 276
Kyushu University 65

LAN 79
language level constraint 367
large system development 370
large system software implementation 354
layered architecture 82
LISP 49
local area network 79
Lockheed 298
LogicalDevice 82, 86
LogicalJob 86
LogicalLayer 82
LogicalManager 86
LogicalObject 84
LogicalProcess 86
LogicalTask 86
loosely coupled processes 40
lunch information 128

MacroMenu 84, 86, 91
manager computer 81
manipulator 276
Man-Vehicle System 196

mask geometry editing software   253
mask layout design   254
match   13
meeting   128
- attendance subsystem   142
- resource subsystem   145
- -visitor subsystem   143
Meiji Gakuin University   109
menu driven system   224
menu generator   86, 96
menu-oriented program   84
menu transition graph   89
menu window   104
mesh generation algorithm   233
mesh generator   231
message passing   40
metrics   366
mighty mouse   115, 123
Mitsubishi Bank   320
Mitsui Bank   308, 311, 317
Mitsui Engineering & Shipbuilding   52, 64
Modula-2   121
modular development   303
modular processing   56
module   56
- level design   357
motor vehicle design   196
MOVIE-BYU   239
multi-media database system   65

multi-window environment   73
mutual financing bank   313

NASA   291
NASTRAN   239
National Aeronautics and Space Administration   291
natural tone   189
NCS   318
nested table model   4
networking   309
NetworkLevel   84
network of processes   41
network protocols   122
NIL   40
Nippon Gakki (YAMAHA)   188, 195, 250, 265
Nippon Steel Corporation   334
Nippon Telegraph and Telephone (NTT)   122
noise level   213
non-procedual specifications   4
non-programmers   168
nonsharable resource   137
NSC   334
NTT   122, 320

on-line testing   180
operability   207
optimistic recovery   48

optimistic transfer   48
ordering   14
outstanding balance   308
ownership transfer   41
Oxford University Computing Laboratory   126

PADL   222
parallel-operation computer   53
part and assembly description language   222
Pascal   156
pentahedron   229
persomal CAD system   253
PHIGS   267
PhysicalDevice   82
PhysicalLayer   82
physical screen   120
piecewise polynomials   229
plan-oriented   348
point-of-sales   312
pool system   142
port   44
POS   312
practice   359, 360
pragmatics   356
procedure flow   54
pre-condition   136
process   359
- control   355
Production Automation Project at the University of Rochester   222
productivity   369
product level design   357

program development 358
program evolution 357
programming environment 356, 362
programming process architecture 355
Programming Research Group at Oxford University 126
PROLOG 30
Prototyping system 88
publication 358

Q 30
QBE 166
QBE/OBE 4
QMF 180
quadratic isoparametric family of finite elements 242
quadratic lagrange shape functions 242
qualifying predicate 136
quality 370
- control 338
- focus 360
- measure 367
query langage 30
query management facility 180
queued asynchronous communication 42
queued synchronous communication 42

reading test 169
relational information manager 293

relational model 166, 174
relational schema 44
remittance 308
remote banking 311
remote devices 81
renault 197
repeationg group 5
resources 130
Ricoh Software Research Center 109
RIM 293
robot 266
robotics 266
run-time environment 48

schedule 348
schema 131
Schottky SITL 259
semantic knowledge 27
semilattice 46
SEQUEL 166
Service Plan 358
Service Training Plan 358
session information 128
SICS 320
simulation 104
sites, models, and trails model 119
SITL 255, 256
SIT LSI 261
software development 52
software human factors 165
software manpower 332
software productivity 332, 371
Sogo banks 313
solid modeling 220

specification document 126
specification library 138
SPIFFE 238
SQL 165
SQL-QBE comparison 179
Standard Telecommunication Laboratories 126, 127
standard man-machine interfaces 114
STAR 4
state 116
- transition diagram 89, 98
static induction transistor logic 256
STD 89, 98
steel industry 334
steel production process 337
step-by-step process flow diagrams 350
strongly typed programming language 44
strong types 27
structured method 355
structured programming 45
superentities 31
supertypes 31
sustain effect 189
system life concept 351

target architecture 357
task planning system 270, 272

teaching method 273
technical awareness 360
Technical University Munich 27, 38
TEL 327
templates 3
testing 357, 358
tetrahedron 229
third nomal form 44
three-demensional model 199
time 128
- slots 130
TOCS 320
Tohoku University 265
tone 189
tools 359
total system development approach 302
Toyota 205
- ESV-1 215
- Motor 196
- SOARER 203
TPS 272
transport reservation 128
- subsystem 141
triangulation 233
two-way mass communications medium 115
type families 44
typestate 45, 46

understanding test 170
UNIVAC 327
University of North Carolina 114
University of Tokyo 78, 109
UNIX 116
unnormalized form 73
unnormalized relations 73
user interface 79, 289
users 130
user's trail 117

validation criteria 355
validation level 355
VAX-780 294
version 72, 364
very high level programming language 40
very high level "specification" language 41
view-port 268
virtual screen 120
visual programming 88
- language 24
VLSI 250

walk-service 358
work cells 266
WorkstationLevel 84
world model 268
writing test 169

Xerox 4
- Alto computer 123
- Star 123
- PARC 27
XS-1 119
- kernel 120
XS-2 119

YAMAHA 188, 250
Yamaha Music Foundation 195
Yamaha Music School 195
Yokosuka Electrical Communication Laboratory of NTT 123

Z 126
Zengin 312
- Telecommunication System 313, 315

RAYMOND H. FOGLER LIBRARY
DATE DUE

BOOKS ARE SUBJECT TO
RECALL AFTER TWO WEEKS

1 5 1987